...NDON BOROUGH OF

The CASSELL

DICTIONARY of
PHYSICS

The CASSELL

DICTIONARY of
PHYSICS

PERCY HARRISON

CASSELL

A CASSELL BOOK

First published in the UK in 1998 by

Cassell
Wellington House
125 Strand
London WC2R 0BB

British Library Cataloguing-in-Publication Data
A catalogue record for this book is available from the British Library

ISBN 0-304-35034-6

Designed, edited and typeset by Book Creation Services, London

Printed and bound in Great Britain by
Mackays of Chatham PLC, Chatham, Kent

Contents

How to use *The Cassell Dictionary of Physics*

Arrangement of the dictionary
Entries are arranged alphabetically on a letter by letter basis, ignoring hyphens and spaces between words. Headwords – or main entries – are shown in **bold** type; ***bold italics*** are used to indicate an alternative form of the main headword.

Cross-references
Words that appear in SMALL CAPITALS in articles have their own entries elsewhere in the dictionary. Certain very common scientific words, such as 'element' or 'atom', are not automatically cross-referenced each time they are mentioned in the text.

See denotes a direct cross-reference to another article. *See also* indicates related articles or entries that contain more information about a particular subject.

Units
SI and metric units are used throughout the dictionary.

Abbreviations
In those cases where the part of speech of a headword is specified, the abbreviations used are as follows:

adj.	adjective
n.	noun
vb.	verb

A

aberration Any DISTORTION in an optical image, in particular one formed as a result of imperfections in a LENS or MIRROR, which mean that rays passing though different parts of the lens or mirror, or which have different WAVE-LENGTHS, are not all brought to a focus at a single point.

The two commonest forms of aberration are chromatic aberration and spherical aberration. In chromatic aberration, different colours of light are refracted to differing degrees by a lens, resulting in coloured fringes. The effect can be corrected by using an ACHROMATIC LENS – two lenses made of different types of glass. Spherical aberration results from the fact that a spherical surface will not bring parallel light to a perfect focus. Aberration can often be reduced by minimizing the diameter of the lens or mirror. *See also* ASTIGMATISM, CAUSTIC, COMA, REFRACTION.

absolute temperature A TEMPERATURE SCALE in which the temperature is proportional to the energy of the random thermal motion of the molecules. This is the same as a temperature scale based on the PRESSURE LAW, which states that temperature is directly proportional to the pressure exerted by an IDEAL GAS held in a fixed volume. The SI UNIT of absolute temperature is the KELVIN. *See also* ABSOLUTE ZERO, INTERNATIONAL PRACTICAL TEMPERATURE SCALE.

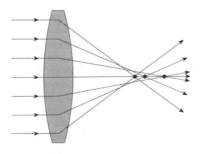

One form of aberration: spherical aberration.

absolute value 1. *See* MODULUS

2. A measurement of any physical quantity that is independent of any arbitrary or variable standard. An example is a pressure measured as the total force exerted by a gas, rather than in terms of a pressure difference relative to ATMOSPHERIC PRESSURE.

absolute zero The lowest TEMPERATURE that is theoretically attainable; zero on the ABSOLUTE TEMPERATURE scale (0 K). Absolute zero is the temperature at which molecules would stop moving and an IDEAL GAS would produce no pressure. It is equivalent to −273.15°C. It is impossible to achieve absolute zero, but temperatures as low as 2×10^{-9} K have been reached.

absorptance, *absorptivity* The fraction of ELECTROMAGNETIC RADIATION falling on a body that is absorbed by that body. The absorptance of any body is equal to its EMISSIVITY. *See also* PRÉVOST'S THEORY OF EXCHANGES.

absorption 1. The taking up of a gas by a solid or a liquid, or of a liquid by a solid. The molecules of the absorbed substance penetrate throughout the whole of the absorbing substance. *Compare* ADSORPTION.

2. The removal of energy from a wave, or particles from a particle beam, as the wave or beam passes through a material. The energy of the wave or beam is usually converted into heat within the absorbing material. *See also* FILTER.

absorption band In an ABSORPTION SPECTRUM of a solid or liquid, the range of wavelengths that are absorbed as electrons gain energy from incoming electromagnetic radiation and are excited from one ENERGY BAND to a higher band.

absorption spectrum A SPECTRUM formed by shining a continuous range of wavelengths of electromagnetic radiation through a sample that absorbs only certain characteristic wavelengths. The absorption spectrum of a solid is made up of one or more absorption bands, where each band represents a range of wavelengths removed from the incoming radiation. In a gas, only narrow ranges of wavelengths are removed and so the bands are much narrower. These are a subset of the lines in the EMISSION SPECTRUM of the same material. *See also* ATOMIC ABSORPTION SPECTROSCOPY, SPECTROSCOPY.

absorptivity *See* ABSORPTANCE.

a.c. *See* ALTERNATING CURRENT.

acceleration (*a*) The change in the VELOCITY of a body divided by the time over which that change takes place.

In physics, acceleration is usually measured in metres per second per second, also called metres per second squared (ms^{-2}). However, the acceleration figure of a car is normally quoted as the time to reach a given speed, usually 60 miles per hour, from rest. Thus an acceleration time of 10 s from 0–60 mph represents an average acceleration of 10 mph per second.

$$acceleration = change in velocity/time$$

See also ACCELEROMETER.

acceleration due to gravity (*g*) The ACCELERATION experienced by an object in FREE FALL, a measure of the strength of the GRAVITATIONAL FIELD. On Earth, *g* is about 9.8 ms^{-2}, but it varies from place to place, being greater at the poles as these are closer to the centre of the Earth than the equator. *See also* GRAVITATIONAL FIELD STRENGTH.

accelerator See PARTICLE ACCELERATOR.

accelerometer A device for the measurement of ACCELERATION. Accelerometers range from simple devices measuring the extension of a spring produced by an accelerating mass and used to check the efficiency of brakes on motor vehicles, to sophisticated devices based on lasers and used in INERTIAL NAVIGATION SYSTEMS.

acceptor atom An atom that accepts an electron to produce a HOLE in a SEMICONDUCTOR, to form a P-TYPE SEMICONDUCTOR.

acceptor impurity An element added to a SEMICONDUCTOR in a DOPING process that accepts an electron, thereby producing a HOLE.

accretion disk A disk of matter gathered around a STAR. Such disks are believed to be the first stage in the formation of PLANETS around stars. Accretion disks are also believed to be formed in some BINARY STARS as one member of the pair draws material out of the other. Accretion disks around BLACK HOLES are believed to be

responsible for the strong X-RAY sources observed in some galaxies.

accumulator 1. An obsolete term for a BATTERY of LEAD-ACID CELLS.

2. The SHIFT REGISTER in which the output from the ARITHMETIC LOGIC UNIT of a MICROPROCESSOR is stored.

achromatic lens A lens designed to correct chromatic aberration (*see* ABERRATION). The simplest type of achromatic lens is a combination of two lenses made of different types of glass, such that their DISPERSIONS neutralize each other.

acoustics The study of SOUND. The term also refers to the behaviour of sound in a particular room, such as a concert hall. The acoustics of a room may be modified by introducing sound-absorbing materials to reduce the REVERBERATION TIME.

action and reaction Two forces that are equal and opposite to each other according to Newto's third law, and which act on different bodies. *See* NEWTON'S LAWS OF MOTION.

activation analysis A technique for detecting small quantities of an element present in a sample. The sample is bombarded with neutrons from a NUCLEAR REACTOR, forming unstable ISOTOPES of the elements present. These decay, emitting GAMMA RADIATION with energies characteristic of the elements present.

activation energy (*E*$_a$) The minimum amount of energy required before a particular process can take place. The term is usually applied to chemical reactions, where the activation energy is the ENERGY BARRIER that must be overcome for the reaction to occur. On the atomic scale, the process may be a NUCLEAR FISSION event or the release of an electron or a photon of light, such as in the THERMOELECTRIC EFFECT or the PHOTOELECTRIC EFFECT. *See* ACTIVATION PROCESS.

activation process Any process in which the particles involved can only take part in the process if they have more than a specified amount of energy, known as the ACTIVATION ENERGY. This acts as an ENERGY BARRIER, which must be overcome for the process to take place. To do this, energy may be supplied externally (for example by light in the case of some chemical reactions). The energy barrier may also be overcome raising the temperature of the material so that more molecules have sufficient energy. Many chemical reactions, the evaporation of liquids and CREEP are all examples of activation processes. *See also* MAXWELL–BOLTZMANN DISTRIBUTION.

active anode An ANODE that is chemically involved in an ELECTROLYSIS process.

active device 1. An electronic device that can act as an AMPLIFIER or SWITCH. The behaviour of active devices, in contrast to devices such as RESISTORS or CAPACITORS, cannot be described in a simple mathematical way. Most active devices are based on SEMICONDUCTORS. A common example is the TRANSISTOR, which can use a small current to control a larger one. Modern active devices are often manufactured as INTEGRATED CIRCUITS. *See also* PASSIVE DEVICE.

2. An artificial SATELLITE that receives and retransmits signals after amplification (*see* AMPLIFIER).

activity The level of IONIZING RADIATION emitted by a radioactive material. Activity is usually measured in BECQUEREL, or becquerel per litre, although other units, for example the CURIE, are sometimes used. *See also* RADIOACTIVITY.

adhesion The intermolecular force of attraction between molecules of one substance and those of another. This effect causes water to spread out and wet glass, for example. The adhesion of the water molecules to the glass molecules is greater than COHESION between the water molecules themselves. In the case of mercury on glass, cohesion between the mercury molecules is greater than the adhesion of mercury to glass. The mercury molecules pull inwards, away from the glass, and small drops of mercury on a glass plate are almost spherical. *See also* CAPILLARY EFFECT, SURFACE TENSION.

adiabatic (*adj.*) Describing a change in which there is no exchange of energy between a system and its surroundings.

adsorption The taking up of a gas by the surface of a solid. Unlike ABSORPTION, the gas does not penetrate the solid material, but is held on the surface either by the formation of chemical bonds or by VAN DER WAALS' FORCES.

advection The transport of energy by the horizontal bulk motion of a fluid.

advection fog Fog produced when warm moist air from above the sea is blown over cooler land. *See also* RADIATION FOG.

aerial, *antenna* A device used to transmit or receive RADIO WAVES. Aerials have many shapes and sizes, but are generally most efficient when they have a size of the same order as the wavelength of the wave concerned. An aerial may be equipped with a PARABOLIC DISH that produces a parallel beam of radio waves from a source of radiation placed at its focus, or that collects and focuses a parallel beam of radio waves. *See also* DIPOLE, YAGI.

aerodynamics The study of the flow of gases (such as air) over solid objects. Aerodynamics has particular applications in the design of vehicles, especially aircraft. It is also used in the study of the flight of birds and insects. The term aerodynamics is also used to refer to the features in the design of a vehicle that smooth the flow of air over it.

The two chief effects of airflow are LIFT and DRAG. Drag is generally undesirable, and designs seek to minimize this. The amount of drag depends on the area of the vehicle and on its shape: flat shapes and rough surfaces produce more drag than smooth STREAMLINED shapes. This effect is measured by a quantity called the COEFFICIENT OF DRAG. Lift, which is desirable in aircraft, but which needs to be suppressed in high speed road vehicles, is described by a quantity called the COEFFICIENT OF LIFT. A shape designed to produce lift is called an AEROFOIL.

Provided the flow over a surface is not TURBULENT, the change of pressure with speed of flow, which is responsible for both lift and drag, is described by Bernoulli's equation (*see* BERNOULLI'S THEOREM). This equation relates the KINETIC ENERGY of the airflow to the POTENTIAL ENERGY represented by the pressure of the air. At speeds that are small compared to the speed of sound, air can be treated as an incompressible fluid, but at higher speeds SHOCK WAVES begin to form and COMPRESSIBILITY effects are important. In considering the flight of insects, the VISCOSITY of the air also needs to be taken into account.

See also REYNOLD'S NUMBER.

aerofoil, *airfoil* A surface designed to produce LIFT. At high speeds, any smooth flat surface meeting the air at an angle will act as an aerofoil. However, the optimum shape of an aerofoil is a lower surface that is flat or nearly so and upper surface that is more strongly curved, with the aerofoil reaching greatest thickness about one third of the way back along the surface. *See also* ANGLE OF ATTACK, COEFFICIENT OF LIFT, STALL.

afterburner *See* REHEAT SYSTEM.

air The mixture of gases forming the Earth's ATMOSPHERE. Dry air contains 78 per cent nitrogen, 21 per cent oxygen, 0.9 per cent argon, 0.03 per cent carbon dioxide and traces of other noble gases. In addition, air usually contains a few per cent water vapour, though the concentration varies widely.

airfoil *See* AEROFOIL.

air mass A volume of air that typically extends for several hundred kilometres over the surface of the Earth. Within such regions temperature and HUMIDITY are fairly constant. Air masses may be classified as tropical or polar depending on whether they originate from regions closer to the equator or to the poles. They are classified as CONTINENTAL or MARITIME depending on whether they travel to the point in question mainly over land or over water. The arrival of an air mass at a point on the Earth's surface will affect the weather at that point, whilst the characteristics of the air mass are only slowly modified by the regions over which they pass.

air resistance The force on an object moving through air that acts to oppose the motion of the object. Air resistance is neglected in many simple calculations, but becomes significant for any object moving at high speeds. At lower speeds, air resistance becomes significant for objects such as feathers, which have a large surface area for their weight. When an object falls, it accelerates until its weight is balanced by the force of air resistance. The speed at which this happens is called the TERMINAL VELOCITY. A simple model of air resistance describes it as being proportional to the area of the object and its speed through the air, though in fact a (speed)^2 dependence is more accurate at higher speeds. *See also* AERODYNAMICS, COEFFICIENT OF DRAG, DRAG.

air speed indicator On an aircraft, an instrument used to display the speed of the aircraft through the air, typically by measuring the difference between HYDROSTATIC PRESSURE and DYNAMIC PRESSURE.

albedo The measure of the amount of light reflected on average by a planet or planetary satellite. An albedo of 1 indicates a perfectly reflecting surface, whilst a body with an albedo of 0 is perfectly black, absorbing all light falling on it.

alloy A material with metallic properties consisting of two or more metals or a metal with a nonmetal. An alloy may be a SOLID SOLUTION, a compound or a mixture of two or more crystalline solids. Alloys are very often used in engineering applications. They are

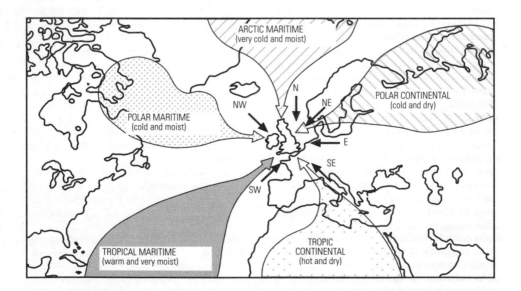

Air masses affecting the British Isles.

often stronger, harder or more resistant to corrosion than their constituent metals.

The most common example of an alloy is steel, which consists of a few per cent of carbon in iron. Many steels also contain other elements, such as chromium and manganese. The carbon atoms are much smaller than the iron atoms, and become INTERSTITIAL atoms, occupying the gaps between the roughly spherical iron atoms. The carbon atoms have relatively little effect on the ELASTIC properties of the iron, which depend on the interatomic forces between the iron atoms. However, they do prevent the onset of PLASTIC behaviour, caused by imperfections or gaps in the lattice, called DISLOCATIONS. These dislocations can move through a metal: as one atom moves to fill a gap in the lattice it leaves a gap in its original location, so the dislocation effectively moves through the lattice in the opposite direction. This makes the metal far softer than it would have been without any dislocations. In small quantities, the carbon in steel 'pins' the dislocations, increasing the strength of the metal without destroying its ductility (*see* DUCTILE). Larger quantities of carbon in steel produce a very hard but brittle material called cast iron. Similar properties apply to other metal alloys: for example, aluminium-magnesium alloy has a high strength for its density.

Although they may have enhanced mechanical properties, alloys are generally poorer conductors of electricity and heat than pure metals. This is because the lattice structure is less highly ordered, making it harder for any FREE ELECTRONS to pass through the metal.

alpha decay The spontaneous disintegration of an unstable atomic nucleus with the emission of an ALPHA PARTICLE. *See* ALPHA RADIATION.

alphanumeric display A system containing a pattern of shapes that display various letters and numbers when the appropriate segments of the display are illuminated. Such displays may be based on LIGHT-EMITTING DIODES, GAS DISCHARGES or LIQUID CRYSTAL DISPLAYS.

alpha particle A helium-4 nucleus (a stable particle consisting of two protons and two neutrons), emitted during ALPHA RADIATION.

alpha radiation The emission of alpha particles, which are helium-4 nuclei (two protons and two neutrons bound together in a stable

entity). Alpha radiation occurs when large nuclei, which are unstable due to the ELECTRO-STATIC repulsion of the protons in the nucleus for one another, spontaneously disintegrate. Alpha particles are highly ionizing (*see* IONIZING RADIATION), and hence lose their energy very quickly. They have a range of only a few centimetres in air and can be stopped by a thin sheet of paper. When a nucleus emits an alpha-particle, it changes into a new nucleus with an ATOMIC NUMBER smaller by 2, and a MASS NUMBER smaller by 4. For example, the metal radium-226 decays to the gas radon-224 with the emission of an alpha particle.

alternating current (a.c.) Electric CURRENT flowing first in one direction, then the opposite one. Alternating current is used for most power supplies as it has the advantage that the voltage can be changed using TRANSFORMERS. *See also* ALTERNATOR, DIRECT CURRENT, OSCILLATOR.

alternator An a.c. GENERATOR: a machine for transferring MECHANICAL ENERGY to ELECTRICAL ENERGY in the form of ALTERNATING CURRENT. An alternator usually consists of a rotating magnet surrounded by coils in which an alternating current is produced. The magnet may be an ELECTROMAGNET, in which case it will be supplied with current by brushes rubbing against slip rings – rotating metal rings connected to the coils of the electromagnet. The electromagnet may be powered by the alternator itself, but in the case of larger machines, such as those used in power stations, a separate DYNAMO, called an EXCITER, is used. Cars are equipped with alternators to recharge their batteries. In such cases the alternating current generated is converted to DIRECT CURRENT (rectified) by a series of DIODES.

altitude 1. The height of an object, such as an aircraft, above some specified surface, usually sea level.

2. (*astronomy*) The angle of a star or similar above the horizon

3. (*mathematics*) The perpendicular distance between the point of a cone or pyramid and its base.

ammeter An instrument for the measurement of electric current. Many ammeters work by using the magnetic field produced when a current flows through a conductor. The most common of this type of ammeter is the MOVING-COIL GALVANOMETER.

Many modern ammeters are actually digital VOLTMETERS connected across a known small resistance; the voltage across the resistance is measured, from which the current can be found.

An ammeter is connected to a circuit by breaking the circuit and inserting the ammeter at the point where the current is to be measured. So as not to alter the size of this current, an ammeter should have as small a resistance as possible. To reduce further the resistance of the ammeter, and to enable it to measure larger currents, a RESISTOR, called a SHUNT, can be connected across the ammeter (in parallel with it).

amorphous (*adj.*) Having no particular shape. The term especially refers to solids such as glass, where the molecules have no regular lattice arrangement. *See also* DISORDERED SOLID.

ampere (*abbrev.* amp; *symbol* A) The SI UNIT of electric CURRENT. Since a current produces a magnetic field and a current in a magnetic field experiences a force, two currents flowing close to one another will produce a force on one another. The size of this force is used to define the ampere: one ampere is equal to that current which, when flowing in two infinitely long parallel wires one metre apart in a vacuum, will produce a force of 2×10^{-7} N on each metre of their length. This force will be attractive if the currents are in the same direction, repulsive if they are in opposite directions. This definition of the ampere determines the strength of the magnetic field produced by a given current.

ampere-hour (Ah) A unit of CHARGE, used to measure the storage capacity of a battery or electrochemical CELL. The capacity in ampere-hours is the current that can be provided by the battery, multiplied by the number of hours for which this current can be supplied. One ampere-hour is equal to 3,600 COULOMBS.

ampere-turns In a SOLENOID or other magnetic system, the magnetizing current multiplied by the number of turns in the coil carrying that current.

amplifier An ANALOGUE electronic system that multiplies a voltage or current. The output is equal to the input signal multiplied by a constant factor called the GAIN. Amplifiers are used to boost small signals, such as the signal from a microphone, to higher levels, to feed to a loudspeaker, for example. As well as having a

high gain, practical amplifiers should be free from DISTORTION and be able to operate over a range of frequencies. Simple amplifiers may be made from TRANSISTORS or VALVES, but more modern devices are complex INTEGRATED CIRCUITS. *See also* DIFFERENTIAL AMPLIFIER, INVERTING AMPLIFIER, OPERATIONAL AMPLIFIER, SUMMING AMPLIFIER.

amplitude The maximum distance from equilibrium reached by a wave or oscillating motion.

amplitude modulation A method of transmitting information in which the amplitude of a CARRIER WAVE is varied at the frequency of the signal. Amplitude modulation is often used in telecommunication systems, to convey information such as a speech signal or the brightness of a particular part of a television picture. Amplitude modulation systems are simple and can be designed to operate in narrower BANDWIDTHS than FREQUENCY MODULATION, but are more prone to INTERFERENCE. *See also* MODULATION, SINGLE SIDEBAND.

analogue (*adj.*) In electronics, describing an electronic signal (voltage or current) where each value of the signal is used to represent a value of some continuously varying quantity. A microphone, for example, will produce a voltage that varies continuously in accordance with the changing pressure of the air. Analogue signals are extremely important in the interfacing of electronic systems to the outside world; that is, in converting information to or from electronic form. This interfacing is done using various TRANSDUCERS. *See also* DIGITAL.

analogue-to-digital converter, *A to D converter* A device used to convert ANALOGUE electronic signals to DIGITAL form for storage or processing. The analogue signal is usually produced by some form of TRANSDUCER, such as a THERMISTOR for temperature, or the amplified output of a microphone for digital storage or transmission of speech or music.

In one form of converter, a series of POTENTIAL DIVIDERS produce fixed reference voltages, which are compared with the input signal in a series of COMPARATORS. The fixed voltage that most closely matches the input voltage is the digital representation of the input signal. Such converters are fast, but require complex circuitry if conversion is to be accurate, with the digital output being many BITS long. A slower but more precise device comprises a single, constantly changing voltage source, from a RAMP GENERATOR, which is again fed into a comparator with the signal being converted. As the ramp generator is started, a COUNTER runs until the output of the comparator changes; the value stored in the counter at this point is the digital representation of the analogue signal.

Recent progress in digital broadcasting and digital music storage systems, such as COMPACT DISCS, has required the development of fast, accurate analogue-to-digital converters for the recording stage.

analysis 1. (*maths*) The branch of mathematics that deals with functions of continuously variable quantities, particularly the behaviour of such functions in the limit of vanishingly small changes, and differential CALCULUS.

2. (*science*) Any process that is used to determine the constituents of a material, particularly in terms of the elements it contains. Analysis can be divided into qualitative, which simply determines which elements are present, and quantitative, which provides a measurement of the amount of each material present. SPECTROSCOPY is an important tool in qualitative analysis, with the sample being vaporized in a flame and the ABSORPTION SPECTRUM or EMISSION SPECTRUM then being examined for features characteristic of the presence of a particular element. *See also* ABSORPTION SPECTRUM, EMISSION SPECTRUM.

analytical (*adj.*) In mathematics, describing a procedure or result that follows the rules of ANALYSIS, and so is a precise solution in terms of algebraic quantities, as opposed to a numerical solution. *See also* GEOMETRY.

AND gate A LOGIC GATE that has a HIGH output only if all its inputs are high.

anechoic (*adj.*) Describing a room or a material that completely absorbs all sound falling on it. Anechoic chambers are used to test the properties of TRANSDUCERS such as microphones and loudspeakers.

anemometer A device for measuring wind speed. An anemometer normally consists of three cups mounted horizontally, which rotate at a rate proportional to the wind speed. This rotation drives a pointer across a scale calibrated in speed units.

aneroid Any container from which all the air has been removed, particularly the pressure sensing element in an ANEROID BAROMETER. In a barometer, an aneroid usually takes the form of a flat cylinder formed from two circular sheets or metal. Ridges provide sufficient rigidity to prevent the structure from collapsing completely under the influence of atmospheric pressure, whilst the aneroid remains flexible enough for changes in ATMOSPHERIC PRESSURE to be recorded as changes in its shape.

aneroid barometer A type of BAROMETER consisting of a sealed metal vessel, called an ANEROID, from which all the air is removed. The vessel has a thin metal lid, which is supported by a spring. Changes in the ATMOSPHERIC PRESSURE cause the lid to move against the spring by varying amounts. This movement is transmitted by the spring to a pointer on a calibrated scale. *See also* BAROGRAPH.

angle A measure of rotation, or a measure of the space between two intersecting lines or planes. Angles are measured in DEGREES or RADIANS. One rotation through a full circle is equal to an angle of 360 degrees (360°) or 2π radians.

angle of attack The angle between an AEROFOIL and the flow of air through which it is moving. The COEFFICIENT OF LIFT of an aerofoil increases with angle of attack for angles up to about 14°, but decreases rapidly at larger angles. *See also* STALL.

angle of depression The angle below the horizontal at which some point appears.

angle of elevation The angle above the horizontal at which some point appears.

angle of friction The angle between the NORMAL and the LINE OF ACTION of the overall force of contact between two surfaces that are sliding over one another. The TANGENT of the angle of friction is equal to the COEFFICIENT OF FRICTION. If θ is the angle of friction and μ is the coefficient of friction, then

$$\mu = \tan\theta$$

angle of incidence The angle at which a ray strikes a surface, measured from the NORMAL to the surface.

angle of reflection The angle at which a ray leaves a surface, having been reflected. The angle of reflection is measured from the NORMAL to the surface. *See* REFLECTION.

angle of refraction The angle, measured from the NORMAL, at which a ray leaves the boundary between two transparent materials, having been refracted (*see* REFRACTION).

angstrom (Å) A non-SI UNIT of length, equal to 10^{-10} m. It is sometimes used to specify WAVE-LENGTHS and intermolecular distances, but has largely been superseded by the nanometre (nm), which is 10^{-9} m.

angular acceleration The rate of change of ANGULAR VELOCITY with time. See ROTATIONAL DYNAMICS.

angular frequency In SIMPLE HARMONIC MOTION, 2π times the frequency measured in HERTZ.

angular magnification See MAGNIFICATION.

angular momentum (L) The ANGULAR VELOCITY of an object multiplied by its MOMENT OF INER-TIA. For a point of mass m moving at a speed v at a distance r from the axis, with the motion at right angles to the line joining the object and the axis

$$L = mvr = m\omega r^2$$

where ω is the angular velocity and L is the angular momentum. See also LAW OF CONSER-VATION OF ANGULAR MOMENTUM, ROTATIONAL DYNAMICS.

angular velocity (ω) The rate of change of the angular position of an object with time. See ROTATIONAL DYNAMICS.

anion A negatively charged ion. So called because it will be attracted to the ANODE in ELECTROLYSIS. Compare CATION.

anisotropic (adj.) Describing a medium, usually a crystalline solid, in which certain physical properties, such as electrical and THERMAL CONDUCTIVITY, are different in different directions. GRAPHITE (a crystalline form of carbon) is an example: electric conduction can take place relatively easily along the planes of carbon atoms, but with much more difficulty across the planes.

annealing A process in which the metal is heated and then allowed to cool slowly. The result is that DISLOCATIONS are formed, under the influence of thermal vibrations. Some of these dislocations disappear as the material cools, but sufficient remain for the material to be soft and easily worked. If the material then undergoes sufficient PLASTIC deforma-tions, it may become hard and brittle (WORK HARDENING), as the dislocations become tangled with one another or run up against the edges of the individual crystals in the POLYCRYSTALLINE metal.

If the material is again annealed, the effects of work hardening are reversed and the material again becomes soft and easily worked. See also QUENCHING.

annihilation The complete destruction of mat-ter, such as takes place when a particle and its antiparticle collide. The energy generated in such a collision is either carried away in the form of photons of GAMMA RADIATION, or as MESONS.

annulus A ring-shaped figure, the surface between two CONCENTRIC CIRCLES. If the circles have radii r_1 and r_2, the area A of the annulus is

$$A = \pi(r_1^2 - r_2^2)$$

anode A positively charged ELECTRODE.

anodize (vb.) To apply a protective coating of aluminium oxide to a piece of aluminium, by making it the ANODE in an ELECTROLYSIS process. A solution of sulphuric acid or chromic acid is normally used as the ELEC-TROLYTE and the oxygen released at the anode reacts with the aluminium to produce a thin oxide layer.

antenna See AERIAL.

anthropic principle Any of a series of state-ments which make deductions about the nature of the Universe with the basic premise that the Universe contains at least one intelli-gent life form that is able to speculate about the nature of the Universe. The weak anthropic principle states simply that the exis-tence of intelligent life places certain con-straints on the Universe; that we must be observing it at a stage when it is several billion years old (for example) since that is the time life takes to evolve. The strong anthropic prin-ciple is more controversial – this states that the Universe was set up with the purpose of creat-ing intelligent life, and is often put forward as an argument for the existence of an intelligent creator. The final anthropic principal takes this one stage further, and states that since the pur-pose of the universe is to support intelligent life, and once intelligent life has emerged in the Universe it must continue to exist in some form.

antibonding orbital The higher energy of the two MOLECULAR ORBITALS formed when two atomic ORBITALS overlap. This orbital tends to push atoms apart, preventing closer bond-ing. The high energy of this orbital means that atoms only form COVALENT BONDS if this orbital is not filled. See also BONDING ORBITAL.

anticyclone A region of high ATMOSPHERIC PRESSURE. Anticyclones are generally slow moving and correspond to a period of settled weather. Winds radiate from a calm centre in a clockwise direction in the northern hemisphere, and an anticlockwise direction in the southern hemisphere.

antiferromagnetic (adj.) Desciding a CRYSTAL, or the arrangements of atoms within a crystal, where the individual atoms behave like magnets but are arranged with adjacent atoms aligned in opposite directions, so that there is no overall magnetic effect. *Compare* FERROMAGNETIC.

antimatter Matter made up of ANTIPARTICLES. Antimatter is not found in nature, as contact with an equivalent amount of matter would result in the complete destruction of both. *See also* ANNIHILATION.

antinode A region in a STANDING WAVE where the oscillations have a maximum amplitude. The distance from one antinode to the next is half the wavelength of the wave. *See also* NODE.

antiparticle A particle with the same mass but an opposite QUANTUM NUMBER as another particle. Every type of particle has an equivalent antiparticle with an opposite charge, or MAGNETIC MOMENT, or SPIN, etc. Thus antiprotons are negatively charged versions of the proton. The antiparticle of the electron is called the POSITRON. *See also* ELEMENTARY PARTICLE.

aperture A gap in an otherwise solid or OPAQUE object; in particular one for the admission of light into a camera or other optical instrument. *See also* DIAPHRAGM.

aphelion For an object in orbit around the Sun, the point in the orbit at which it is furthest from the Sun.

apogee For an object in orbit around the Earth, the point in the orbit at which it is furthest from the Earth.

apparent depth The depth that a transparent medium, such as water, appears to have when viewed from above. The apparent depth is less than the true depth by a factor of the REFRACTIVE INDEX of the material concerned. The effect is caused by light that has passed through the transparent material being refracted on entering the air between this material and the eye. *See also* REFRACTION.

apparent luminosity The amount of light energy received from a star per second per square metre. The apparent luminosity depends on the distance of the star from the Earth and on the true LUMINOSITY of the star.

approximation Any result that is only roughly correct, having been obtained either on the basis of approximate measurements, or on the basis of a calculation that gives only approximate results. Such calculations are often far simpler than completely correct solutions, which in any case are often not available.

arc, electric *See* ELECTRIC ARC.

Archimedes' principle The UPTHRUST on any object immersed (partially or totally) in a fluid is equal to the weight of fluid displaced. *See also* BUOYANCY, FLOTATION.

area The measure of the size of a surface. For a rectangular surface, the area is defined as the product of the lengths of the two sides of the surface. Areas of other surfaces are described in terms of the total amount of rectangular area into which they can be divided.

argument Any number used as the starting point for the calculation of some function. For example, in the equation $\sin 30° = \frac{1}{2}$, 30° is the argument of the SINE function.

arithmetic logic unit (ALU) The part of a MICROPROCESSOR or computer that performs actual manipulations on numbers fed into it from some form of memory, with the output also being stored in memory.

arithmetic mean *See* MEAN.

arithmetic progression A SERIES of numbers in which each number is greater than the previous number in the series by a fixed amount. Thus if the nth member of the series is a_n, the $(n+1)$th member will be

$$a_{n+1} = a_n + k$$

where k is a constant called the common difference. For an arithmetic progression of N terms, the sum of the series will be

$$Na_1/2 + \frac{1}{2}N(N-1)k$$

armature The moving part of a MOTOR, DYNAMO or other electromagnetic system.

arrow of time A concept arising from the idea of irreversibility in the SECOND LAW OF THERMODYNAMICS. Although the individual interactions between molecules could equally well happen backwards, the overall direction of the universe is from order to chaos. The fact that this feature is present only on the macroscopic scale (when large numbers of interacting particles are considered) and not on the microscopic scale is a key area of philosophical interest.

arteriogram A RADIOGRAPH of the arteries, made with an injection of a CONTRAST ENHANCING MEDIUM.

associated production Any process in which a QUARK-antiquark pair of one of the heavier quarks (for example, the STRANGE quark) are produced, with the two members of the pair ending up in different HADRONS.

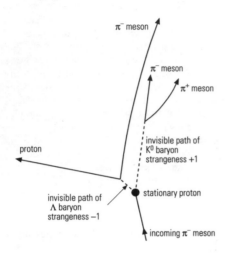

Associated production as seen in a bubble chamber.

astable A digital electronic device that changes its output from HIGH to LOW and back at a rate determined either by the charging of a CAPACITOR or by a resonant device, such as a PIEZOELECTRIC quartz crystal.

asteroid, *minor planet* Any one of a large number of small rocky bodies, less than 1,000 km in diameter, that orbit the Sun. Many of the smaller asteroids are irregularly shaped. Most asteroids orbit in a region of the Solar System called the asteroid belt, which lies between the orbits of Mars and Jupiter. However, a number of asteroids have been found in other orbits, including some that cross the orbit of the Earth. It has been proposed that collisions of such asteroids with the Earth may have caused major climatic changes in the past, leading to the sudden extinction of many species, particularly the dinosaurs.

asteroid belt The region between the orbits of Mars and Jupiter where most ASTEROIDS are found.

asthenosphere A layer of the Earth's MANTLE that extends from about 70 km to 260 km below the surface. It is a soft, probably partially molten, layer on which TECTONIC PLATES move. SEISMIC WAVES slow down in this zone.

astigmatism An ABERRATION in mirrors and lenses (including the CORNEA of the eye) that arises when the surface is more strongly curved in one direction than another. Parallel rays of light are not brought to a focus at the same point, leading to DISTORTIONS in the image. Astigmatism in the eye can be corrected with spectacles that have a cylindrically curved surface.

astrometric binary *See* BINARY STAR.

astrometry The branch of astronomy concerned with the measurement of the position of astronomical objects.

astronomical unit (AU) A unit of distance used in astronomy, the mean distance between the Earth and the Sun. One AU is equivalent to 1.50×10^8 km.

astronomy The study of celestial bodies, their positions, motions, nature and evolution. The main branches of astronomy are ASTROMETRY, CELESTIAL MECHANICS and ASTROPHYSICS and COSMOLOGY.

Historically, astronomy began with the study of the motions of the PLANETS. The invention of the TELESCOPE enabled Galileo (1564–1642) to observe SATELLITES in ORBIT around JUPITER and the PHASES of VENUS. These observations lead to the rejection of the PTOLEMAIC MODEL of the Universe, which placed the Earth at the centre, with the Sun and all the planets in orbits around it. This was replaced by the COPERNICAN MODEL, in which the Sun was at the centre. Further refinements led Johannes Kepler (1571–1630) to propose his three laws of planetary motion (*see* KEPLER'S LAWS), which were subsequently explained by Isaac Newton's (1642–1727) theory of GRAVITY. Applications of the theory of gravity subsequently led to the discovery of the previously unknown planet NEPTUNE and an understanding of the motion of COMETS, which had previously been regarded as atmospheric effects.

In the twentieth century, an understanding of NUCLEAR FUSION led to a model for the life cycle of STARS, and the discovery of the red shift of GALAXIES to a cosmological theory based around the BIG BANG model of the Universe.

astrophysics The branch of ASTRONOMY concerned with the study of the physical and chemical properties of astronomical objects and phenomena. Astrophysics deals with the structure and evolution of STARS, CLUSTERS and GALAXIES, and the properties of interstellar matter. Astrophysicists generally study the composition of celestial objects by analysing the ELECTROMAGNETIC RADIATION (such as light, radio waves and X-rays) they emit, although NEUTRINOS from the Sun and SUPERNOVAE have been observed. *See also* COSMOLOGY, SPECTROSCOPY.

asymptote The line towards which a curve approaches more and more closely, but never reaches. For example, the graph of the equation $y = 1/x$ is asymptotic to $y = 0$ as x tends to infinity.

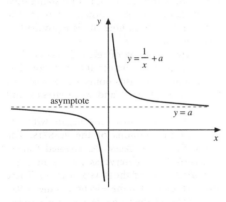

Asymptote.

atmosphere A layer of gas surrounding the surface of the Earth and some other planets. The Earth's atmosphere has a thickness of approximately 400 km but its density decreases with height.

The lowest layer of the atmosphere, about 10 km thick, is called the TROPOSPHERE. All CLOUDS and WEATHER SYSTEMS occur within this region. All the movement of the atmosphere is driven by the energy absorbed

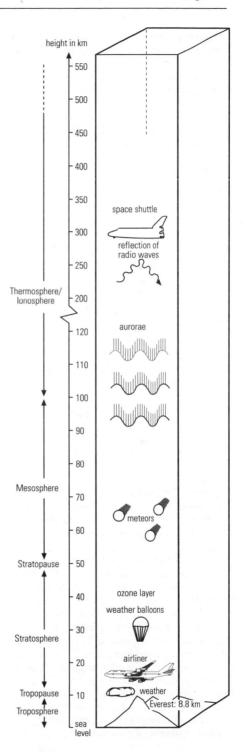

The structure of the Earth's atmosphere.

from solar radiation. The Earth also radiates energy into space, and clouds in the troposphere play a major part in the energy balance, reflecting both incoming and outgoing radiation. By the top of the troposphere, called the TROPOPAUSE, the ATMOSPHERIC PRESSURE has fallen to about 1 per cent of its surface value and the temperature is typically –50°C.

The next layer, from 25 km to 100 km, is called the STRATOSPHERE. Within this layer the temperature rises, heated by energy absorbed from ultraviolet light emitted by the Sun. At this temperature 'normal' oxygen, O_2, is converted into ozone, O_3, and the OZONE LAYER is thus formed. Recently, fears have been expressed over the depletion of this layer, the so-called OZONE HOLES, which have been observed over Antarctica and parts of Europe. Man-made chemicals, particularly chlorofluorocarbons, which act as catalysts for the decomposition of ozone, are believed to be responsible. This has led to a reduction in the use of these materials in refrigerators and as the propellant in aerosol sprays. It is feared that reductions in the concentration of ozone over populated areas may allow ultraviolet light to reach the surface, resulting in an increase in the incidence of skin cancers.

The top of the stratosphere is called the STRATOPAUSE. Here the temperature is about 0°C and the pressure only about one thousandth of the surface pressure. Above the stratopause, from 50 km to 100 km is the MESOSPHERE, where temperature falls with increasing height to about –100°C and the pressure falls to 10^{-5} times the surface pressure. Above this lies the THERMOSPHERE, the lower region of which is called the IONOSPHERE as the gas here is ionized by the absorption of X-rays from the Sun. The energy absorbed causes temperatures to rise, but the density and pressure of the material in this layer is very low.

See also CLIMATE, GREENHOUSE EFFECT.

atmosphere, *standard atmosphere* (atm.) A unit of pressure equivalent to 101,325 Pa. This is equal to the pressure that will support a column of mercury 760 mm high, at 0°C, at sea level and latitude 45°.

atmospheric pressure The pressure produced by gravity acting on the ATMOSPHERE. At the surface of the Earth this produces a pressure of about 1×10^5 Pa (1 ATMOSPHERE), though the exact value varies from day to day. Changes in atmospheric pressure are responsible for WEATHER SYSTEMS. *See also* BAROMETER.

A to D converter *See* ANALOGUE-TO-DIGITAL CONVERTER.

atom The smallest particle into which an element can be divided without losing its chemical identity. Atoms were originally believed to be indivisible objects, but it is now known that they comprise a small dense positive NUCLEUS of PROTONS and NEUTRONS, surrounded by ELECTRONS.

Almost all the mass of an atom is contained in its nucleus: protons and neutrons have similar masses, while electrons are about 1/1,836 the mass of protons. In a neutral atom, the number of electrons is equal to the number of protons in the nucleus (called the ATOMIC NUMBER). The number and arrangement of the electrons in an atom is what gives each element its distinct chemical properties.

In a simple model of the atom, electrons are regarded as orbiting the nucleus in SHELLS. However, a more sophisticated treatment recognizes the influence of QUANTUM MECHANICS on the atom, and places the electrons in ORBITALS, each having a defined energy. The sequence in which these orbitals are filled, and the energies involved are crucial in explaining the chemical properties of each element.

Chemistry generally concerns itself with the reactions of neutral atoms or of IONS that contain a few electrons more or less than the number needed to make the atom neutral.

See also ATOMIC THEORY, BOHR THEORY, LIQUID DROP MODEL, MOLECULE, QUANTUM THEORY, RUTHERFORD–BOHR ATOM, RUTHERFORD SCATTERING EXPERIMENT, RYDBERG EQUATION, SHELL MODEL, WAVE NATURE OF PARTICLES.

atomic absorption spectroscopy An ANALYSIS technique based on the ABSORPTION SPECTRUM formed when white light is shone through a sample in the form of a vapour. The wavelengths of light absorbed are characteristic of the elements present.

atomic clock A clock that measures time by some periodic process occurring in atoms or molecules, such as atomic vibrations or the frequency of emission or absorption of electromagnetic radiation. The caesium clock is based on the very precise frequency of radiation produced or absorbed by transitions between ENERGY LEVELS in caesium atoms.

This frequency is used to define the second, and so caesium clocks are used in international timekeeping. Experiments with atomic clocks have been used to confirm the predictions of TIME DILATION in the SPECIAL THEORY OF RELATIVITY.

atomic emission spectroscopy An ANALYSIS technique in which a sample is IONIZED using an ELECTRIC ARC or flame and the wavelengths in the EMISSION SPECTRUM so produced are measured. These wavelengths are characteristic of the elements present.

atomic force microscope A microscope that produces an image using a diamond-tipped probe, which is moved over the surface of a sample and responds to the interatomic forces between the probe and the sample. The probe in effect 'feels' its way over the contours of the surface, and its up-and-down movements are transmitted to a computer that produces a profile of the sample. The atomic force microscope can resolve (*see* RESOLUTION) single molecules, and is useful for biological specimens as the sample does not have to be electrically conducting.

atomic mass unit (amu) The unit in which nuclear masses are usually measured. The mass of one atom of carbon–12 (the ISOTOPE of carbon with MASS NUMBER 12) is defined to be 12 amu exactly. Nuclear masses can be measured using a MASS SPECTROMETER.

atomic number The number of PROTONS in an atomic NUCLEUS.

atomic theory The idea that all materials are made up of small particles called ATOMS. The motion of these particles leads to KINETIC THEORY in physics, whilst the way in which they combine to form MOLECULES is the foundation of chemistry.

atomic volume The RELATIVE ATOMIC MASS of an element divided by its volume, usually expressed in $cm^3 mol^{-1}$.

atomic weapon Any weapon that derives its energy from NUCLEAR FISSION. *See also* NUCLEAR FALLOUT.

atomic weight *See* RELATIVE ATOMIC MASS.

attenuation The factor by which the INTENSITY of a wave or the size of an electronic signal is reduced when passed through some device. In electronics, attenuation is often measured in DECIBELS.

attenuator A device designed to produce a certain ATTENUATION, especially of an electronic signal.

aurora A luminous glow in the atmosphere caused by the IONIZATION of gases by high-energy charged particles in the SOLAR WIND. The MAGNETIC FIELD OF THE EARTH causes these particles to enter the atmosphere only in regions close to the magnetic poles. An aurora in the northern hemisphere is called aurora borealis, or northern lights. An aurora in the southern hemisphere is called aurora australis.

auxiliary circle A mathematical construction for the study of SIMPLE HARMONIC MOTION. In this construction, a point is considered to move around a circle of radius A at a constant ANGULAR VELOCITY ω. The mathematical projection of this point onto the x-axis then moves with simple harmonic motion,

$$x = A\cos \omega t$$

avalanche breakdown The mechanism by which a gas conducts electricity. An avalanche breakdown results when a few stray ions of a gas in an electric field are accelerated so violently by the field that they collide with other molecules with enough energy to break them apart. This produces further IONIZATION and leads to a rapid increase in the number of ions present and a decrease in the resistance of the gas.

At low pressures, the molecules in a gas are more widely spaced, and a GAS DISCHARGE, as the phenomenon is then called, can be produced relatively easily. Each ion has space in which to accelerate before hitting another atom, so weaker electric fields are required. At higher pressures, the result is a spark once the electric field reaches the required value. In each case light is given off from the energy released as ions of opposite charges recombine and release energy.

average A term used in STATISTICS to indicate the typical member of a set of data. It usually refers to the arithmetic MEAN, which is calculated by adding together a group of numbers and dividing the total sum by the number of samples. A mean value is often expressed plus or minus the STANDARD DEVIATION. The term average is also used to refer to the MEDIAN value, which is the middle number in a set of numbers arranged in increasing or decreasing order, or the MODE, which is the most frequently occurring number in a group.

Avogadro constant, *Avogadro number* (L or N_A) The number of atoms in one MOLE of atoms, molecules in one mole of molecules, etc. It is equal to 6.022×10^{23}. The mass of this number of carbon–12 atoms is 12 g.

Avogadro number *See* AVOGADRO CONSTANT.

Avogadro's hypothesis A given number of molecules of any gas at a given temperature and pressure will occupy the same volume, regardless of the nature of the gas. In particular, one MOLE of any gas occupies a volume of 22.4 dm^3 under conditions of STANDARD TEMPERATURE AND PRESSURE (atmospheric pressure and a temperature of 0°C). This result was originally based on EMPIRICAL observations but is now seen to be a consequence of KINETIC THEORY.

axiom A mathematical statement from which other statements may be deduced, but which itself cannot be proved but rather is accepted as self-evident.

axis (*pl. axes*) **1.** One of a set of lines from which CO-ORDINATES are measured in co-ordinate geometry. For example, in CARTESIAN CO-ORDINATES, the x-axis is the line along which the x co-ordinate increases.

2. The line about which an object rotates (an axis of rotation), or about which an object can be rotated, possibly only through certain angles, and appear unchanged (an axis of symmetry).

B

back e.m.f. A voltage that opposes the current flowing in a circuit. The back e.m.f. of a coil is produced as a result of a change in the magnetic field in that coil, caused by a change in the current flowing. *See* SELF-INDUCTANCE. *See also* ELECTROMOTIVE FORCE, LENZ'S LAW.

background radiation The collective name for the many sources of IONIZING RADIATION. The most important of these are naturally occurring radioactive materials in rocks, soil and atmosphere. The other main source of background radiation is COSMIC RADIATION – high-energy charged particles, mostly protons, that enter the atmosphere from space.

Compared to these two sources of radiation, the radiation present from NUCLEAR REACTORS and NUCLEAR WEAPON tests represents only 1 or 2 per cent of the total exposure to ionizing radiation for the average human. In addition, individuals often experience significant doses of ionizing radiation from medical sources, mainly X-RAYS. The level of medical exposure can vary widely from one individual to another, though in the West it typically accounts for about 13 per cent of the lifetime dose. *See also* RADIOACTIVITY.

balance A device for weighing. A beam balance consists of a lever balanced on a central pivot, with known and unknown masses being suspended on opposite sides of the pivot, usually at equal distances. The two masses are compared by using the pull of gravity on the masses to produce turning MOMENTS about the pivot: when the masses are equal, the beam is exactly horizontal. A top-pan balance does not compare one mass with another, but uses the pull of gravity on an unknown mass to deform a STRAIN GAUGE, with the mass being presented directly as a digital reading. *See also* CURRENT BALANCE.

ball bearing *See* BEARING.

ballistic (*adj.*) Describing an object that is moving solely under the force of gravity, especially one moving at high speed. An example is a spacecraft when its rocket motors are produc-ing no thrust, or a military aircraft flown in such a way that its wings are producing very little lift.

ballistics The study of PROJECTILES. In particular ballistics is the study of bullets or shells fired from guns, but also of any object that is set moving by a sudden IMPULSE and then moves under the influence of forces, such as gravity and the force of resistance exerted by the air.

balloon A roughly spherical envelope made of flexible material and containing a gas, usually of lower density than the surrounding air. Balloons fly because their weight is less than the UPTHRUST produced by the air they displace. The balloon may be filled with a low density gas, such as hydrogen or helium, or it may be filled with air, which is then heated to reduce its density.

Helium balloons are used to carry meteorological packages to high altitude. Hydrogen is sometimes also used, but poses problems because it is inflammable and also tends to diffuse rapidly through the material of the envelope. Hot air balloons are used for recreational purposes and usually carry propane gas which is burnt to heat the air inside an envelope that is open at the base. *See also* ARCHIMEDES' PRINCIPLE.

Balmer series A series of lines in the HYDROGEN SPECTRUM. The wavelengths are mostly in the visible part of the ELECTROMAGNETIC SPECTRUM, though the series extends slightly into the ultraviolet. Each line corresponds to a transition between the second ENERGY LEVEL and some higher level. *See also* BOHR THEORY.

band gap In the BAND THEORY of solids, the gap between one ENERGY BAND and the next, particularly between a full VALENCE BAND and an empty CONDUCTION BAND.

band theory The branch of QUANTUM MECHANICS that explains the properties of solids in terms of ENERGY LEVELS. The electrons in a single atom exist in discrete, sharply defined energy levels. When the atoms come together to form a solid, these sharply defined energy levels become bands of allowed energies. These bands may be filled with electrons, or may be partially full, or empty. Between the allowed bands are 'forbidden' bands.

The VALENCE ELECTRONS, those involved in chemical bonding, form the VALENCE BAND of a solid. The only electrons free to move through the solid, carrying heat and electricity, are those in partially full bands. The PAULI EXCLUSION PRINCIPLE forbids an electron from gaining energy unless there is an empty energy level for it to move to. If the valence band is full, the electrons must move to an unfilled band, called the CONDUCTION BAND. Conductors are those materials whose valence band and conduction band are unfilled, or whose conduction and valence bands overlap – in either case there are vacant energy levels.

In INSULATORS, the valence band and conduction band are separated by a wide forbidden band. Such materials do not conduct because the electrons do not have enough energy to cross from one band to another. In SEMICONDUCTORS, the energy difference between the valence band and the next band is sufficiently small that thermal vibrations may give electrons enough energy to enter the empty conduction band. The energy required is called the BAND GAP. If the band gap has an energy corresponding to a PHOTON of visible light, light will be absorbed or emitted as electrons cross the band gap. This effect is exploited in PHOTODIODES and LIGHT-EMITTING DIODES.

bandwidth The spread of frequencies needed to convey information in any TELECOMMUNICATIONS system, particularly those using radio waves. An important feature of any telecommunications system is the rate at which information is transmitted (for digital systems this is called the BAUD RATE). The faster the information is transmitted, the greater the bandwidth needed to transmit it and so the more widely spaced must be the chosen CARRIER WAVE frequencies.

Speech and music have a bandwidth of only a few kilohertz (*see* HERTZ), depending on how high a quality of reproduction is required. Television, on the other hand, requires a bandwidth of many megahertz if a complete picture is to be transmitted enough times per second for the impression of movement to be produced. *See also* SIDEBAND.

bar A unit of pressure in the C.G.S. SYSTEM. One bar is equal to 10^5 PASCAL, or approximately one ATMOSPHERE. The millibar (100 Pa) is used as the unit of pressure in METEOROLOGY.

barium meal A drink containing barium in the form of barium sulphate, given to a patient as a CONTRAST ENHANCING MEDIUM before taking a RADIOGRAPH of the stomach or intestines.

bar magnet A magnet in the shape of a bar or rod with a POLE at each end.

barn A unit of area used in particle physics to measure cross-sections of atomic nuclei in interactions. One barn is equivalent to 10^{-28} m^2.

barograph An instrument that gives a record of changes in ATMOSPHERIC PRESSURE over time. It is similar in construction to the ANEROID BAROMETER, but has an inked pointer that leaves a trace on a sheet of paper moving under it. Falling pressures indicate the arrival of a DEPRESSION, which usually means deteriorating weather, whilst rising pressures generally mean improving weather.

barometer Any instrument used to measure ATMOSPHERIC PRESSURE. *See* ANEROID BAROMETER, BAROGRAPH, FORTIN BAROMETER, MERCURY BAROMETER.

baryon Any HADRON with half integral SPIN. Baryons are composed of three QUARKS and carry a QUANTUM NUMBER called the BARYON NUMBER, which is believed to be conserved in all interactions. Baryons are either nucleons (PROTONS or NEUTRONS) or short-lived particles known as HYPERONS that include nucleons in their decay products. *See also* LAMBDA PARTICLE, OMEGA-MINUS PARTICLE, SIGMA PARTICLE.

baryon number A QUANTUM NUMBER carried by all BARYONS and the QUARKS from which they are made. Each quark has baryon number ⅓, whilst antiquarks all have baryon number −⅓. Thus a baryon (which comprises three quarks) has baryon number 1 and a MESON (which comprises a quark and an antiquark) has baryon number 0. It is thought that baryon number is exactly conserved in all interactions, but some GRAND UNIFIED THEORIES, predict a very slow rate of PROTON DECAY, leading to a violation of baryon number conservation.

base 1. (*mathematics*) The flat surface at the bottom of a geometrical figure, such as the flat surface of a cylinder or cone, or the rectangular surface of a pyramid.

2. (*mathematics*) A description of a number system, the base of the system being the number of different values that can be represented by a single digit. Thus in the decimal (base–10) system there are 10 number symbols, the digits 0 to 9. A binary (base–2) system has only 2 digits, for which the symbols 0 and 1 are normally used. Hexadecimal (base–16) numbers are widely used in computing, the digits are given the symbols 0 to 9 and A to F, with A to F corresponding to 11 to 15 in the decimal system.

3. (*electronics*) The central ELECTRODE in a JUNCTION TRANSISTOR.

base unit, *fundamental unit* Any unit whose size is fixed by reference to some experimental measurement, as opposed to a DERIVED UNIT. In the SI system, the base units are the KILOGRAM (mass), METRE (length), SECOND (time), AMPERE (current), KELVIN (temperature) and MOLE (amount of substance).

battery Several electrochemical CELLS connected together, usually in series to produce a larger voltage.

baud rate In a digital TELECOMMUNICATIONS system, the number of BITS sent per second. *See also* BANDWIDTH.

beam balance *See* BALANCE.

beam tube In a PARTICLE ACCELERATOR, the evacuated tube through which the particle beam passes.

bearing 1. A direction measured as an angle clockwise from some reference direction, usually north. Thus a bearing of 150° represents an angle of 150° around from north, which can be written as S 30° E; that is, 30° east of south. Bearings are often expressed as three digit numbers, such as 090° for East.

2. (*mechanics*) A point at which a LOAD is supported, particularly if the load is able to rotate and steps have been taken to reduce friction by supporting the load on balls (a ball bearing) or rollers (a roller bearing).

beats An INTERFERENCE effect, often observed in sound, that occurs when two waves of slightly different frequency interfere. The two waves drift in and out of PHASE producing a periodic increase and decrease in loudness.

beauty In particle physics, *see* BOTTOM.

becquerel (Bq) The SI UNIT of radioactive ACTIVITY, one becquerel being an activity of one ionizing particle per second.

Bernoulli effect The fall in pressure in a fluid as it accelerates. In an aircraft wing, for example, the curved upper surface forces air to travel more quickly than over the flat under surface. This acceleration results in a fall in pressure and the aircraft flies because the upward pressure on the underside of the wing is greater than the downward pressure on the upper surface. *See also* BERNOULLI'S THEOREM.

Bernoulli's theorem The total energy per unit volume in a fluid remains constant, so changes in pressure (POTENTIAL ENERGY) and speed (KINETIC ENERGY) compensate for one another. If the density of the fluid is ρ and it moves at speed v, with pressure p, then

$$\tfrac{1}{2}\rho v^2 + p = \text{constant}$$

See also BERNOULLI EFFECT.

beta decay The emission of BETA PARTICLES, which may be either positive or negative, from an unstable atomic nucleus. Negative beta particles are fast-moving electrons, produced by atomic nuclei with too many neutrons; a neutron turns into a proton, and an electron is emitted together with an antineutrino (*see* NEUTRINO). Positive beta particles are POSITRONS, produced when a nucleus with too few neutrons converts a proton into a neutron emitting a positron and a neutrino.

In beta decay, the MASS NUMBER is unchanged, but the ATOMIC NUMBER is increased by 1 in negative beta decay and decreased by 1 in positive beta decay.

beta particle An electron or POSITRON produced in an atomic nucleus by the process of BETA DECAY. *See also* BETA RADIATION.

beta radiation A stream of BETA PARTICLES, emitted from unstable nuclei by the process of BETA DECAY. Beta radiation is less ionizing (*see* IONIZATION) than ALPHA RADIATION, but more ionizing than GAMMA RADIATION. Beta particles have a range of several metres in air but can be stopped by a layer of aluminium a few millimetres thick.

bifocal (*adj.*) Describing a lens that is effectively two lenses with different powers made from a single piece of glass.

big bang The currently accepted model of the early Universe. According to the big bang theory, the Universe was initially vastly hotter and denser than it is today. In the very early stages the temperature was high enough to create a 'soup' of QUARKS and antiquarks, electrons and POSITRONS and NEUTRINOS. As the Universe expanded and cooled, the quarks began to fuse into protons and neutrons. Some of these then fused into simple nuclei: first into deuterium nuclei (1 proton, 1 neutron) and then into helium nuclei (2 protons, 2 neutrons). Helium now makes up about 25 per cent of the matter in the Universe, the rest being mostly hydrogen. This matter was initially in the form of a PLASMA, but eventually the Universe was cool enough for electrons and nuclei to combine into atoms faster than these atoms were ionized by collisions. Once the Universe contained mostly neutral atoms ELECTROMAGNETIC RADIATION was able to pass through it without ABSORPTION. At this stage the Universe became transparent. The radiation from this time is still visible today, stretched to longer wavelengths by the expansion of the Universe, as the COSMIC MICROWAVE BACKGROUND. Much later, the primordial hydrogen and helium began to clump into regions that eventually became the GALAXIES which we see today.

One problem with theories of the early Universe is that little is known about how matter behaves under such extreme conditions. Theories also tend to have problems with predicting the correct degree of 'lumpiness'. On a small scale, stars are dense objects separated by large regions of space, and galaxies form CLUSTERS that appear to lie along string-like paths enclosing voids (empty regions of space). On a larger scale, the distribution of matter in the Universe seems very uniform, and there are only very small variations in the intensity of the cosmic microwave background. One solution to these problems is to suggest that the very early Universe underwent a period of INFLATION – very rapid growth in which different regions of space became separated from one another. *See also* BIG CRUNCH, COSMOLOGY, HUBBLE'S LAW.

big crunch The hypothetical end of the Universe if its density exceeds the CRITICAL DENSITY. *See* COSMOLOGY. *See also* BIG BANG.

bimetallic strip A device made from strips of two metals of different EXPANSIVITIES securely fastened together. As the strip is heated or cooled, the two metals will expand or contract by different amounts, causing the strip to bend. A bimetallic strip can be used as a temperature controlled switch, or THERMOSTAT, in which the bending of the strip makes or breaks an electric circuit.

binary (*adj.*) Describing a system that works in BASE–2; that is, all numbers contain only the digits 1 and 0. *See also* DECIMAL.

binary star A system in which two (or sometimes more) stars orbit around one another. Binary stars that are directly visible as two stars are called physical binaries to distinguish them from optical binaries, which are not linked but appear close together due to line-of-sight effects. Some binaries are observed as spectroscopic binaries, where the two components are inferred from the SPECTRUM received from what appears to be a single star. For astrometric binaries, the existence of a faint companion star is deduced from its influence on the motion of a brighter star. Eclipsing binaries are stars whose light output falls regularly as one member of the pair passes in front of (eclipses) its partner.

Binary stars are important in ASTROPHYSICS because their observation enables the masses of the stars in the system to be measured using Kepler's third law (*see* KEPLER'S LAWS).

binding energy *See* NUCLEAR BINDING ENERGY.

binoculars Low-powered magnifying devices for viewing distant objects. They consist of two small telescopes mounted side by side (one for each eye). Inside each telescope is a pair of PRISMS which reverse the direction of travel of the light beam, so it passes along the length of the binoculars three times, producing an upright image and enabling the binoculars to be more compact.

birefringence, *double refraction* The property of some crystals, such as the mineral calcite, by which two refracted rays of light are formed from a single unpolarized ray (*see* POLARIZATION). The two refracted rays are polarized at right angles to each other: one (the ordinary ray) follows the normal laws of REFRACTION; the other (the extraordinary ray) follows different laws. *See also* PHOTOELASTICITY.

bistable A LOGIC GATE with two outputs, one of which is HIGH and the other LOW. It also has two inputs, usually called set and reset. Making one of these inputs low puts the bistable into one of its two stable states, fixing one output as high and the other as low. When the inputs are both high, the bistable does not change states but remains in the state to which it was set. This is the basis of electronic MEMORY, which is composed of one bistable for each BIT which is stored. These memory bistables are set into one of their two stable states to store a binary 1 or 0, and this state can be observed at a later time to recall the data stored in the memory. *See also* CLOCKED BISTABLE, SHIFT REGISTER.

bit A BINARY digit, a 1 or a 0, or a signal representing a binary digit.

black body A hypothetical object that absorbs and emits all wavelengths of ELECTROMAGNETIC RADIATION perfectly. *See also* BLACK-BODY RADIATION.

black-body radiation The ELECTROMAGNETIC RADIATION emitted by a BLACK BODY at any temperature above ABSOLUTE ZERO. It extends over the entire range of wavelengths. A graph of the distribution of energies has a characteristic shape, with a maximum at a wavelength dependent on the temperature. Black body radiation is used as an approximation for the radiation given off by any hot object. *See also* STEFAN'S LAW, WIEN'S LAW.

black dwarf The final state of any STAR once it has consumed all its nuclear fuel and cooled down. Black dwarves are formed by the cooling of WHITE DWARVES and NEUTRON STARS. The nature of the material in a black dwarf depends on the mass of the original MAIN SEQUENCE STAR, as the higher pressures in more massive stars mean that the NUCLEAR FUSION processes can continue further. Because of the small surface area and large HEAT CAPACITY of white dwarves and neutron stars, black dwarves form relatively slowly, and it is believed that very few stars have reached this stage at the current point in the evolution of the Universe. However, all but the most massive stars, which form BLACK HOLES, will end their lives as black dwarves if the Universe is not first destroyed in a BIG CRUNCH.

black hole An astronomical object so dense that not even light can escape from its GRAVITATIONAL FIELD. Stars with cores of mass greater than about 2.5 SOLAR MASSES are believed to collapse without limit at the end of their lives, forming black holes. The density of matter at the centre of a black hole is believed to increase without limit, forming a SINGULARITY in SPACE-TIME. Any material which falls into a black hole will be unable to escape once it crosses a surface, called the event horizon, inside which the ESCAPE VELOCITY is greater than the speed of light. In practice, any object falling towards a black hole will be torn apart by TIDAL FORCES long before it reaches the event horizon.

Whilst black holes were originally put forward as hypothetical consequences of the GENERAL THEORY OF RELATIVITY, there is increasing experimental evidence for their existence in the form of intense X-RAY sources and jets of excited matter that have been observed near the centre of several GALAXIES. This radiation is believed to be formed as a result of intense heating of matter as it falls into the black hole. *See also* HAWKING RADIATION, SCHWARZSCHILD RADIUS.

blind spot The point in an eye where the OPTIC NERVE leaves the RETINA. At the blind spot, there are no RODS or CONES, so no visual images are transmitted.

blue-shift *See* DOPPLER EFFECT.

bob The mass in a SIMPLE PENDULUM.

body centred cubic A crystalline structure in which the UNIT CELL is a cube with an atom at the centre surrounded by one eighth of an atom at each corner. This structure has a CO-ORDINATION NUMBER of 8.

All the alkali metals (elements in Group I of the PERIODIC TABLE) form crystals with this structure, which accounts for their low density compared to other metals of comparable atomic mass. The structure is also found in IONIC compounds where the two ions have equal charges and similar radii. If the ions differ much in radius, such as in sodium chloride, a different cubic structure is favoured with a co-ordination number of 6.

Bohr atom *See* BOHR THEORY.

Bohr magneton A unit of MAGNETIC MOMENT equal to 9.27×10^{-24} Am². *See* MAGNETON.

Bohr theory, *Bohr atom* A simple model of ATOMS put forward in 1913 by Niels Bohr (1885–1962) to explain the LINE SPECTRUM of hydrogen. It has three basic postulates: (i) electrons revolve around the nucleus in fixed orbits without the emission of ELECTROMAGNETIC RADIATION; (ii) electrons can only occupy orbits in which the ANGULAR MOMENTUM of the electron is a whole number times $h/2\pi$, where h is PLANCK'S CONSTANT – in other words, within each orbit an electron has a fixed amount of energy; and (iii) when an electron jumps from one orbit to another, energy is emitted or absorbed as a PHOTON of electromagnetic radiation.

Whilst this model explains the broad features of the HYDROGEN SPECTRUM, it does not account for the fine detail and has been superseded by more complete quantum mechanical descriptions. It was, however, an important early step in the development of QUANTUM THEORY.

boil (*vb.*) To turn from a liquid to a vapour, with bubbles of gas being formed in the liquid.

boiling point The temperature at which a liquid boils, usually quoted at a pressure of one ATMOSPHERE. More technically, the boiling point of a liquid is defined as the one temperature for a given VAPOUR PRESSURE at which the liquid and its vapour can exist in equilibrium together. Boiling points increase with pressure, as it is harder for a bubble to form if the liquid is at high pressure. All materials expand on boiling and have boiling points that increase with increasing pressure. *See also* SATURATED VAPOUR PRESSURE.

bolometer A device for the detection and measurement of INFRARED radiation. It consists of a blackened strip of platinum metal, the resistance of which increases as it heats up in response to absorbed radiation. *See also* THERMOPILE.

Boltzmann constant (*k*) The constant in the IDEAL GAS EQUATION

$$pV = NkT$$

where N is the number of molecules, p the pressure, V the volume and T the ABSOLUTE TEMPERATURE. The Boltzmann constant is equal to $1.38 \times 10^{-23}\,\text{JK}^{-1}$.

Boltzmann factor The factor $e^{-E/kT}$ in the MAXWELL–BOLTZMANN DISTRIBUTION. It describes the number of particles with an energy greater than E when the ABSOLUTE TEMPERATURE is T, where k is the BOLTZMANN CONSTANT.

bonding orbital The lower energy of the two MOLECULAR ORBITALS formed when two atomic ORBITALS overlap. The bonding orbital is the structure that holds covalently bonded atoms together. For the bond to be stable, the bonding orbital must be full and the ANTIBONDING ORBITAL empty. Since each orbital can hold two electrons, this explains why COVALENT BONDS involve the sharing of pairs of electrons in an orbital.

Bose–Einstein condensation The accumulation of large numbers of BOSONS in the lowest available ENERGY LEVEL at low temperatures. The

effect accounts for SUPERFLUIDITY in liquid helium and SUPERCONDUCTIVITY. Superfluidity is only observed in liquid helium-4, where the atoms are bosons. At low temperatures, close to ABSOLUTE ZERO, large numbers of bosons occupy the lowest energy level, and behave as a single entity, called a superatom. Under the influence of a force, the particles all gain in momentum, and so flow without any of the VISCOSITY associated with intermolecular collisions. In some metals, electrons form pairs, called Cooper pairs, with equal and opposite momenta and opposing SPIN. These behave like bosons, and again can move without opposition at low temperatures. This flow of charge without resistance is observed as superconductivity. The effect only occurs with bosons, as FERMIONS are kept in higher energy levels even at low temperatures by the PAULI EXCLUSION PRINCIPLE.

boson Any particle with integral SPIN; that is, a spin which is an even number times $h/4\pi$, where h is PLANCK'S CONSTANT. *See also* BOSE–EINSTEIN CONDENSATION. *Compare* FERMION.

bottom, beauty A FLAVOUR of QUARK.

boundary layer A thin layer of fluid (liquid or gas) that moves with an object, such as an aircraft, as it passes through the fluid. In TURBULENT flow, the boundary layer becomes detached from the surface. By delaying the point at which the boundary layer becomes detached, it is possible to reduce the amount of turbulent flow and thus reduce DRAG.

Bourdon gauge A pressure measuring device in which a curved metal tube straightens out slightly under the influence of the applied pressure. A system of gears converts this small motion into the larger motion of a pointer across a scale.

Boyle's law For a fixed mass of an IDEAL GAS at constant temperature, the pressure is inversely proportional to the volume, i.e. the pressure multiplied by the volume is a constant. For a fixed mass of gas in a volume V at a pressure p, at a constant temperature,

$$pV = \text{constant}$$

See also CHARLES' LAW, GAS CONSTANT, IDEAL GAS EQUATION, PRESSURE LAW.

Bragg's law In X-RAY DIFFRACTION, the relationship between the angles at which CONSTRUCTIVE INTERFERENCE is observed (measured from the surface of a crystal) and the separation of the planes of atoms in the crystal. For a beam of X-rays of wavelength λ, striking a crystal at an angle θ to the planes of the crystal, which have a separation d between one plane and the next, constructive interference will occur if

$$2d \sin\theta = n\lambda$$

where n is a whole number.

brake horsepower (bhp) The output of a machine, especially an INTERNAL COMBUSTION ENGINE, measured in HORSEPOWER using a DYNAMOMETER.

brass An ALLOY of copper with around 20 per cent zinc. Harder than copper but still soft enough to be worked easily, brass is used for small metal parts but is too expensive and too weak to find widespread use.

bremsstrahlung X-RAYS produced by rapidly decelerating electrons, as in an X-RAY TUBE, for example.

Brewster angle The ANGLE OF INCIDENCE at which light falls on a transparent material such that the reflected and REFRACTED RAY are at right angles to one another. At this angle, the reflected light will be completely POLARIZED, with its ELECTRIC FIELD parallel to the reflecting surface. *See also* BREWSTER'S LAW.

Brewster's law Light reflected from a transparent material is completely POLARIZED when the reflected and refracted rays are at right angles to one another. *See also* BREWSTER ANGLE.

bridge rectifier A RECTIFIER circuit containing four DIODES in an arrangement similar to the WHEATSTONE BRIDGE. It has the advantage that current flows from the ALTERNATING CURRENT (a.c.) supply to the LOAD throughout the a.c. cycle. *See also* PN JUNCTION DIODE.

British thermal unit (Btu) An obsolete unit of energy, originally defined as the energy required to heat one POUND of water by one degree FAHRENHEIT. One Btu is equivalent to 1,055 JOULES.

broadcasting The transmission of signals in all directions for general reception. *See also* RADIO.

brown dwarf An astronomical object that is formed when a gas cloud in space collapses to form a PROTOSTAR, which is of too small a mass to reach a high enough temperature to begin the processes of NUCLEAR FUSION that release energy in a STAR. Such objects have been observed by the INFRARED radiation they emit as a result of the heating that takes place when

they collapse. Brown dwarves have masses less than 0.08 SOLAR MASSES and there is evidence that they exist in orbit around many nearby MAIN SEQUENCE STARS. In some senses the planet JUPITER can be regarded as a brown dwarf, though most brown dwarves are larger than this.

Brownian motion The rapid, random motion of small particles suspended in a fluid, which is seen when such particles are viewed through a microscope. Brownian motion provides evidence for KINETIC THEORY, as it is caused by the bombardment of the particles by far smaller, but fast moving, molecules in the suspending fluid. Brownian motion was first observed in pollen grains suspended in water, but is now usually demonstrated with smoke particles in air.

brush An electrical contact between the moving and stationary parts of a motor or generator. Brushes are usually made from carbon and kept in place with a spring.

bubble chamber A PARTICLE DETECTOR comprising a vessel containing liquid, often liquid hydrogen, held under pressure by a piston. When the piston is withdrawn, reducing the pressure, the liquid starts to boil, but it does so first on any CONDENSATION NUCLEI (irregularities in the liquid). The trail of IONIZATION left by any charged particle that has recently passed through the chamber acts as a source of condensation nuclei, giving rise to a trail of bubbles, which are photographed. The piston is then forced back into the chamber, causing the bubbles to collapse, ready for the next event.

bulk modulus A measure of the COMPRESSIBILITY of a material when subjected to an increase in external pressure tending to decrease its volume. If a pressure change of Δp causes a change in volume of ΔV in a sample of volume V, then the bulk modulus is $V\Delta p/\Delta V$.

bumping The violent boiling of a SUPERHEATED liquid.

buoyancy The UPTHRUST on a body immersed in a fluid, equal to the weight of the fluid displaced by the body. If the object is of a sufficiently low density, the upthrust will exceed the weight and the object will float. *See also* ARCHIMEDES' PRINCIPLE, FLOTATION.

byte, *word* A number made up of several (usually 4) BINARY digits.

C

calculus The branch of mathematics that deals with functions that change in a continuous way, particularly the rate of change of a function with a change in an independent variable.

If y is a function of x, then the rate of change of y with respect to a change in x is defined as the DERIVATIVE of y, denoted by the symbol y' or dy/dx. The process of finding a derivative is called DIFFERENTIATION. The reverse process of finding a function with a particular derivative is called INTEGRATION.

Graphically, the process of differentiation may be regarded as equivalent to finding the gradient of the TANGENT to a curve at a given point, and integration as finding the area bounded by the curve and other specified lines. *See also* ANALYSIS, INTEGRAL.

calibration 1. The process of using a measuring device to measure known quantities in order to check or adjust the instrument concerned.

2. A marking on the scale of an instrument representing a specified numerical value for the quantity being measured.

caloric *See* HEAT.

calorie (cal) A unit of quantity of heat in the C.G.S. SYSTEM of units. One calorie is the amount of heat needed to raise the temperature of one GRAM of water by 1°C. The calorie has been largely replaced by the JOULE; it is roughly equivalent to 4.2 J.

Calorie (kcal) A unit sometimes used to specify the energy value of foods. One Calorie (capital c) is equivalent to 1,000 CALORIES (small c).

calorific value The energy content of a fuel or food. It is the amount of heat generated by completely burning a given mass of fuel in a piece of apparatus called a bomb CALORIMETER. Calorific value is measured in JOULES per kilogram.

calorimeter A container for performing experiments related to heat transfer and temperature changes, such as the measurement of SPECIFIC HEAT CAPACITY and LATENT HEAT. Calorimeters are generally made of a metal (a good conductor of heat, so the entire vessel reaches the same temperature), of known heat capacity. *See also* CALORIFIC VALUE.

camera A device used to record an optical image of an object. The object to be photographed is placed in front of a lens with PHOTOGRAPHIC FILM behind the lens. A shutter controls how long the film is exposed to the light, while an adjustable hole called an APERTURE, or DIAPHRAGM, controls the amount of light reaching the film. The distance from the lens to the film can be varied to produce a sharp image on the film – a process called focusing.

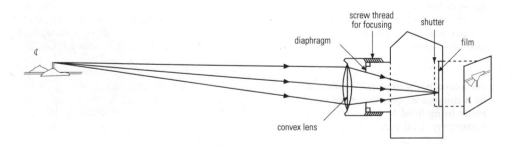

Principle of the camera.

For a distant object, the lens to film distance is equal to the FOCAL LENGTH of the lens; for closer objects this distance will be greater. By changing the focal length of the lens the image size can be altered: the longer the focal length, the larger the image for a given object distance.

When the aperture size is decreased, the outermost rays of light are stopped from reaching the film. This has the effect of making those parts of the image that would not be perfectly sharp on the film appear sharper. The range of distances over which the image on the film is acceptably sharp is called the DEPTH OF FIELD.

candela (Cd) The SI UNIT of LUMINOUS INTENSITY. One candela is equal to the luminous intensity in a given direction of a source of MONOCHROMATIC radiation of frequency 5.4×10^{14} Hz and has a radiant intensity in that direction of 1/683 WATT per STERADIAN.

capacitance The charge storage ability of a CAPACITOR; that is, the electric charge stored divided by the voltage between the plates. The unit of capacitance is the FARAD.

capacitor A device used to store electric charge. In its simplest form, a capacitor consists of two metal plates separated either by a vacuum or by some insulating material, called a DIELECTRIC. The amount of charge stored is directly proportional to the POTENTIAL DIFFERENCE (p.d.) between the two plates. The charge stored for each volt of potential difference is called the CAPACITANCE. The capacitance of a capacitor is directly proportional to the area of overlap of the plates and inversely proportional to their separation, thus the greatest capacitance is obtained by having large, closely-spaced plates. In practice these plates are made of metal foil separated by an insulator and folded or rolled into a many-layered construction.

If a capacitor stores a charge Q when there is a potential V between its plates, the capacitance is C, where

$$C = Q/V$$

For a capacitor with plates of area A separated by a distance d, with the space between the plates being filled by a dielectric of RELATIVE PERMITTIVITY ε_r, the capacitance C is

$$C = \varepsilon_0 \varepsilon_r A/d$$

where ε_0 is the PERMITTIVITY of free space (8.85×10^{-12} Fm^{-1}).

Capacitors are often distinguished by the material used for their dielectric. Various types of plastic are often used, but for large capacitances in small packages, ELECTROLYTIC CAPACITORS are used.

When two or more capacitors are connected in SERIES, the charge that flows off one plate of one capacitor flows onto the first plate of the next capacitor and so on; thus each capacitor has the same charge. The total capacitance of this combination is the reciprocal of the sum of the reciprocals of the individual capacitances. When two or more capacitors are connected in PARALLEL, the total charge stored is just the sum of the charges that each capacitor alone would store if charged to the same voltage. Thus the total capacitance is simply the sum of the individual capacitances.

For capacitors C_1, C_2 and C_3 connected in series, the total capacitance C is

$$1/C = 1/C_1 + 1/C_2 + 1/C_3$$

For the same capacitors connected in parallel,

$$C = C_1 + C_2 + C_3$$

The rate of flow of charge on or off a capacitor is directly proportional to the difference between the potential difference across the capacitor and the voltage of any supply of charge to which it is connected, and is inversely proportional to the resistance in the circuit. The charge on a capacitor rises or falls exponentially, with a TIME CONSTANT equal to RC, where R is the resistance in the circuit and C the capacitance. For a capacitor of capacitance C, initially charged to a charge Q_0, discharging through a resistance R, the charge Q after time t is

$$Q = Q_0 e^{-t/RC}$$

For a capacitor of capacitance C, initially uncharged, connected to a supply of voltage V by a resistance R, the charge Q after a time t is

$$Q = CV(1 - e^{-t/RC})$$

As well as storing charge, a capacitor stores energy, in that energy is needed to force the charges onto the plates. This energy will be released when the opposite charges on the two plates are provided with a conducting path from one plate to the other. The energy stored in a capacitor is equal to the average potential difference between the plates (which is half the final p.d.) multiplied by the charge on the plates.

Capacitors do not store much energy for their size, so are not an alternative to CELLS, but they can be used for short-term energy storage, for example in photographic flash guns, or in SMOOTHING in power supply circuits.

For a capacitor of capacitance C charged to a p.d. V, storing a charge Q, the energy stored E is

$$E = \tfrac{1}{2}QV = \tfrac{1}{2}CV^2 = \tfrac{1}{2}Q^2/C$$

See also VARIABLE CAPACITOR.

capillary effect The effect that causes most liquids, including water, to rise up a glass tube with a narrow bore (a capillary tube). This arises due to the intermolecular attraction between the water and the glass (called ADHESION) being stronger than the forces between the water molecules, called COHESION in this context. The same effect causes water to spread out on a glass surface, provided it is free of oil or grease.

It is the capillary effect that causes water to form a curved surface, called a MENISCUS, in a tube, and which allows materials with fine cracks (POROUS materials) to draw up water. Many plants rely on this mechanism to draw water from their roots into their leaves, though the process of OSMOSIS is also important. With some liquids, such as mercury, the cohesion is so strong that the liquid will form nearly spherical droplets on almost any surface. This results in the outward curving surface on mercury in a glass tube.

In a capillary tube, the height h by which the liquid inside the tube rises above that outside is

$$h = 2T/r\rho g$$

where T is the SURFACE TENSION, r the radius of the tube, ρ the density of the liquid and g the ACCELERATION DUE TO GRAVITY.

capillary tube A glass tube with a narrow bore. See CAPILLARY EFFECT.

Carnot cycle In physics, an ideal cycle of operations, leading to the greatest efficiency of conversion of heat energy into WORK attainable by any reversible HEAT ENGINE.

The Carnot cycle has four reversible stages, operating between two HEAT RESERVOIRS.

An ideal gas starts at a high temperature and pressure, and expands isothermally (at a constant temperature, see ISOTHERMAL), converting energy from a high temperature reservoir into mechanical work. The gas then expands further adiabatically (with no exchange of heat energy to its surroundings, see ADIABATIC) cooling to a lower temperature. In the third stage, the gas is compressed isothermally, releasing heat energy. Finally the gas is compressed adiabatically to its original state. Heat energy is taken from the high temperature reservoir and a smaller amount given up to the lower temperature. The difference between these two energies is the amount of mechanical work done.

Since every stage in the Carnot cycle is reversible, there is no overall change in ENTROPY and the Carnot engine must be the most efficient heat engine permitted by the SECOND LAW OF THERMODYNAMICS. Realistic heat engines are far less efficient, but are designed to approach the Carnot cycles as closely as possible.

See also INTERNAL COMBUSTION ENGINE.

Carnot engine A hypothetical HEAT ENGINE operating on the CARNOT CYCLE.

carrier wave A radio wave used to carry some information, such as speech or a picture, or which may be switched between two similar frequencies to convey BINARY information. The process of imparting information on a carrier wave is called MODULATION. See also AMPLITUDE MODULATION, FREQUENCY MODULATION, PHASE MODULATION, PULSE-CODE MODULATION, SIDEBAND.

Cartesian co-ordinates A mathematical system for locating a point in space by measuring its distance from three fixed lines, called axes. The axes are at right angles to one another and intersect at a fixed point, called the origin. These distances are called the co-ordinates of the point, and are usually written in the form (x,y,z). On a two-dimensional surface only the x and y co-ordinates are needed.

cathode A negatively charged ELECTRODE.

cathode ray A stream of electrons emitted by a CATHODE in a vacuum, such as in a CATHODE RAY TUBE. See also CATHODE RAY OSCILLOSCOPE.

cathode ray oscilloscope A device based on the CATHODE RAY TUBE and used to display how voltages change with time. The voltage to be measured is connected, via an AMPLIFIER, to the

Y-PLATES (*see* CATHODE RAY TUBE), whilst a voltage that increases steadily with time, called the TIMEBASE, is connected to the X-PLATES. The timebase causes the spot on the screen to sweep at a steady rate from left to right, whilst the voltage being measured moves the spot up or down, thus effectively producing a graph of the voltage against time. When the spot reaches the right-hand end of the screen, the CONTROL GRID switches off the electron beam and the spot is returned to the left-hand edge of the screen, a process called flyback.

cathode ray tube The display device used in televisions and the CATHODE RAY OSCILLOSCOPE. In the neck of a cathode ray tube there is an ELECTRON GUN comprising a heated filament together with two or more cylindrical ANODES. Electrons are released from the filament by THERMIONIC EMISSION. A beam of electrons is thus formed that can be focused to a narrow point on a PHOSPHOR-coated screen at the far end of the tube by varying the potential on the anodes. The electron gun also contains a CONTROL GRID, an ELECTRODE that can be made negative to reduce the number of electrons leaving the gun and thus control the brightness of the spot of light produced on the screen. The electron beam can be steered to any point on the screen, either electrostatically, by two pairs of plates called X-PLATES and Y-PLATES that steer the beam horizontally and vertically, or magnetically by coils placed around the neck of the tube.

Cathode ray tubes have the disadvantages of requiring high voltage power supplies and being rather large in terms of the amount of space required between the electron gun and the screen. They are gradually being replaced by LIQUID CRYSTAL DISPLAYS, but at present these are too expensive to replace large cathode ray tubes.

cation A positively charged ion. So called because it will be attracted to the CATHODE in ELECTROLYSIS. *Compare* ANION.

caustic The curve that contains all the rays from a point source of light as they approach and depart from the focus of a lens or mirror that suffers ABERRATION.

Cavendish's experiment An experiment to determine the GRAVITATIONAL CONSTANT. Two small lead spheres are fastened to the ends of a rod, which is suspended from its midpoint by a fine fibre. The suspended spheres are allowed to rotate to-and-fro on this fibre and their period of oscillation is timed, enabling the stiffness of the fibre to be calculated. Two larger lead spheres are then placed near to the smaller ones in such a way that the gravitational attraction between two pairs of spheres twists the fibre. By measuring the deflection of the rod for different configurations of the four spheres and from the known stiffness of the fibre, the gravitational constant can be determined using NEWTON'S LAW OF GRAVITATION.

CCD (charge coupled device) An array of small detectors based on the PHOTOELECTRIC EFFECT. CCD's are built as the top layer of an INTEGRATED CIRCUIT, which allows the build up of charge on each line of detectors to be 'read out' in turn. This is done by using a series of electrical PULSES to transfer charge, which has accumulated in each element as a result of exposure to light, from one line of detectors to the next. As each line of charges reaches the edge of the detector it is transferred out of the CCD, giving a series of outputs representing the brightness of each part of the image. CCD's can be made very small and produced at a reasonable cost, allowing the production of small, cheap, television cameras. CCD's can also be used at very low light levels, which has made them very popular in astronomy.

celestial equator The extension of the Earth's equator onto the sky. Points on the celestial equator appear overhead when standing on the Earth's equator and lie on the horizon when standing at the poles.

celestial mechanics The study of the motions of astronomical objects and the forces between them, based on NEWTON'S LAWS OF MOTION and NEWTON'S LAW OF GRAVITATION.

cell, *electrochemical cell* A device that connects chemical reagents to an electrical circuit. A cell consists of two ELECTRODES in contact with an ELECTROLYTE – a conducting liquid or jelly. An electrolytic cell is one in which a chemical change takes place when a current is passed through the electrolyte (*see* ELECTROLYSIS). In a voltaic cell, a chemical reaction between the electrolyte and the electrodes produces an ELECTROMOTIVE FORCE, which drives a current around a circuit.

Some voltaic cells can be used once only: examples are the ZINC-CARBON CELL, which is cheap but stores relatively little energy, and the more expensive but longer lasting MANGANESE-

ALKALINE CELL. Others are rechargeable by passing a current through them in the reverse direction, the chemical reactions can be reversed and electrical energy converted back to chemical energy. Examples of rechargeable cells are the LEAD-ACID CELL, used to provide high currents in cars for example, and the NICKEL-CADMIUM (or Nicad) cell, often used in electronic appliances. The amount of charge that a voltaic cell can drive around a circuit is measured in AMPERE-HOURS.

See also DRY CELL, HALF-CELL, WET CELL.

Celsius A TEMPERATURE SCALE in which the freezing point of water is defined as zero degrees Celsius (0°C), whilst the boiling point of water is 100°C, both temperatures being measured at ATMOSPHERIC PRESSURE. The degree Celsius is the same size as the KELVIN. *See also* ABSOLUTE TEMPERATURE.

centi- Prefix used to denote one hundredth. For example, one centimetre is one hundredth of a metre (0.01 m).

centre of gravity *See* CENTRE OF MASS.

centre of mass, *centre of gravity* The single point in an object at which the whole of the gravitational force on that object can be considered to act. When an object is suspended, the centre of mass will always be directly below the point where the object is suspended. By suspending an object from several different points in turn, its centre of mass can be found.

centrifugal force An apparent (but not real) force used in some descriptions of CIRCULAR MOTION to balance the CENTRIPETAL FORCE as if the object moving in a circle were in equilibrium. This is sometimes a useful way of thinking about the motion of an object moving in a circle, but it must be remembered that there is no real equilibrium. Inside an object accelerating in a circular path, such as an orbiting spacecraft, it is convenient to think of the centripetal force, which acts towards the centre of the circle, as being balanced by an equal and opposite centrifugal force. However, this will not provide a full description of the situation unless a second imaginary force called the CORIOLIS FORCE is also included.

centrifuge A machine for separating two different materials on the basis of their relative densities. A centrifuge may be used to separate the components in a mixture of two liquids or a solid suspended in a liquid. The mixture is placed in a tube and rotated very rapidly in a horizontal circle. The CENTRIPETAL FORCE needed to make the mixture rotate forces the denser component outwards along the tube, displacing the less dense component, and collecting at the bottom of the tube. Centrifugation is used, for example, for separating the different constituents of blood.

centripetal acceleration The acceleration of an object moving in a circle. *See* CIRCULAR MOTION.

centripetal force The force needed to make an object move in a circular path. Note that this is not a particular type of force, but a force doing a particular job: for a car going around a bend, the centripetal force is provided by the FRICTION between the wheels and the road. For a satellite in orbit, the centripetal force is provided by gravity. *See* CIRCULAR MOTION. *See also* CENTRIFUGAL FORCE.

centroid The geometric centre of a body: the point from which the average displacement of all other points is zero. This is the CENTRE OF MASS of a shape or body of uniform density.

Cepheid variable A type of RED GIANT that varies in brightness over a few days. There is a definite link between the period of this variation and the LUMINOSITY of these stars, which enables their distance to be found. Once the distance to a Cepheid variable within a CLUSTER or galaxy is known, the distances of all the other members of that cluster or galaxy can be taken as being roughly equal.

CERN (European Centre for Nuclear Research) A multinational institution, located in Geneva, Switzerland, home to many of the world's largest PARTICLE ACCELERATORS. *See also* LEP, SPS.

c.g.s. system A system of physical units derived from the metric system and based on the centimetre, gram and second. It has been replaced by SI UNITS.

chain reaction Any reaction in which the products of the reaction initiate further reactions of the same type, leading to an exponential growth in the rate of reaction. In physics, a NUCLEAR FISSION process may lead to a chain reaction, where neutrons initiate fission, which produces further neutrons for more fission, and so on. Such reactions can be controlled by using control rods to absorb excess neutrons, but if there is no control this can lead to a nuclear explosion. *See also* CRITICAL MASS, THERMONUCLEAR REACTION.

change of state Any process in which substance changes from one of the STATES OF MATTER to another. In any change of state, LATENT HEAT is taken in or given out. *See* BOILING POINT, MELTING POINT, SUBLIMATION.

channel 1. A radio frequency allocated for a particular purpose. The spacing between channels is arranged to allow for the BANDWIDTH of the transmitted signal.

2. The semiconducting material through which CHARGE CARRIERS flow in a FIELD EFFECT TRANSISTOR.

chaos The state of a system whereby a small change in the initial state of the system leads to large changes in the final state. Since initial conditions can never be known precisely, it is impossible to predict the behaviour of chaotic systems over long periods of time, although they may obey essentially simple laws. An important example is in meteorology, where the pattern of air movement is so complex that it is difficult to confidently predict the weather more than a day or so ahead. It has been said that a butterfly flapping its wings in one part of the world may eventually lead to a hurricane somewhere else. Surprisingly, even quite simple systems can exhibit chaotic behaviour.

charge A fundamental property of matter. Charges are of two types, positive and negative. Charged particles exert forces on one another: charges of the same type repel one another whilst opposite charges attract. The SI UNIT of charge is the COULOMB (C).

Electrons are negatively charged, whilst protons have an equal and opposite positive charge of 1.6×10^{-19} C. These are the basic units of electrically charged matter. If an object has equal numbers of electrons and protons it is electrically neutral, if it has an excess of electrons it has an overall negative charge, while if it has an excess of protons it has a positive charge. The flow of charged particles, in particular the flow of electrons, is what constitutes an electric CURRENT. The removal of a charge from an object is called discharging.

See also CAPACITOR, CHARGING BY FRICTION, CHARGING BY INDUCTION, COULOMB'S LAW, INDUCED CHARGE, MILLIKAN'S OIL DROP EXPERIMENT, STATIC ELECTRICITY.

charge carrier A charged particle that moves through a substance to carry a current. In metals, the charge carriers are electrons. In SEMICONDUCTORS they are electrons or HOLES. In IONIC liquids and IONIZED gases the charge carriers are ions.

charge coupled device *See* CCD.

charged (*adj.*) The state of a CAPACITOR or electrochemical CELL when it is storing the greatest possible amount of charge. In the case of a capacitor, this is when the POTENTIAL DIFFERENCE across its plates is equal to the ELECTROMOTIVE FORCE of the charging supply. The opposite state, when the charge stored is zero, is described as discharged.

charge density The amount of charge per unit volume or per unit area.

charging by friction The mechanism by which a pair of objects gain an equal and opposite charge when they are rubbed against one another. Electrons are transferred from one of the objects, leaving it with a positive charge, to the other, which gains a negative charge. *See also* LIGHTNING, VAN DE GRAAF GENERATOR.

charging by induction The process by which an object can be given a charge by placing it near an electrically charged object. If an uncharged conductor is placed near a positively charged rod, for example, electrons in the conductor are attracted towards the positive rod. The conductor remains electrically neutral overall, but the influence of the charged rod redistributes the charges in the conductor. The surface of the conductor closest to the rod will become negatively charged, leaving the opposite surface positively charged. If the conductor is then separated into two parts, they will have equal and opposite charges. Such charges are called induced charges.

Charles' law For a fixed mass of an IDEAL GAS at constant pressure, the volume is proportional to the ABSOLUTE TEMPERATURE; that is, the volume divided by the absolute temperature is a constant. For a fixed mass of an ideal gas with an absolute temperature T in volume V, at constant pressure

$$V/T = \text{constant}$$

See also BOYLE'S LAW, GAS LAWS, IDEAL GAS EQUATION, PRESSURE LAW.

charm A FLAVOUR of QUARK or the QUANTUM NUMBER associated with that quark. Charm is conserved in interactions involving the STRONG NUCLEAR FORCE but not in those involving the WEAK NUCLEAR FORCE. *See also* J/Ψ.

chemical energy The energy that can be released in forming or breaking chemical bonds. As the forces involved in chemical reactions are ELECTROSTATIC in nature, this can be thought of as a form of ELECTRICAL POTENTIAL ENERGY.

CHP *See* COMBINED HEAT AND POWER.

chromatic aberration *See* ABERRATION.

circuit, electric Any closed conducting path around which an electric current can flow. *See also* CONTACT BREAKER, OPEN CIRCUIT, SHORT-CIRCUIT.

circuit breaker A device for introducing a gap into a CIRCUIT. The term circuit breaker may refer (especially in the US) to any form of on/off switch or RELAY, but refers more specifically to a device that operates when a CURRENT exceeds a certain value. Some circuit breakers are thermally operated, with the current flowing through a heating coil, which warms a BIMETALLIC STRIP to operate a spring-loaded switch. Others are operated by the MAGNETIC FIELD produced by the current flowing in a coil. The thermal type of circuit breaker operates only when the current exceeds a certain value for a few seconds, whilst the magnetic type operates more rapidly. *See also* EARTH-LEAKAGE CIRCUIT BREAKER.

circularly polarized (*adj.*) Describing a form of polarized light (*see* POLARIZATION) in which the vector describing the electric field rotates as the wave propagates, tracing out a spiral path. Circularly polarized light can be thought of as the superposition of two plane polarizations with a 90° phase difference. *See also* PLANE-POLARIZED.

circular motion An object moving along a circular path at a constant speed is accelerating due to the change in its direction. In other words, its velocity is changing though its speed may be constant. This acceleration is directed towards the centre of the circle and is called CENTRIPETAL ACCELERATION. To produce this acceleration, a force is needed; this force is called a CENTRIPETAL FORCE. For an object moving at a constant speed v around a circle of radius r, the centripetal acceleration is v^2/r. *See also* CENTRIFUGAL FORCE, CORIOLIS FORCE.

circumpolar orbit A satellite ORBIT that passes over the Earth's North and South Poles. For an orbit just above the atmosphere, the period of orbit is about 90 minutes, whilst the Earth rotates underneath the satellite every 24 hours.

Thus the satellite has a view of every point on the Earth at least once every 12 hours. Such orbits are used by some weather satellites and also to view the Earth for military and environmental monitoring. *See also* GEOSTATIONARY ORBIT. ·

cirrus High-level clouds, composed of ice crystals. They appear as feathery white wisps. *See also* WEATHER SYSTEMS.

Clausius statement of the second law of thermodynamics The SECOND LAW OF THERMODYNAMICS forbids any system in which heat energy flows from a region of low temperature to one of higher temperature with no WORK being done. For example, a refrigerator cannot work without a supply of extra energy, usually electrical, which is also deposited in the warm part of the system.

cleavage plane In a non-metallic crystal, any surface along which the crystal will break cleanly and relatively easily. If an attempt is made to break the crystal in other directions, it may shatter into smaller fragments rather than breaking cleanly. A knowledge of the cleavage planes of diamond is important for breaking naturally occurring diamonds, which usually have no particular shape, into gemstones for jewellery. *See also* SLIP PLANE.

climate The general long-term pattern of weather as it varies from place to place on the planet or from year to year. The temperature of the Earth's surface is greater near the equator than near the poles, where the Sun hits the Earth at an angle (so the intensity of radiation reaching the surface is less). Since the axis about which the Earth rotates is not at right angles to the Earth's orbit around the Sun, the northern and southern hemispheres each receive more heat in one half of the year (called summer) than in the other (winter).

Heat energy absorbed from the Sun sets up major CONVECTION CURRENTS within the ATMOSPHERE, which form closed loops of convection called HADLEY CELLS. The convection currents set up by solar heating produce areas of high and low pressure. Air does not flow directly from high to low pressure, but tends to rotate. In the northern hemisphere the rotation is clockwise around areas of high pressure and anticlockwise around areas of low pressure (which are called DEPRESSIONS). This rotation (which is in the opposite direction in the southern hemisphere) is caused by the

CORIOLIS FORCE: the Earth is spherical so the surface rotates faster near the equator than at the poles. Air moving from the equator towards the poles thus appears to be moving quickly relative to its surroundings, as if there was a force on it.

In addition to these global patterns of air flow, the climate of a region is determined by the distribution of land and sea (affecting humidity), by altitude (affecting temperature and rainfall) and ocean currents (affecting temperature).

See also WEATHER SYSTEMS.

clocked bistable, *flip flop* In electronics, a BISTABLE with an additional 'clock' input. Pulses on the clock input change the output from one state to another. A COUNTER can be made by connecting the output of one bistable to the clock input of another, as the second bistable will receive clock pulses at half the rate of the first. This is the basis of digital counting and timing circuits. See also SHIFT REGISTER.

close packed (*adj.*) Describing a pattern in a crystal that maximizes the numbers of atoms that will fit into a given volume. The CO-ORDINATION NUMBER of a close packed structure is 12. See also CUBIC CLOSE PACKED, HEXAGONAL CLOSE PACKED.

cloud A mass of minute water droplets or ice crystals suspended in the atmosphere. The droplets or crystals are formed when water vapour condenses on tiny dust or salt particles.

cloud chamber A device that shows the path taken by IONIZING RADIATION, particularly ALPHA PARTICLES. The chamber contains a vapour (usually alcohol), which is cooled to the point where it is about to condense. Droplets of liquid will form first on any CONDENSATION NUCLEI present, such as trails of ionized particles left by an alpha or BETA PARTI-CLE (beta particles leave only very faint tracks).

cluster In astronomy, a group of stars (star cluster) or galaxies (galaxy cluster) held together by GRAVITY. See also GLOBULAR CLUSTER.

coalesce (*vb.*) To stick together. The term is used in mechanics to describe objects that combine and move with a common velocity after an INELASTIC COLLISION.

coefficient Any number that multiplies some other number or variable in an equation.

coefficient of drag A mathematical represen-tation of the amount of DRAG produced by a given shape. It is lower for more STREAMLINED shapes. The coefficient of drag does not depend on the size or speed of motion of the object. For TURBULENT flow of an object of cross-sectional area A through a fluid of density ρ at a speed v, the force F is

$$F = \tfrac{1}{2} C_D \rho v^2$$

where C_D is the coefficient of drag.

coefficient of expansion See EXPANSIVITY.

coefficient of friction (μ) The ratio of the DYNAMIC FRICTION to the NORMAL REACTION. It is a measure of the roughness of a surface. Values for the coefficient of friction generally vary between about 0.01 for smooth surfaces to about 0.8 for rough surfaces. Much larger coefficients of friction can occur in situations where there is a strong force of ADHESION – such as very smooth metal or glass surfaces. For a normal reaction of N, the frictional force F is given by

$$F = \mu N$$

See also FRICTION.

coefficient of lift A measure of how effective a wing, or other AEROFOIL, is at generating LIFT. The coefficient of lift is given by

$$L = \tfrac{1}{2} C_L A \rho v^2$$

where C_L is the coefficient of lift and L is the lift generated by an aerofoil of area A moving at velocity v through air of density ρ. The value of the coefficient of lift varies depending on the angle at which the wing meets the air-flow, called the ANGLE OF ATTACK, but a typical aerofoil has a maximum coefficient of lift of 1.2. See also AERODYNAMICS.

coefficient of restitution A quantity used to describe the degree of elasticity in partially ELASTIC COLLISIONS. The coefficient of resti-tution is the RELATIVE VELOCITY of the colliding objects immediately after impact divided by their relative velocity before the collision. For a perfectly elastic collision, the coefficient of restitution is 1, whilst in a totally INELASTIC COLLISION (where as much KINETIC ENERGY is lost as is allowed by the LAW OF CONSERVATION OF MOMENTUM), the coefficient of restitution is 0. See also HYPERELASTIC COLLISION.

coercive force The magnetic field that must be applied in a reverse direction to remove the magnetism of a PERMANENT MAGNET.

coherent (*adj.*) Describing a wave that has a steadily varying PHASE, or two or more waves that have a constant phase difference. Light from a laser has a steadily varying phase, but light from other sources has jumps in phase, due to the light being emitted by one atom after another with no link between the phases produced by each atom. *See also* DIVISION OF AMPLITUDE, DIVISION OF WAVEFRONT.

cohesion The attractive force between molecules of the same type. Cohesion is responsible for SURFACE TENSION. *See also* ADHESION, CAPILLARY EFFECT.

coil A wire helix, usually used to exploit some property of the wire. Springs, RESISTORS and ELECTROMAGNETS are all often made in the form of a coil. *See also* SOLENOID.

cold front A region where cold air attempts to force its way under warmer air, producing rain showers and possibly thunderstorms. *See* WEATHER SYSTEMS. *See also* WARM FRONT.

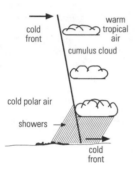

Formation of a cold front.

cold fusion NUCLEAR FUSION at room temperatures. A process by which DEUTERIUM nuclei undergo fusion by absorption into palladium atoms was claimed to have been observed in 1989. Subsequent work failed to confirm this discovery and the process is now generally believed not to take place.

collector In a JUNCTION TRANSISTOR, the ELECTRODE at which CHARGE CARRIERS arrive having passed through from the EMITTER and the BASE.

colliding beam experiment Any particle physics experiment in which two beams collide head-on, as opposed to a beam colliding with a fixed target. It has the advantage that the total MOMENTUM is zero, so all the energy of the beams can be harnessed for the production of new particles, rather than being used to give

those particles the KINETIC ENERGY they need to conserve momentum.

colligative properties Those properties of a solution that are dependent on the concentration of particles in the solution, but not on the nature of those particles. Examples are the LOWERING OF VAPOUR PRESSURE, the DEPRESSION OF FREEZING POINT, the ELEVATION OF BOILING POINT and OSMOSIS.

collimator A tube with an adjustable slit at one end and a CONVERGING LENS at the other, with adjustments to allow the slit to be placed in the FOCAL PLANE of the lens. A collimator is used to produce a parallel beam from the light that DIFFRACTS from the slit, which provides a COHERENT source of light for SPECTROSCOPY.

collision cross-section The cross-sectional area of the volume swept out by a molecule or some other particle as it moves through space. If another particle enters this volume, there may be a collision. Collision cross-sections are used in the calculation of collision rates in particle physics, and for the calculation of the MEAN FREE PATH of a molecule in a gas.

colour 1. (*optics*) The name given to the physiological sensation caused by different wavelengths of visible light. The eye can distinguish three wavelength bands which produce the sensations of the three primary colours: red (the longest wavelengths), green (intermediate wavelengths) and blue (short wavelengths). When the eye receives wavelengths between these bands, or a mixture of two of these colours, the secondary colours are seen. These are yellow (red plus green), cyan (green plus blue) and magenta (blue plus red). The mixture of all three primary colours produces the sensation of white, whilst the sensation of black is produced by an absence of any visible light.

Objects appear coloured because they reflect the different colours of light falling on them by differing amounts. A red object is red because when viewed under white light only red light is reflected. Viewed under green light, for example, it will appear dark as there is no red light falling on it to be reflected. Colour filters work in a similar way, modifying the light by absorbing certain colours. A yellow filter, for example, absorbs blue light whilst transmitting red and green. For this reason yellow and blue are known as complementary colours

– added together they will produce white light. The other pairs of complementary colours are green/magenta and red/cyan. *See also* RETINA.

2. (*particle physics*) A quantity possessed by QUARKS and GLUONS in QUANTUM CHROMODYNAMICS, analogous to the electrical charge in ELECTROMAGNETISM.

colour filter A device that absorbs some wavelengths of visible light whilst allowing others to pass. A yellow filter, for example, is transparent to the red and green parts of the SPECTRUM, but absorbs blue. *See also* COLOUR.

colour television A television system in which three ELECTRON GUNS are used. Electrons are sent through a SHADOWMASK – a metal screen with holes just behind the PHOSPHOR coating. This ensures that electrons from each gun can only reach certain parts of the phosphor screen. These regions are coated with different phosphors, which glow red, green and blue when struck by electrons. In this way, each electron gun can be used to build up a picture in each of the PRIMARY COLOURS.

coma 1. (*physics*) An ABERRATION in images formed by lenses and CURVED MIRRORS when the rays forming the image are at a large angle to the PRINCIPAL AXIS. Coma is so called because the image of a point is shaped like a comet.

2. (*astronomy*) The cloud of luminous gas around a comet when it is close to the Sun.

combined heat and power (**CHP**) The combined production of heat and electricity in a power station. The steam produced to drive turbines in the power station is afterwards used to heat nearby buildings. The high costs of CHP plants have meant that few such systems have been built, despite the greater EFFICIENCY in energy usage.

comet A small body of rock and ice that orbits the Sun. Most comets are believed to spend most of their time in very distant orbits that make up a region called the OORT CLOUD. Occasionally, a gravitational disturbance will cause a comet to leave this cloud and fall into a highly elliptical orbit, bringing it very close to the Sun. As it heats up, some of the comet's material is vaporized and then IONIZED by solar radiation, producing a glowing tail, which always points away from the Sun. As the comet moves away from the Sun it cools and fades from view. Comets in short orbits lose a lot of their material on each pass and

rapidly become fainter. Some comets make only one pass of the Sun and are never seen again.

commutator A device for reversing electrical connections. An example is the SPLIT-RING COMMUTATOR used in many ELECTRIC MOTORS and DYNAMOS.

compact disc (*see also following page*) A system for the digital storage of information. A compact disc contains music, visual images or computer data encoded as a series of BINARY 1's and 0's recorded as small hollows or pits etched into a thin layer of aluminium sandwiched between transparent plastic layers. The disc is read by the reflection of a low power laser beam. The digital nature of the data and the error correction systems built into the way the data is encoded mean that the data can be read perfectly unless the disc is badly damaged. There is also no wear as there is in a conventional vinyl record, as there is no mechanical contact between the compact disc and the reading device.

comparator A circuit that compares two voltages and gives a digital output depending on whichever voltage is the larger. Comparator circuits are often based on OPERATIONAL AMPLIFIERS. Provided there is no FEEDBACK, the output of the operational amplifier will be high if the NON-INVERTING INPUT is at a higher potential than the INVERTING INPUT and low (or negative) if the non-inverting input is at the lower potential.

compass *See* MAGNETIC COMPASS.

complement The complement of an angle is 90° minus the angle concerned. In other words, an angle and its complement add up to 90°.

complementary angles Any pair of angles that add up to 90°; the two other angles in a right-angled triangle.

complementary colour The colour that must be added to a specified colour to make white light. Thus blue and yellow, for example, are complementary colours.

complementary pair A pair of JUNCTION TRANSISTORS, one NPN and the other PNP, used together.

complex conjugate *See* CONJUGATE.

complex number Any number formed by adding a REAL NUMBER and an IMAGINARY NUMBER. A complex number can be expressed as $x + iy$, where x is the real part and iy is the

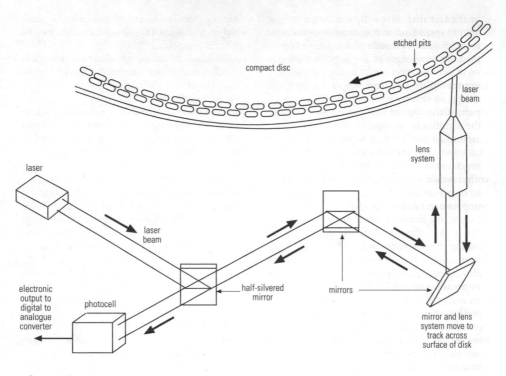

Compact disc player.

imaginary part. Both x and y are real numbers and i is the square root of -1.

Complex numbers are widely used in physics. This may be a matter of convenience, when dealing with ALTERNATING CURRENT, for example, but equations of QUANTUM PHYSICS, such as SCHRÖDINGER'S EQUATION, have complex COEFFICIENTS and require that the WAVEFUNCTION be represented by complex numbers.

See also CONJUGATE.

component 1. (*mathematics*) Any one of a set of VECTORS that add together to give some other vector. The term is particularly used in reference to any one of three vectors parallel to the axes of a CARTESIAN CO-ORDINATE system, described as the components of the vector formed by adding them together. The process of finding the components of a vector in specified directions is called resolving (*see* RESOLVE).

2. (*mechanics*) That part of a FORCE that acts in a particular direction. The component of a force F in a direction at an angle θ to the direction of the force is $F\cos\theta$.

composite material Any engineering material made from two or more different materials, designed to exploit the advantages of each without suffering their weaknesses.

compound A substance made up of two or more ELEMENTS that cannot be separated by physical means. In a compound, the quantity of the elements present is fixed and is a simple ratio, though more than one compound may exist containing the same elements, for example, water (H_2O) and hydrogen peroxide (H_2O_2).

compound lens An optical lens made from several pieces of glass.

compound microscope A MICROSCOPE that uses two lenses. The first lens, called the object lens, is used to form an enlarged REAL IMAGE of the object. This image is then viewed through a second lens, called the EYEPIECE, which acts as a magnifying glass, further enlarging the object. The MAGNIFICATION of the microscope can be altered by adjusting the power of the two lenses, but the useful magnification is limited by the fact that DIFFRACTION effects cause the light to spread out as it enters the objective. It is therefore not possible to see details smaller than the wavelength of the light used to view them. *See also* SIMPLE MICROSCOPE.

compressibility The tendency of a material to change its density with pressure. The term is particularly applied to the effects arising from the significant changes in the density of air around an aircraft as it approaches the SPEED OF SOUND. The compressibility of a material is measured by its BULK MODULUS.

compression A pushing force, tending to make an object smaller.

compression ratio In an INTERNAL COMBUSTION ENGINE, the factor by which the volume of the gas in the cylinder of the engine is reduced during the COMPRESSION STROKE. Higher compression ratios lead to greater efficiency but also increase problems of PRE-IGNITION.

compression stroke The motion of a PISTON in an INTERNAL COMBUSTION ENGINE to compress and heat the air or fuel/air mixture prior to the POWER STROKE. *See also* COMPRESSION RATIO.

concave (*adj.*) Describing an inwardly curving surface, particularly in a lens or mirror.

concentric circles Two or more circles with a common centre.

condensation 1. The process by which a vapour turns into a liquid.

    2. Droplets, usually of water, formed by this process.

condensation nucleus A small imperfection or piece of dirt that may act as a centre for a change of state, such as the condensation of a SUPERSATURATED VAPOUR or the boiling of a SUPERHEATED liquid.

condense (*vb.*) To turn from a gas or vapour into a liquid.

condensed matter Solids and liquids, those states of matter where the spacing between the molecules is of the same order as their size.

condenser 1. *condensing lens* (*optics*) A lens of short FOCAL LENGTH designed to concentrate as much light as possible onto an object, to illuminate it as brightly as possible.

    2. (*electronics*) An obsolete word for CAPACITOR.

condensing lens *See* CONDENSER.

conductance In electricity, the reciprocal of resistance. A material that has a large resistance has a low conductance and vice versa. This means it is sometimes more convenient

to think in terms of conductance, or CONDUCTIVITY, the reciprocal of RESISTIVITY. The unit of conductance is the SIEMENS.

conduction The flow of electricity or heat through a material. *See* CONDUCTOR, THERMAL CONDUCTION.

conduction band In the BAND THEORY of solids, an energy band that is not completely occupied by electrons, so an electron in that band can gain a small amount of energy and move through the material, carrying heat or electricity. *See also* VALENCE BAND.

conductivity The reciprocal of RESISTIVITY. The unit of conductivity is SIEMENS per metre. *See also* CONDUCTANCE, THERMAL CONDUCTIVITY.

conductor, electrical Any material through which current can flow. All metals are conductors, as their structure contains VALENCE ELECTRONS that are free to move through a lattice of positive IONS. Metals conduct less well as their temperature increases, since thermal vibrations make the lattice less regular, making collisions between the electrons and the lattice more likely. For similar reasons, pure metals, which have highly regular lattice structures, usually conduct electricity (and heat – *see* THERMAL CONDUCTION) better than alloys. Graphite is one of the few nonmetallic solids that conducts electricity.

    Molten IONIC materials and solutions of ionic salts all conduct electricity. In all these cases there are both positive and negative ions free to move through the liquid and carry the charge. Once these ions reach an ELECTRODE, the ions gain or lose electrons and chemical reactions may take place (*see* ELECTROLYSIS). In IONIC SOLIDS, these ions are locked into a rigid lattice and so are not free to carry charge. Molten metals also conduct electricity.

    Gases do not normally conduct electricity, but may do so if they are IONIZED – that is, if they contain a number of ions as well as uncharged atoms or molecules. Sources of IONIZATION include IONIZING RADIATION from radioactive materials, flames or very high voltages, which may produce an AVALANCHE BREAKDOWN.

    See also BAND THEORY, INSULATOR, SEMICONDUCTOR.

cone 1. A solid shape produced by the set of all lines that pass through a fixed point, called the vertex of the cone, and a circle (called the base).

2. Light-sensitive cells in the RETINA of the human eye. Cones are sensitive to colour and are used mostly for day vision.

conical pendulum A variation on the SIMPLE PENDULUM in which the BOB moves in a circle rather than from side-to-side.

conic section Any of the curves that can be formed by the intersection of a plane and a cone. The four main types of conic sections are the circle (if the plane is perpendicular to the axis of a right circular cone); the PARABOLA (if the plane is parallel to the edge of the cone); the ELLIPSE (if the intersection follows a closed curve that is not a circle); and the HYPERBOLA (if an open curve is produced that is not a parabola).

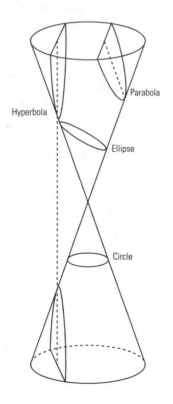

Conic section.

conjugate, *complex conjugate* One of a pair of COMPLEX NUMBERS with the same real part but opposite imaginary parts: that is, $a + ib$ and $a - ib$.

constant A quantity that does not change or depend on other quantities.

constant flow method A way of measuring the HEAT CAPACITY of a liquid. The power of an electric heater is calculated from the current through it and the voltage across it whilst the fluid passes the heater at a measured rate. The heat capacity can then be found from the temperature difference in the fluid produced by the heater.

constant volume gas thermometer A thermometer that uses the changes in pressure of a fixed amount of gas held in a constant volume to measure temperature. The gas (generally hydrogen) is usually held in a glass or metal bulb connected by a CAPILLARY TUBE to a mercury-filled MANOMETER. The manometer incorporates a flexible section, so the open arm can be raised or lowered. The gas is allowed to come to equilibrium at the temperature being measured and the open end of the manometer is raised or lowered to restore the mercury level on the closed side of the manometer to its original position. The volume of gas thus remains constant, whilst its pressure can be measured as being equal to ATMOSPHERIC PRESSURE plus an amount represented by the height difference in the mercury level on the two sides of the manometer. For very precise work, a series of readings are taken at different pressures and extrapolated to zero pressure to compensate for the effects of intermolecular forces. A correction can also be made for the volume of gas in the capillary tube, which is at intermediate temperatures. *See also* PRESSURE LAW.

constructive interference The effect produced when two or more waves arrive IN PHASE, in which the waves add to produce a wave with an amplitude equal to the sum of the amplitudes of the original waves. *Compare* DESTRUCTIVE INTERFERENCE. *See also* INTERFERENCE.

contact breaker A switch, often operated by the rotation of some machinery, such as an INTERNAL COMBUSTION ENGINE, designed to break the flow of current in a device such as an INDUCTION COIL.

contact force Any force that results from two solid objects touching each other. The contact force is usually a combination of the NORMAL REACTION, which acts at right angles to the surface and prevents one object from entering the other, and friction, which acts along the surfaces and prevents them from sliding along one another.

contact lens A small MENISCUS lens designed to have the same effect as spectacles but resting directly on the cornea (the curved front surface of the eye).

continent A large mass of land rising above the ocean floor. The boundary of a continent lies at the edge of the continental shelf, so offshore islands are considered part of the nearby continent. There are seven continents: Africa, Antarctica, Asia, Australia, Europe, North America and South America. *See also* CONTINENTAL DRIFT, PANGAEA.

continental (*adj.*) Describing an AIR MASS that has travelled mostly over land.

continental drift The slow motion of the continents from one part of the globe to another. It is thought that the Earth's continents once formed a single large land mass, given the name Pangaea, and then moved relative to each other. The phenomenon of continental drift may be explained by the motion of TECTONIC PLATES.

continental shelf The sea bed surrounding the land mass of a CONTINENT at an average depth of 200 metres. At the edge of the shelf, the sea bed drops steeply to the ocean floor.

continuous data In statistics, measured numbers that can take on any numerical value so cannot be represented fully by INTEGERS. Most physical data is continuous. *Compare* DISCRETE DATA.

continuous spectrum *See* SPECTRUM.

contrast enhancing medium An X-ray-absorbing material introduced into a patient to show up more clearly the differences between various organs or structures. *See also* ARTERIOGRAM, BARIUM MEAL, RADIOGRAPH.

control grid In a CATHODE RAY TUBE, or any other device that produces a beam of electrons by the process of THERMIONIC EMISSION, an ELECTRODE used to control the flow of electrons. If the control grid is more negative than the heated CATHODE, it limits the number of electrons able to reach the ANODE or form a beam.

control rod In a NUCLEAR REACTOR, a rod made of a neutron-absorbing material, usually boron or cadmium, which can be inserted into the reactor to control the CHAIN REACTION by absorbing excess neutrons.

convection The motion of a fluid (usually air or water) in a GRAVITATIONAL FIELD driven by changes in density resulting from temperature changes. Fluids expand and become less dense as they are heated, so a source of heat near the base of a volume of fluid will set up convection currents carrying energy throughout the fluid. If the heat is applied near the top of the fluid there will be no convection. An important exception to the rule that fluids expand on heating is water in the temperature range 0°C to 4°C. For this reason, lakes and ponds freeze from the surface down rather than freezing solid as soon as the air temperature falls below freezing point.

convection current The movement of a volume of fluid caused by CONVECTION. *See also* WEATHER SYSTEMS.

conventional current A current imagined to flow in the direction that would be taken by positive charges – that is, from the positive terminal of the battery around the circuit to the negative terminal.

converging lens An optical lens that is thicker at the centre than at the edges and brings parallel light rays together at a point. The distance between the lens and the point at which parallel rays of light are brought together is called the FOCAL LENGTH.

An object placed more than two focal lengths from a converging lens will form an inverted (upside down), diminished REAL IMAGE located at a distance between one and two times the focal length of the lens. This is the arrangement in a camera. If an object is placed in front of a converging lens at a distance between one and two focal lengths, a real image will be formed which is inverted and enlarged (is larger than the original object). This is the system used in a PROJECTOR.

If an object is placed in front of a converging lens at a distance of less than one focal length, the lens will behave as a magnifying glass, producing an enlarged, upright, VIRTUAL IMAGE, which can be seen by looking at the object through the lens. *Compare* DIVERGING LENS.

convex (*adj.*) Describing an outward curving surface, particularly in a lens or mirror.

cooling by evaporation When a liquid evaporates, the fastest molecules escape from the surface, resulting in a reduction of the average energy of the remaining molecules and so producing a lower temperature in the liquid.

cooling correction A calculation designed to compensate for any heat losses, such as in an experiment to measure HEAT CAPACITY.

Cooper pair *See* SUPERCONDUCTIVITY.

co-ordinate geometry *See* GEOMETRY.

co-ordinates A set of numbers that represent the position of a point measured from some fixed point called the ORIGIN. To describe the position of a point in an *n*-dimensional space, *n* co-ordinates are needed. CARTESIAN CO-ORDINATES or POLAR CO-ORDINATES are generally used, though other systems are possible.

co-ordination number In a crystalline structure, the number of nearest neighbours surrounding any atom or ion. In the sodium chloride structure, for example, the co-ordination number is 6: each sodium ion has six nearby chlorine ions and vice versa. In a BODY CENTRED CUBIC structure, the co-ordination number is 8, whilst CLOSE PACKED structures have a co-ordination number of 12.

Copernican model The modern model of the SOLAR SYSTEM, which places the Sun at the centre with all the PLANETS, including the Earth, in ORBIT around it (*compare* PTOLEMAIC MODEL). The model was first put forward by Nicolaus Copernicus (1473–1543). In his original model, the orbits of the planets were circular, but more accurate measurements have shown that the planets move in elliptical orbits.

copper losses In electromagnetic systems, particularly TRANSFORMERS, energy losses in the electrical rather than the magnetic parts of the system, such as the coils of a transformer, which are normally made of copper. *Compare* IRON LOSSES.

core 1. (*geology*) The central part of a star or planet. The Earth's core is believed to be composed of nickel and iron, and to be partly liquid with a temperature in excess of 6,000°C.

  2. (*electromagnetism*) The iron centre of a SOLENOID or other electromagnetic device.

Coriolis force An imaginary force that makes a moving object appear to follow a curved path when viewed against a rotating background treated as stationary. The Coriolis force is used to simplify calculations involving rotating systems, which can then be treated as INERTIAL REFERENCE FRAMES. It is the Coriolis force that creates the rotating flow of air in WEATHER SYSTEMS as air flows from high to low pressure. It is also responsible for the rotation of water flowing down a plug hole. In both cases the motion is described with reference to the Earth as being at rest, although it is in fact rotating. *See also* CENTRIFUGAL FORCE, CENTRIPETAL FORCE, CIRCULAR MOTION.

cornea The outer layer of the human eye. The cornea is curved and acts as a fixed lens carrying out most of the REFRACTION of the light entering the eye, before it reaches the lens.

corona The region of high temperature, low density PLASMA around the Sun. The corona is too faint to be seen normally, but is visible during a total ECLIPSE of the Sun. It is distorted by the Sun's magnetic field.

corona discharge An area of IONIZATION around a conducting object in an electric field that allows charge to leak away but is not strong enough to produce a spark.

corpuscle *See* LIGHT.

correlation A measure of the extent to which two quantities are related. For example, in any given population of people there will be a strong correlation between their height and their weight.

  The degree of correlation can be observed by plotting the two quantities on a SCATTER DIAGRAM. The greater the correlation, the closer to a single straight line the points will lie. If the line has a positive gradient the correlation is described as positive, with a perfect correlation being given the value of +1. For a negative gradient the correlation is negative.

  Correlation can be measured numerically by a correlation coefficient. This can be calculated in a number of ways depending on the nature of the data.

correlation coefficient *See* CORRELATION.

cosecant A function of angle. In a right-angled triangle, the cosecant of an angle is defined as the length of the HYPOTENUSE divided by the length of the side opposite to the angle. The cosecant of an angle is the reciprocal of its SINE.

cosine A function of angle. In a right-angled triangle, the cosine of an angle is defined as the length of the side adjacent to the angle divided by the length of the HYPOTENUSE. The cosine of an angle is equal to the SINE of the COMPLEMENTARY ANGLE.

cosmic microwave background ELECTROMAGNETIC RADIATION received from all directions in space with a SPECTRUM characteristic of a BLACK BODY at a temperature of 2.7 K. This radiation is interpreted as the BLACK-BODY RADIATION produced by the hot early Universe at a time when most of the atoms were IONIZED. As the Universe cooled, neutral atoms formed, which no longer interacted with the radiation. This radiation was then stretched to longer wavelengths by the continued expansion of the Universe. The observed distribution of wavelengths matches this theory very well and provides strong evidence for the BIG BANG model of the Universe. The intensity of the radiation arriving from different directions varies very little, but very small differences that have been observed provide evidence of the first stages of clumping which eventually led to the formation of GALAXIES. *See also* COSMOLOGY.

cosmic radiation High-energy charged particles arriving at the Earth from space. Cosmic radiation consists mostly of protons, but also includes electrons and other atomic nuclei. When these so-called 'primary rays' enter the Earth's atmosphere, collisions with oxygen and nitrogen nuclei generate other ELEMENTARY PARTICLES and GAMMA RADIATION. The origin of cosmic radiation is uncertain, though SUPERNOVAE or other violent processes are probably involved, given the very high energies of some of the particles.

cosmological principle The idea that there is no privileged central position for observing the UNIVERSE and that the Universe would appear basically the same viewed from any other point. HUBBLE'S LAW satisfies this principle, as an observer standing in some distant galaxy would also believe it to be true. *See also* COSMOLOGY.

cosmology The study of the Universe as a whole, particularly the early stages in the evolution of the Universe.

The currently accepted model is called the BIG BANG model, in which the Universe expanded from an initial state of very high density and temperature at a fixed time about 12 billion years ago (exact estimates for the current age of the Universe vary quite widely). Evidence for the big bang model comes from the observed motion of the GALAXIES away from one another (*see* HUBBLE'S LAW) and the COSMIC MICROWAVE BACKGROUND. The model is also consistent with the GENERAL THEORY OF RELATIVITY, and is able to explain the observed ratio of helium to hydrogen in the Universe. The big bang model displaced the previous

steady-state theory, which suggested that the Universe has always had its present form. The steady-state theory did not adequately account for the constant supply of hydrogen needed to fuel STARS and had problems in avoiding the tendency of the Universe to collapse under the influence of GRAVITY.

Cosmology also concerns itself with the end of the Universe. If the density of the matter observed is greater than the CRITICAL DENSITY, the expansion will eventually be overcome by gravity and the Universe will end in a BIG CRUNCH, like the big bang in reverse. If the density is less than critical it will expand forever. Theories that predict a period of INFLATION suggest that the Universe is just poised between these two possibilities, with the positive KINETIC ENERGY of the expansion exactly balancing the negative GRAVITATIONAL POTENTIAL ENERGY.

The observed matter in the Universe accounts for a few per cent of the mass needed to reach the critical density, and astronomers are currently engaged in searches for the 'MISSING MASS'. Some believe that this may be in the form of WIMP'S (weakly interacting massive particles) not yet discovered by particle physics. Others believe that the matter is in the form of MACHO'S (Massive Compact Halo Objects), possibly burnt-out stars orbiting in the regions where GLOBULAR CLUSTERS are found (GALACTIC HALOS).

See also COSMOLOGICAL PRINCIPLE.

cotangent A function of angle. In a right-angled triangle, the cotangent of an angle is defined as the length of the side adjacent to the angle divided by the length of the side opposite to the angle. The cotangent of an angle is the reciprocal of its TANGENT.

coulomb The SI UNIT of electric charge. One coulomb is the amount of charge carried in one second by a current of one AMPERE. *See also* FARADAY.

Coulomb's law For point charges, the size of the attractive or repulsive force between two charged particles is proportional to the size of each charge and inversely proportional to the square of the distance between them. Algebraically, the force F between two point charges, q_1 and q_2 is

$$F = q_1 q_2 / 4\pi\varepsilon_0 r^2$$

where ε_0 is the PERMITTIVITY of free space

(equal to $8.85 \times 10^{-12}\,Fm^{-1}$), and r is the separation of the two charges.

counter In physics, any device for detecting and counting IONIZING RADIATION, PHOTONS, etc. They generally work by electronically counting the current or voltage pulse created when a charged particle or photon causes IONIZATION. *See also* CLOCKED BISTABLE.

countercurrent system Any system in which two fluids flow in opposite directions along vessels close to one another, so that exchange, for example of heat or contents, can occur. The level of substance or heat drops in one fluid and rises in the other.

couple A pair of parallel forces that are equal in size and opposite in direction, but do not act along the same line. Whilst a couple will not cause the CENTRE OF MASS of an object to accelerate, it will cause an ANGULAR ACCELERATION (that is, cause the object to turn). The MOMENT of a couple is equal to the size of the forces multiplied by the distance between their lines of action.

covalent bond A bond between two atoms in which electrons in the outer ORBITALS are shared to give each atom a share in a full SHELL of electrons. Covalent bonds are directed in space and so tend to lead to the formation of molecules that are electrically neutral and have only small interactions between one another. Thus covalent materials generally have low melting and boiling points. Examples include gases such as hydrogen, oxygen and carbon dioxide. Covalent bonds also lead to the formation of giant molecules such as diamond, where carbon atoms are bound to one another in a giant lattice. Such structures are very hard and have high melting and boiling points. The rigid nature of the bonds also leads to such solids having unusually high THERMAL CONDUCTIVITIES for non-metals.

covalent crystal A crystalline MACROMOLECULE with all molecules being attached to their neighbours by a regular pattern of COVALENT BONDS. Covalent crystals are hard and have high melting points. *See also* IONIC SOLID.

creep The slow flow shown by some plastics and metals close to breaking point when subject to a LOAD. Creep occurs when atoms in a solid move from one position to another in a lattice. The extra energy they need to do this comes from their thermal vibrations. Creep is thus an ACTIVATION PROCESS, which happens more rapidly at high temperatures. Conversely, materials that are ductile at ordinary temperatures can become far more brittle when cold, as more time is needed for flow to take place.

critical angle The ANGLE OF INCIDENCE at a boundary at which TOTAL INTERNAL REFLECTION first takes place. When light passes from a more dense to a less dense medium, some of the light does not pass the boundary, but is internally reflected. As the angle of incidence increases, the intensity of the reflected beam increases, until, at the critical angle, none of the light passes through the boundary and the whole beam is internally reflected.

critical density The density above which there is enough matter in the Universe for it eventually to collapse on itself as a result of gravity (the 'BIG CRUNCH'). *See* COSMOLOGY.

critical mass The mass of a piece of FISSILE material above which a CHAIN REACTION is no longer prevented by the escape of neutrons from the surface. Many NUCLEAR WEAPONS are triggered by a chemical explosive forcing together two smaller masses to form a lump of greater than critical mass.

critical point *See* CRITICAL TEMPERATURE.

critical temperature, *critical point* The temperature at which there is no distinction between the liquid and gas states of a substance. Above the critical temperature a gas cannot be liquefied by pressure alone. *See also* GAS.

cross-section The surface formed by cutting through an object at right angles to its axis of symmetry. The area of this surface is also called the cross-section. *See also* COLLISION CROSS-SECTION.

crust The solid surface layer of the Earth. *See* GEOLOGY.

cryogenics The study of materials and processes at temperatures close to ABSOLUTE ZERO. *See* SUPERCONDUCTIVITY, SUPERFLUIDITY.

cryostat A vessel for storing a material at low temperatures, usually immersed in a liquefied gas such as liquid helium or liquid nitrogen. Cryostats are generally many-walled vessels with a vacuum between the walls to prevent heat entering by conduction or CONVECTION, and also usually incorporate reflective coatings to reflect THERMAL RADIATION.

crystal A piece of solid material throughout which the atoms are arranged in a single regular arrangement called a lattice. This arrangement is apparent in many naturally occurring crystals, which have symmetrical shapes reflecting the long-range ordering of their atoms.

Crystals are grouped into seven crystal systems, each one based on a different geometric shape. The classes are: cubic, hexagonal, MONOCLINIC, ORTHORHOMBIC, TETRAGONAL, TRICLINIC and trigonal. A given mineral will always crystallize in the same system, although the crystals may not always be the same shape.

Compare DISORDERED SOLID. *See also* BODY CENTRED CUBIC, CLEAVAGE PLANE, CLOSE PACKED, COVALENT CRYSTAL, CRYSTALLOGRAPHY, CUBIC CLOSE PACKED, HEXAGONAL CLOSE PACKED, IONIC SOLID, ISOMORPHIC, SLIP PLANE.

crystalline (*adj.*) Describing a material having the structure of a crystal, though the material itself may be POLYCRYSTALLINE; that is, made of many small crystals with irregular shapes.

crystallography The study of crystals. Early crystallography was aimed at discovering the arrangement of atoms or molecules within a crystal, but more recently similar techniques have been used to map the density of electrons within the molecules forming organic crystals. X-RAY DIFFRACTION and NEUTRON DIFFRACTION are important techniques in this study. *See also* X-RAY CRYSTALLOGRAPHY.

crystal oscillator An OSCILLATOR that has its frequency determined by the RESONANCE of a PIEZOELECTRIC crystal, usually of quartz. Such oscillators are very stable, particularly if the temperature of the crystal is thermostatically controlled. They are commonly used in clocks and watches.

cube 1. A three-dimensional figure having 12 sides all of equal length and at right angles to one another. The volume of a cube is equal to the third power of its length.

2. The third power of any number, thus the cube of a (written a^3) is $a \times a \times a$.

cubic close packed, *face-centred cubic* (*adj.*) Describing a crystal structure in which each layer of atoms is CLOSE PACKED, with each atom surrounded by six others in that layer. The next layer is placed so that it lies above gaps in the previous layer with the third layer again lying in gaps in the second layer, but also above the gaps in the first layer. The atoms in the fourth layer lie directly above those in the first layer, so by labelling the layers of atoms as A, B and C, the packing can be described as ABCABC...

Seen from an angle, this structure can be seen as being cubic, with the UNIT CELL having an eighth of an atom at each corner and half an atom in the middle of each face. This accounts for the alternative name for this structure: face-centred cubic. Many metals occur with this structure, including calcium and copper, though the HEXAGONAL CLOSE PACKED structure is also common.

cumulative frequency In STATISTICS, a measure of the total number of objects in the sample for which some quantity is less than or equal to a certain value. *See also* OGIVE.

cumulonimbus A dense, heavy cloud that appears as a mountain or a huge vertical tower. They are usually associated with heavy rain and sometimes thunderstorms. *See also* WEATHER SYSTEMS.

cumulus A heaped, fluffy cloud, produced by CONVECTION or a COLD FRONT. *See also* WEATHER SYSTEMS.

curie (Ci) A unit of ACTIVITY, now superseded by the BECQUEREL. One curie is equivalent to 3.7×10^{10} Bq.

Curie point The temperature above which a material loses its FERROMAGNETIC properties. *See* PERMANENT MAGNET.

currant-bun model A model of the ATOM, subsequently superseded by the RUTHERFORD-BOHR ATOM model. In the currant-bun model, negative ELECTRONS are dotted around within a large diffuse sphere of positive charge, rather like currants in a currant bun. This model was shown to be incorrect by the RUTHERFORD SCATTERING EXPERIMENT, in which the deflection of ALPHA PARTICLES through large angles showed the existence of a far denser centre of charge in the atom.

current The flow of electric charge through a CONDUCTOR. For a current to flow continuously, there must be a closed conducting path, called a circuit, together with some source of electrical energy, an ELECTROMOTIVE FORCE (e.m.f.). The size of the current through a certain cross-section of conductor is equal to the rate of flow of charge. The unit of current is the AMPERE.

CHARGE CARRIERS may be ELECTRONS, HOLES, or IONS. The direction in which these particles flow around a circuit depends on their charge. In a metal wire, electrons will flow from the negative terminal of the battery (or other source of e.m.f.) around the rest of the circuit to the positive terminal. In many cases it is more convenient to think in terms of CONVENTIONAL CURRENT, the hypothetical flow of positive charge around the circuit from the positive terminal of the battery around the circuit to the negative terminal. The amount of charge is measured in a unit called the COULOMB (C) and the size of the charge on one electron or PROTON is 1.6×10^{-19} C. The current is the number of coulombs passing a given point in one second. When a current flows through a conductor, the material becomes hot; the charge carriers collide with one another or with the surrounding lattice, producing random thermal vibrations.

See also AMMETER, CONDUCTION, CURRENT DENSITY, KIRCHHOFF'S LAWS.

current balance A device in which two parallel wires carry currents, with one of the wires pivoted so it is free to move. The force from the interaction between the currents is then balanced against known weights. See also BALANCE.

current density The amount of electric current flowing per unit area. This is a useful concept, as a conductor of a certain material can generally handle a fixed current density without overheating, regardless of the thickness of the conductor in question.

curvature of space The effect of gravity in the GENERAL THEORY OF RELATIVITY, which accounts for change in the motion of light and of massive objects under the influence of a GRAVITATIONAL FIELD. The path followed by light in curved space is called a GEODESIC LINE. Within a curved space, the usual rules of geometry do not apply: the angles of a triangle do not add up to 180°, for example. A space that is only weakly curved may appear flat over short distances, just as we can imagine small regions of the Earth to be flat, though we know it to be spherical. Depending on whether the Universe is closed or open (see COSMOLOGY), the curvature of space may be positive, like a sphere, where all paths are of finite length, or negative, like a saddle, where all paths are infinitely long.

curve In co-ordinate geometry, a line on a graph that is not straight; that is, where a given change in the x co-ordinate does not always produce the same change in the y co-ordinate.

curved mirror Any mirror with a surface that is not flat, although in many applications the amount of curvature is very slight. Curved mirrors are usually of one of two types, parabolic or spherical.

Rays of light leaving a point in front of a curved mirror called the PRINCIPAL FOCUS of the mirror will be reflected to form a parallel beam. This effect is used in car headlights – a bulb is placed at the principal focus and all the light rays that hit the mirror leave it travelling in a single direction. The same effect is used in reverse in a REFLECTING TELESCOPE – parallel rays of light are brought together at the focus if they are travelling parallel to the PRINCIPAL AXIS of the mirror (the line joining the focus to the centre of the mirror). If the parallel rays of light are at a small angle to the principal focus they will be brought together (or focused) at a point in the FOCAL PLANE. Thus a REAL IMAGE is formed in the focal plane. See also PARABOLIC MIRROR, SPHERICAL MIRROR.

cyclone Alternative name for a DEPRESSION, a region of low ATMOSPHERIC PRESSURE. Severe cyclones in the tropics (tropical cyclones) are accompanied by strong winds and can develop into hurricanes or typhoons.

cyclotron A PARTICLE ACCELERATOR in which charged particles are accelerated in a spiral path inside two hollow D-shaped electrodes, called dees. A magnetic field is applied at right angles to the plane of the dees, which keeps the particles moving in a circular path. The particles are accelerated by an electric field in the evacuated gap between the dees, which then reverses direction whilst the particles are within the dees. As the particles move faster the radius of their path increases, so the time taken to complete one orbit remains constant; thus there is no need to alter the frequency of the electric field or the strength of the magnetic field. After several thousand revolutions, the particles reach the perimeter of the dees, where they are deflected onto the target. The energies that can be achieved by the cyclotron are limited by the RELATIVISTIC increase in mass, which upsets the balance between increasing speed and path radius. See also SYNCHROCYCLOTRON, SYNCHROTRON.

cylinder A space, straight sided and circular in cross-section, in which a PISTON can move.

cylindrical co-ordinates See POLAR CO-ORDINATES.

D

Dalton's law of partial pressures In any mixture of gases, the total PRESSURE exerted by the mixture is equal to the sum of the PARTIAL PRESSURE of each gas on its own (that is, the pressure that each gas would exert if it were present alone).

damping Any force that always opposes motion and increases with the speed of the motion, particularly in the context of oscillating systems. If no other forces acted other than those which caused a system to return to its EQUILIBRIUM position, oscillation would continue for ever. In practice, however, this is never the case. Provided the damping forces are small (that is, the system is underdamped) they will produce an exponential decrease in the amplitude and a small reduction in the frequency. Larger damping forces – systems which are overdamped – will result in no oscillation at all, but the system slowly returns to equilibrium after it has been displaced. An example of overdamping is that of a pendulum immersed in treacle or thick oil – it will no longer swing to and fro, but simply return to equilibrium. The borderline state between underdamping and overdamping is called critical damping, and gives the most rapid approach to equilibrium. *See also* Q-FACTOR.

Darlington pair In electronics, a pair of TRANSISTORS connected together in such a way that the EMITTER current of the first transistor in the pair provides the BASE current of the second, producing a much larger GAIN.

day The time taken for the Earth to revolve once on its axis. The solar day is the time taken for the Earth to revolve once relative to the Sun. The mean solar day is the average value of the solar day for one year. It is divided into 24 hours and is the basis of our civil day.

The sidereal day is the time take for the Earth to revolve once relative to the stars, i.e. the time after which the position of the stars in the sky will be the same as it was one day earlier. Because of the Earth's orbit around the Sun, the Sun appears to move relative to the stars, returning to its original position after one year. This makes the sidereal day four minutes shorter than the mean solar day. Since the Earth's orbit around the Sun is not quite circular, this apparent motion of the Sun is not quite uniform. Thus the time told by a sundial can be in error by as much as fifteen minutes, when compared to a clock that keeps time at a constant rate.

DBS *See* DIRECT BROADCAST SATELLITE.

d.c. motor An electric motor designed to operate from a DIRECT CURRENT (d.c.) supply. Most d.c. motors comprise a set of rotating coils, called an ARMATURE, wound on an iron CORE. The ends of the coils are connected to a series of metal segments arranged in a ring – an assembly known as a SPLIT-RING COMMUTATOR. Brushes, fixed contacts usually made of graphite, rub against the segments of the commutator and carry current in and out of the coils in the armature. The armature assembly rotates in a magnetic field provided either by PERMANENT MAGNETS or by ELECTROMAGNETS. The current in the magnetic field produces a force, which causes the coils to rotate (this is sometimes called the MOTOR EFFECT). The commutator reverses the current direction in each coil every 180° so the force produced always makes the coil rotate in a single direction.

If the coils are electromagnets, the motor can also be made to run off an ALTERNATING CURRENT supply – the reversal of the magnetic field caused by a change in polarity of the supply is compensated by a simultaneous reversal in the direction of the current in the armature.

D.C. motors with electromagnets may have the FIELD COILS (coils that provide the magnetic field in which the armature rotates) connected either in series (series-wound motors) or in parallel with the armature (shunt-wound motors). As the motor speed increases, the movement of the armature through the magnetic field produces a BACK E.M.F. in the armature, reducing the current through it. In a series-wound motor, this also reduces the magnetic field and thus the TORQUE produced by the motor. Series-wound motors thus produce very high torque at low speeds, and are therefore used in electric vehicles. Shunt-wound motors are used where it is important that the speed does not change too much with the mechanical load on the motor.

de Broglie wavelength The wavelength that a particle appears to have when it is exhibiting wave-like properties (*see* WAVE NATURE OF PARTICLES). The de Broglie wavelength λ is given by

$$\lambda = h/p$$

where h is PLANCK'S CONSTANT and p the MOMENTUM of the particle.

debye A measurement of DIPOLE MOMENT used to state the degree to which a molecule is POLAR. One debye is equal to 3.34×10^{-30} Cm. Highly POLAR MOLECULES, such as caesium chloride, have dipole moments of about 10 debye, whilst a more typical polar molecule, such as hydrogen chloride, has a dipole moment of about 1 debye.

decay constant In radioactivity, the probability per second of an atomic nucleus decaying. In a sample of radioactive nuclei, which originally contains N_0 nuclei with decay constant λ, after a time t, the number of nuclei which remain in their undecayed state will be N, where

$$N = N_0 e^{-\lambda t}.$$

decay series *See* RADIOACTIVE SERIES.

deci- Prefix denoting one tenth. For example, a decimetre is one tenth of a metre (0.1 m).

decibel (dB, dBA) A unit for measuring the ratio of two signal levels on a logarithmic scale, commonly used in ACOUSTICS and electronics. A difference of signal strength of 10 dB represents a factor of 10, 20 dB represents a factor of 100, etc, while 3 dB represents a factor of (roughly) 2. Sound levels are often measured in decibels (dBA), with 0 dBA being the quietest sound that can be heard, 10 dBA being 10 times louder, 20 dBA 100 times louder and so on. The 0 dBA level is called the THRESHOLD OF HEARING. *See also* PHON.

decomposition voltage The voltage that must be applied to make an electric current flow in ELECTROLYSIS.

dee The hollow D-shaped ELECTRODE in a CYCLOTRON.

degassing The process of removing dissolved gases from a liquid or gases absorbed into a solid, usually by heating. In particular, degassing is the removal of gas from the walls of a vacuum vessel.

degaussing A term for the removal of any permanent magnetism. *See* PERMANENT MAGNET.

degree (°) A unit based on the division of an interval into equal parts. In particular a measure of ANGLE, with a full circle being 360 degrees (written as 360°) or TEMPERATURE, based on two fixed points, for example the CELSIUS scale, which divides the interval between the freezing and boiling points of water into 100 degrees.

degree of freedom 1. Any one of a set of independent variables needed to define the state of a physical system (such as pressure, temperature, etc.)

2. A way in which a molecule can possess energy independent of any other degree of freedom. Thus a MONATOMIC gas – one composed of molecules each containing only one atom – has three degrees of freedom, motion in each of three independent directions, and a MOLAR HEAT CAPACITY $3R/2$, where R is the MOLAR GAS CONSTANT. A molecule containing two or more atoms in a line has an extra two degrees of freedom, due to rotational motion about two axes at right angles to one another and to the line of the molecule. Molecules with three or more atoms not arranged in a straight line have a further degree of freedom. The molar heat capacity of an IDEAL GAS at constant pressure is $R/2$ times the number of degrees of freedom active in the material.

In solids, there are potentially six degrees of freedom: three due to KINETIC ENERGY from motion in three dimensions and three due to POTENTIAL ENERGY from departure from EQUILIBRIUM in three dimensions. QUANTUM effects mean that these six degrees of freedom are not fully active in most solids at ordinary temperatures – the molecules can have only certain quantized energies. This makes the molar heat capacity less than the expected value of $3R$. *See also* PHASE RULE, DULONG AND PETIT'S LAW.

delocalized orbital One of a number of MOLECULAR ORBITALS that overlap to effectively produce a single large orbital that can hold as many electrons as the original orbitals. This overlap increases the stability of the molecule or crystal lattice. In metals, the VALENCE

ELECTRON orbitals overlap to form a CONDUCTION BAND.

demodulation The reverse process to MODULATION, extracting information from a CARRIER WAVE.

dense (*adj.*) Having a high DENSITY.

density The MASS of a substance contained in a given VOLUME. This depends both on the mass of the molecules or atoms from which the material is made and on their separation. Most solids or liquids have densities of thousands of kilograms per metre cubed, whilst gases under ATMOSPHERIC PRESSURE have densities of just a few kilograms per metre cubed. This difference is due to the fact that in solids and liquids, which are not easily compressed, molecules are pretty closely packed, whilst in gases, which are much more easily compressed, the spacing between the molecules is far greater, typically ten times as great. *See also* DENSITY CAN, RELATIVE DENSITY, SPECIFIC VOLUME.

density can, *eureka vessel* A container with an overflow used to measure the volume of an object of known mass in order to find its DENSITY. The density can is filled with water to the level of the overflow and the object is lowered into it. The volume of water that then overflows is equal to the volume of the object.

dependent variable *See* VARIABLE.

depleted uranium Uranium from which the uranium–235 ISOTOPE has been removed. Depleted uranium is a dense and fairly hard material, and has been used in the manufacture of armour for military vehicles and for the manufacture of bullets and shells. *See also* ENRICHED URANIUM.

depletion layer An area close to a junction between P-TYPE SEMICONDUCTORS and N-TYPE SEMICONDUCTORS, in which there are few free CHARGE CARRIERS as FREE ELECTRONS, and HOLES diffuse across the junction to cancel one another out. *See also* PN JUNCTION DIODE.

depression, *cyclone* In meteorology, a region of lower than average ATMOSPHERIC PRESSURE. As air enters a depression it rotates anticlockwise in the northern hemisphere, and clockwise in the southern hemisphere, as a result of the CORIOLIS FORCE. Depressions typically bring about a mixing of POLAR and TROPICAL AIR MASSES, leading to the arrival of a WARM FRONT, bringing lowering cloud then rain, followed by a COLD FRONT bringing showers. *See also* ANTICYCLONE, WEATHER SYSTEMS.

depression of freezing point The amount by which the freezing point of a solvent is reduced by the presence of dissolved molecules or ions. It is a COLLIGATIVE PROPERTY, independent of the nature of the dissolved particles, but proportional to their molar concentration, with a constant of proportionality known as the cryoscopic constant. Measurements of freezing point depression for measured masses of a material dissolved in a solvent can be used to estimate RELATIVE MOLECULAR MASSES.

depth of field The range of distances away from a camera at which an object can be placed and still produce an acceptably sharp image on the film.

derivative The function that measures the rate of change of some other function. It is found by differentiation (*see* CALCULUS). If $y = x^n$, the derivative of y with respect to x, usually written as dy/dx, is nx^{n-1}.

derived unit Any UNIT that is not a BASE UNIT. For example, the SI UNIT of speed, the metre per second, is derived from the metre and second. Some derived units are given special names – the SI unit of charge, for example, is called the COULOMB, but the definition of the coulomb is derived from those for the AMPERE and the second.

destructive interference The situation in which two waves arrive at the same point exactly OUT OF PHASE. The waves tend to cancel one another out and the resulting wave has an amplitude equal to the difference in amplitude between the two original waves, or zero if the two waves had equal amplitudes.

In order to produce destructive interference, TRANSVERSE WAVES must either be POLARIZED in the same direction, or as is often the case with light, unpolarized (really a mixture of all polarizations). The waves must also be COHERENT if destructive interference is to occur constantly at a given point.

See also CONSTRUCTIVE INTERFERENCE, INTERFERENCE.

deuterium The naturally occurring heavy ISOTOPE of hydrogen, hydrogen–2 (one proton, one neutron).

dew Water vapour from the atmosphere condensed on the ground as the air cools at night. Dew tends to be thicker when the sky is clear, since THERMAL RADIATION is not reflected back to the ground. *See also* WEATHER SYSTEMS.

dew point The temperature at which air becomes SATURATED with water vapour. *See also* HUMIDITY.

diamagnetism The property of all substances that have a RELATIVE PERMEABILITY slightly less than 1. An applied magnetic field is slightly weaker inside the material than it would be if the material were not present. A bar of diamagnetic material will align itself at right angles to a magnetic field. Diamagnetism is caused by the interaction between the electrons in the atom and the external magnetic field. The phenomenon occurs in all substances, although in many cases it is masked by the much greater effects of PARAMAGNETISM or ferromagnetism (*see* FERROMAGNETIC).

diaphragm 1. In general, a thin sheet separating off a region of space, but able to move.
   2. In a CAMERA, an adjustable circular APERTURE made from overlapping sheets of metal.

diatomic (*adj.*) Describing a molecule comprising just two atoms, such as chlorine, Cl_2, or hydrogen chloride, HCl.

dielectric An electrically insulating material, in particular one that experiences a displacement of charge – but not a flow of charge – in an applied electric field. The inclusion of a dielectric between the plates of a CAPACITOR increases the CAPACITANCE over that in a vacuum by a factor known as the RELATIVE PERMITTIVITY. The molecules in a dielectric are POLARIZED by the electric field. This produces surface charges on the dielectric that are of the opposite sign to the charges on the adjacent capacitor plates. The result of this is to weaken the electric field in the dielectric, so that the same charge on the capacitor plates now leads to a lower POTENTIAL DIFFERENCE between them, i.e. to a greater capacitance. *See also* PERMITTIVITY.

dielectric constant *See* RELATIVE PERMITTIVITY.

diesel engine A type of INTERNAL COMBUSTION ENGINE that burns a lightweight fuel oil. It is similar in principle to the four-stroke PETROL ENGINE, but only air is drawn into the engine during the INDUCTION STROKE. The air is then compressed to a greater extent than in a petrol engine, and at the top of the COMPRESSION STROKE fuel is injected directly into the engine. Here, the compressed air is at a high enough temperature for the fuel to ignite without the need for a SPARK PLUG.

   Diesel engines tend to be used on larger vehicles such as trucks and trains. The greater compression of the air means they are more efficient, however the need for a stronger construction makes them heavier and slower to accelerate.

differential amplifier An AMPLIFIER whose output depends on the difference between two inputs, called the INVERTING INPUT and NON-INVERTING INPUT. If the voltages at these inputs are V_- and V_+, the output voltage will be $A(V_+ - V_-)$, where A is the GAIN. *See also* OPERATIONAL AMPLIFIER.

differential equation An EQUATION that contains one or more DERIVATIVES of a variable. In physics, differential equations are vital because they show the rate at which a particular quantity changes as a function of another variable. In mechanics, for example, the rate of change of the momentum of a particle is equal to the force on the particle.

differentiation The mathematical process of finding the DERIVATIVE of a function. *See* CALCULUS.

diffract (*vb.*) Of a wave, to spread out by DIFFRACTION.

diffraction The spreading out of any wave motion as it passes through an APERTURE, or when obstructed by a narrow barrier. The waves do not pass through with a sharp edge, like a shadow, but spread out, or diffract. The amount of diffraction is greatest for small apertures and for long wavelengths and depends on the ratio of wavelength to aperture size. For this reason diffraction is not normally noticed on an everyday basis with light: the wavelength is too short.

   One consequence of diffraction is that no system, such as a telescope for example, can ever form a perfect image of an object – the light entering the telescope has WAVEFRONTS that are limited by the lens or the mirror of the telescope. The larger the aperture compared to the wavelength, the less the image will be affected by diffraction (the RESOLUTION is higher). RADIO TELESCOPES, which operate at longer wavelengths than optical telescopes, have relatively poor resolution, so are less able to distinguish as separate two objects that appear close to one another in the sky.

   On a simple level, diffraction can be explained by recognizing the fact that it is not possible to have a wavefront with a perfectly sharp edge. A more sophisticated explanation relies on HUYGENS' CONSTRUCTION, which provides a detailed description of the distribution of the AMPLITUDE of a wave that has passed through an aperture. In the case of FRAUN-HOFER DIFFRACTION of light from a single slit, there is a broad central peak in the amplitude before it falls to zero, but further out it increases again, reaching about one third of its maximum value (a subsidiary maximum). There are therefore a succession of points where the amplitude reaches zero, with smaller and smaller peaks in amplitude between each minimum. The zero points are evenly spaced. As the slit is narrowed, the pattern becomes wider and less bright. The first minimum is explained by Huygens' construction in terms of light from each point in one half of the slit interfering destructively (see DESTRUCTIVE INTERFERENCE) with light from the corresponding point in the bottom half of the slit.

For Fraunhofer diffraction from a single slit, of width a, provided that the wavelength λ is much less than the width of the slit, the points of zero amplitude will be at angles θ to the straight ahead direction, where

$$\theta = n\lambda/a$$

where n is a whole number.

See also DIFFRACTION GRATING, FRESNEL DIFFRACTION, INTERFERENCE, YOUNG'S DOUBLE SLIT EXPERIMENT.

diffraction envelope A gradual overall variation in intensity of some patterns of light, such as YOUNG'S FRINGES, caused by the DIFFRACTION of the light from the individual slits through which light passes to make the pattern.

diffraction grating A mirror or transparent piece of material on which many regularly spaced lines are ruled to form a series of slits. Light diffracts from each slit, but only interferes constructively (see CONSTRUCTIVE INTER-FERENCE) in those directions in which the PATH DIFFERENCE between light from one slit and the next is a whole number of wavelengths.

For a grating illuminated by light striking the grating at 90°, the light will leave the grating at angles θ such that

$$d\sin\theta = n\lambda$$

where λ is the wavelength of the light, d is the grating spacing (the distance between adjacent slits) and n is a whole number called the order of the diffraction maximum. Thus different wavelengths leave the grating at different

angles. If the light does not hit the grating square on, but at an angle ϕ, this introduces a PHASE difference between light leaving one slit and the next, modifying the formula to

$$d(\sin\phi + \sin\theta) = \lambda$$

The RESOLUTION of a grating is its ability to distinguish between two similar wavelengths and is proportional to the number of slits in the grating. If there are N slits, the smallest wavelength separation detectable will be $\delta\lambda$, where

$$\delta\lambda = \lambda/N$$

The ability of a diffraction grating to split light up according to its wavelength has led to its use in many SPECTROMETERS. PRISMS were originally used, but diffraction gratings can be made to give a greater dispersion for resolving fine detail in a SPECTRUM, and also disperse light in a way which can be calculated with a precise mathematical formula, leading to more accurate measurement of wavelengths.

diffuse (vb.) To spread out by DIFFUSION.

diffusion The spontaneous and random movement of molecules or particles in a fluid (gas or liquid) from a region where they are at a high concentration to one where they are at a low concentration, until a uniform concentration or dynamic EQUILIBRIUM is achieved. Once at a uniform concentration, the molecules will continue to move in random motion, but there is no net diffusion. The concentration gradient is the difference in concentration of a substance between two regions. The rate at which one material diffuses into another is determined by the average speed of its molecules and by the MEAN FREE PATH. Compare ACTIVE TRANSPORT. See also GRAHAM'S LAW OF DIFFUSION.

digital (adj.) Describing a system in which only whole numbers are handled, normally only two values, corresponding to BINARY 1 and 0 are used. Compare ANALOGUE.

digital signal An electrical signal represented by two values only: high and low (or 1 and 0). High is a voltage close to the supply voltage, while low is a voltage close to zero. A digital system has the advantage that any small change in the signal, the addition of electrical NOISE for example, will not prevent the signal from being recognized as a one or a zero. Digital systems are therefore much less prone to interference than systems based on ANALOGUE signals.

The disadvantage of this technique is that most signals that we wish to handle are analogue in nature, and must be converted into a binary number several digits long to give a reasonably continuous range of values. Modern digital electronic circuits can provide complex processing power at high speed in a small space with low cost and low power consumption. This, combined with the other advantages of digital systems, means that they are increasingly used.

digital-to-analogue converter A device that converts DIGITAL SIGNALS into ANALOGUE signals. Such devices are used to convert information that has been stored or processed digitally back to its original analogue form. In playing a COMPACT DISC, for example, the digital signal that represents the stored music must be converted to an analogue signal which, after amplification, is used to drive a loudspeaker. With a suitable choice of resistors and each input connected to either a HIGH or LOW voltage, a SUMMING AMPLIFIER, which adds these signals together in appropriate proportions, can be used as the basis of a digital-to-analogue converter.

dimensionless (*adj.*) Describing a quantity, or combination of quantities, that has no DIMENSIONS, so is expressed without a UNIT and has the same numerical value whatever units system is used to calculate it. A simple example is the COEFFICIENT OF FRICTION, which will take on the same value whatever units of FORCE are used in its calculation, provided they are used consistently.

dimensions The fundamental physical quantities that describe any system, and the powers to which these quantities are raised to obtain a derived physical quantity. In a mechanical system, the basic physical quantities are usually taken to be mass (M), length (L) and time (T). Using these dimensions, the derived physical quantity of volume will have the dimension L^3, velocity is L/T (or LT^{-1}), and MOMENTUM, which is a product of mass and velocity, will have the dimensions MLT^{-1}.

diminished (*adj.*) Describing an IMAGE that is smaller than the original object.

diode An electronic device that allows current to pass in one direction only. *See* PN JUNCTION DIODE, THERMIONIC DIODE, VARICAP DIODE. *See also* BRIDGE RECTIFIER, RECTIFIER.

dioptre (D) A measure of the refractive power of an optical lens, equal to the reciprocal of the FOCAL LENGTH in metres. Thus, for example, a 10 D lens has a focal length of $\frac{1}{10}$ m or 0.1 m. A negative power represents a DIVERGING LENS.

dipole A magnetic system containing two POLES, such as a BAR MAGNET, or an ELECTROSTATIC system containing two equal but opposite charges. At a distance from the dipole that is large compared to the separation of the charges or poles, the field produced decreases with increasing distance by a factor equal to the inverse third power of the distance from the dipole. Thus if the distance is doubled, the field falls to one eighth of its original strength. *See also* MAGNETIC DIPOLE.

dipole aerial An aerial made up of two lengths of wire, or metal rods, end to end, with a SIGNAL fed into, or taken from, a small central gap between the two conductors. Such aerials are directional, best able to transmit and receive in a plane at right angles to length of the aerial. The aerial transmits and receives ELECTROMAGNETIC WAVES that are polarized with the ELECTRIC FIELD parallel to the length of the aerial. The dipole aerial has greatest efficiency when the total length of the aerial is half the wavelength of the signal being transmitted or received. Important variations are the folded dipole, in which a second conductor connects the two ends of the dipole, and the quarter-wavelength vertical in which a single metal rod is used above a metal surface, such as the roof of a vehicle. The reflection of the rod in this surface acts as the second part of the dipole. *See also* YAGI.

dipole moment The strength of a dipole. For an electric dipole, the dipole moment is equal to the size of the charges multiplied by the distance between them. The electrical dipole moment of a MOLECULE is a measure of the extent to which that molecule is POLAR. *See also* DEBYE.

direct broadcast satellite (DBS) A television system in which individual homes are equipped with a receiving AERIAL and PARABOLIC DISH reflector (satellite dish) for the reception of television signals from a satellite in GEOSTATIONARY ORBIT.

direct current (d.c.) A steady electric current flowing always in the same direction. *Compare* ALTERNATING CURRENT.

discharge (*vb.*) To remove the electric charge from an object. This is most easily done by connecting it to the EARTH (a process called earthing or grounding). Electrons flow either to or from the earth until the earthed object has no overall charge.

discharged (*adj.*) The state of a CAPACITOR or CELL when it has no stored charge. *See also* CHARGED.

discrete data In statistics, quantities that can only take on particular values, with values between these having no meaning. For example the number of children in a family can only ever be a whole number. *Compare* CONTINUOUS DATA.

disintegration A term sometimes applied to the decay of a radioactive nucleus. *See* RADIOACTIVITY.

dislocation An imperfection in the lattice structure of a solid, particularly a metal. A dislocation arises as a result of a misalignment of atomic planes as the crystals of the material are forming, with a plane of atoms coming to a halt in the middle of the atom or forming a 'spiral staircase' structure. *See also* ALLOY.

disordered solid A solid in which the atoms are not arranged in a regular crystalline lattice. Some materials, such as glass, are naturally disordered; others form disordered solids only if they are cooled so rapidly that crystals do not have time to form.

dispersion The variation of the REFRACTIVE INDEX of a medium with the wavelength of the light passing through it. Thus the amount of REFRACTION will be slightly different for different colours of light. This effect is used in a PRISM to split white light into the colours from which it is made up.

dispersive (*adj.*) Describing a material that shows DISPERSION.

displace (*vb.*) To push out of the way.

displacement The volume, mass or weight of fluid pushed out of the way by an immersed object. In particular, the water displaced by a ship, where the displacement is equal to the weight of water contained in the volume of the hull below the water level. By ARCHIMEDES' PRINCIPLE, this will be equal to the weight of the ship.

dissipate (*vb.*) In physics, to turn energy, in particular electrical energy, into heat.

dissipative force A force, such as friction, that always opposes the direction of motion, and converts KINETIC ENERGY to heat. Dissipative forces produce DAMPING in SIMPLE HARMONIC MOTION.

distortion An unwanted change introduced in an optical image or an electronic signal. The term particularly refers to the measure of the extent to which the output from an AMPLIFIER is not simply a constant times the input. *See also* ABERRATION.

distribution, *frequency distribution* In statistics, a set of measurements that show how many times a quantity takes a particular value, or a value that lies in a particular range.

diverging lens A lens that is thinner in the middle than at the edges and causes parallel rays of light to spread out as if they had come from a point behind the lens called the VIRTUAL FOCUS. The FOCAL LENGTH of a diverging lens is usually taken as being negative and is the distance of the virtual focus from the lens. An object viewed through a diverging lens always appears diminished (smaller in size than the original object) and is a VIRTUAL IMAGE. *Compare* CONVERGING LENS.

division of amplitude A technique for obtaining two or more COHERENT light beams from a non-coherent source by using partial REFLECTION to split a single beam into two. The beams are then brought together again to interfere after travelling along different paths. *See also* INTERFERENCE.

division of wavefront A technique for obtaining two or more COHERENT light beams from a non-coherent source by using DIFFRACTION to spread a small section of a wavefront over a large angle. These sections are then brought together again to interfere after travelling along different paths. *See also* INTERFERENCE.

dodecahedron A POLYHEDRON with 12 plane faces.

domain **1.** A region within a FERROMAGNETIC material in which all the atomic magnets are aligned in the same direction. *See also* PERMANENT MAGNET.

2. (*mathematics*) The set of values that a quantity may take, particularly the independent variable in a FUNCTION. For example, the domain of the inverse sine function is -1 to $+1$, and the function is meaningless for values outside this range.

domain wall In a FERROMAGNETIC material, the boundary between one DOMAIN and the next, which is usually magnetized in the opposite direction, with the direction of magnetization varying gradually through the thickness of the wall.

donor atom An atom that provides a conduction electron in the DOPING of a SEMICONDUCTOR to form an N-TYPE SEMICONDUCTOR.

donor impurity An element added to a SEMICONDUCTOR in a DOPING process that releases FREE ELECTRONS, producing an N-TYPE SEMICONDUCTOR.

doping The addition of small quantities of impurity to a SEMICONDUCTOR to alter its electric properties. By doping different parts of a silicon wafer in different ways, complex INTEGRATED CIRCUITS can be built up. *See also* N-TYPE SEMICONDUCTOR, P-TYPE SEMICONDUCTOR.

Doppler effect The apparent change in observed frequency (or wavelength) of a wave due to relative motion between the source of the wave and the observer. An example is a police car siren that increases in PITCH (frequency) as it moves towards a stationary observer, and decreases as it moves away. As the car approaches, the WAVEFRONTS emitted in the direction of the observer are 'bunched' more closely together, since the source is chasing the wavefronts in that direction. As a result, the wavelength is reduced and the frequency increased. On moving away, the wavefronts are further apart, leading to a longer wavelength and lower frequency.

With visible light, and other ELECTROMAGNETIC WAVES, the same effects are observed, though the details are modified by the SPECIAL THEORY OF RELATIVITY. If a light source moves towards the observer, the light appears more blue, blue being of a shorter wavelength than the other colours. This phenomenon is called a blue-shift. If the source and the observer are moving away from one another, the light appears more red (a RED-SHIFT).

For a wave source moving towards the observer with a speed u and the observer moving towards the source with a speed v, then if the speed of the waves is c and the frequency of the source is f, the observer will hear a frequency f', where

$$f' = f(c + v)/(c - u)$$

For light, if the relative motion of the source towards the observer is v,

$$f' = f\{(1 + v/c)/(1 - v/c)\}^{-1}$$

The Doppler effect has many uses in detecting and measuring motion. In astron-omy, for example it is widely used to measure the speed of stars and galaxies (*see* HUBBLE'S LAW). It can also be used to infer the temperature of a gas as a result of Doppler shifts in the wavelengths in the ABSORPTION SPECTRUM or EMISSION SPECTRUM as a result of thermal motion.

A signal reflected by a moving object will also be Doppler shifted; this phenomenon is used to measure speeds in some police speed traps, and to distinguish moving targets from stationary clutter in RADAR systems. Doppler shift in ULTRASOUND has also been used to measure the rate of fluid flow, for example in human blood vessels.

Doppler shift The apparent change in WAVELENGTH and FREQUENCY as a result of relative motion between a wave source and the observer of the wave. *See* DOPPLER EFFECT.

d-orbital The third lowest energy ORBITAL for a given PRINCIPAL QUANTUM NUMBER. They exist only for principal quantum number of 3 or greater. There are five d-orbitals for each principal quantum number. Four of these consist of four lobes in an X-shape, each lying in a single plane, whilst the fifth d-orbital consists of two lobes surrounded by a TORUS.

dose The amount of IONIZING RADIATION absorbed by a living organism, usually human, over a specified period of time. It is usually specified in terms of the amount of energy deposited by the radiation, measured in GRAY. Some forms of ionizing radiation are more biologically harmful than others, and to take account of this, the DOSE EQUIVALENT is often considered.

dose equivalent The DOSE of IONIZING RADIATION absorbed by a human, multiplied by a factor that takes account of the relative damage done by different types of ionizing radiation. This factor is 20 for ALPHA PARTICLES and 1 for BETA PARTICLES. For GAMMA RADIATION, the factor varies between 0.60 and 0.28, depending on the energy of the gamma rays concerned. The SI UNIT of dose equivalent is the SIEVERT.

double glazing The fitting of a window with two layers of glass. This reduces heat loss from a room by providing extra glass-air boundaries and so increasing the THERMAL CONTACT RESISTANCE.

double refraction *See* BIREFRINGENCE.

down A FLAVOUR of QUARK.

drag A DISSIPATIVE FORCE caused by a solid object

moving through a fluid, such as air or water. Provided the speed of the motion is slow, and the solid has a smoothly curved or STREAMLINED shape, there will be a smooth flow of the fluid around the solid. This type of flow is called LAMINAR FLOW – there is a gradual change in the speed of the fluid around the moving object. In this case the drag will be proportional to the speed of motion through the fluid and to the VISCOSITY of the fluid.

At higher speeds, or for less streamlined objects, the flow will become TURBULENT, with vortices – irregular spiralling patterns of fluid in the wake of the moving object. Usually there is still a layer of laminar flow, called the BOUNDARY LAYER, close to the surface of at least part of the object. Turbulent flow consumes much more energy than laminar flow, so the force increases much more rapidly with velocity, being roughly proportional to the square of the velocity. The size of the force is proportional to the area of the object moving through the fluid, and also depends on the shape of the object.

See also COEFFICIENT OF DRAG, STOKES' LAW.

drain In electronics, the ELECTRODE at one end of the CHANNEL in a FIELD EFFECT TRANSISTOR, from which CHARGE CARRIERS leave the channel.

drift chamber A development of the SPARK CHAMBER that enables the trails of ionizing particles to be measured more precisely. The wires are placed further apart and low voltages are used, so that the IONS created by the passage of the charged particle move more slowly and do not cause the further ionization that leads to a spark. The voltage and the spacing of the wires are chosen so that the electric field just compensates for the energy lost by the ions in collisions with gas molecules. In this way, the time of arrival of the ions gives a measure of the position at which they were produced, enabling the path of the particles to be reconstructed.

drift tube A hollow metal ELECTRODE through which a charged particle can move without influence from external electric fields, such as in a LINEAR ACCELERATOR.

drift velocity The imbalance in speed that causes CHARGE CARRIERS to produce a net flow of charge though a conductor.

Drift velocities are very small compared to the speed of random motion of charge carriers. In a typical metal, for example, electrons have random speeds of around $10^5\,ms^{-1}$, whilst the largest current that a wire made of such a metal could carry without melting requires the charges to move with an average speed of only a few millimetres per second. For a wire of cross-sectional area A made of a material with n charge carriers per metre cubed, carrying a current I, the drift velocity v is

$$v = I/nAe$$

where e is the charge on an electron.

driving force The external force that drives a FORCED OSCILLATION.

driving frequency The frequency at which a system is forced to oscillate by some external force. See FORCED OSCILLATION.

dry cell Any VOLTAIC CELL in which the ELECTROLYTE is in the form of a paste. An example is the ZINC-CARBON CELL. Such cells can be used at any angle without any risk of the electrolyte being spilt. Compare WET CELL.

D-T reaction A NUCLEAR FUSION process involving deuterium (hydrogen–2, which is naturally occurring) and tritium (hydrogen–3, which is radioactive with a HALF-LIFE of about 12 years, so must be manufactured artificially). At high enough temperatures (many million kelvin), these nuclei collide violently enough to fuse, producing a helium–4 nucleus and a neutron. The neutron released can be captured by lithium to produce further tritium.

ductile (adj.) Able to be drawn out into a long strand, such as a wire. Ductility is an important property of many metals.

Dulong and Petit's law An experimental rule that states that the MOLAR HEAT CAPACITY of a solid tends to $3R$ at high temperatures, where R is the MOLAR GAS CONSTANT. The theoretical basis for this lies in EQUIPARTITION OF ENERGY between six DEGREES OF FREEDOM, ELASTIC and KINETIC ENERGY in each of three dimensions.

dummy leads In electrical engineering, a pair of wires that do not carry a signal, but which follow the same path as the wires that do carry a signal. In this way, any effect on the signal caused by changes in the environment can be measured. In a RESISTANCE THERMOMETER, for example, dummy leads are used to compensate for any changes in resistance caused by changes in temperature of the wires connecting the temperature sensing element to the rest of the instrument.

duplex (*adj.*) Describing a TELECOMMUNICATIONS path along which messages can be sent in both directions simultaneously. *Compare* SIMPLEX.

dynamic equilibrium *See* EQUILIBRIUM.

dynamic friction The FRICTION between two objects moving relative to one another. *See also* LIMITING FRICTION, STATIC FRICTION.

dynamic pressure The pressure an object experiences due to its motion through a fluid, caused by there being more molecular collisions per second on the side that is moving forwards than on the other side. This extra pressure can be used to measure the speed of an object through the fluid – an example of this is the PITOT-STATIC system used to measure the speed of an aircraft through the air by comparing the pressure on a forward facing pipe (the pitot head) with that on a sideways opening (the static vent) where there is no dynamic pressure. The difference in pressures is displayed as a speed on an instrument called an air speed indicator. *See also* HYDROSTATIC PRESSURE.

dynamics The branch of MECHANICS concerned with the motion and acceleration of objects under the action of forces. *Compare* STATICS.

dynamo A device that converts mechanical energy into electrical energy. It is based on the principle that if a conductor moves in a magnetic field, a current will flow in the conductor. The simplest dynamo comprises a coil of conducting wire (called an ARMATURE) rotated between the poles of an ELECTROMAGNET. The mechanical energy of rotation is converted into electrical energy in the form of a current flowing in the armature. The term dynamo is usually taken to mean a device that generates a DIRECT CURRENT. An ALTERNATING CURRENT generator is usually called an ALTERNATOR.

dynamometer A machine for measuring the power output of an engine, particularly of an INTERNAL COMBUSTION ENGINE. The engine is made to do WORK, by driving a paddle wheel through a fluid for example, and the power delivered is measured, usually from the force applied by the engine.

dyne A unit of force in the C.G.S. SYSTEM. One dyne is the force needed to make a MASS of 1 g accelerate at $1\,\mathrm{cm\,s^{-2}}$. It is equivalent to 10^{-5} NEWTON.

E

e, *exponential* The BASE of NATURAL LOGARITHMS, approximately 2.713. e is an irrational number, but it can be represented by the SERIES

$$1 + 1/2! + 1/3! + 1/4! + \ldots.$$

Earth The third planet from the Sun. It has a mean diameter of 13,000 km, a mass of 6.0×10^{24} kg, and mean density of about 5,500 kg m^{-3}. The Earth's orbit lies between those of Venus and Mars, and it has a mean orbital radius of 1 AU (149,500,000 km). It has an orbital period of one year and a rotational period of 24 hours. *See also* ATMOSPHERE, GEOLOGY.

earth A connection to the Earth to DISCHARGE an object; the act of making this connection.

As well as shielding the equipment from electric fields, earthing an object provides protection against ELECTRIC SHOCK. If a LIVE wire touches the case of the object, a large current will flow to Earth and this should blow any FUSE or CIRCUIT BREAKER connected to the power supply. If the earth wire were not present a smaller, but still dangerous, current would flow through any person touching the equipment whilst in contact with the earth. This current may not be sufficient to blow a fuse.

earth-leakage circuit breaker, *residual current device (RCD)* A CIRCUIT BREAKER that operates when it detects a difference in the CURRENTS flowing along the LIVE and NEUTRAL wires of an ALTERNATING CURRENT power supply. This acts as a protective device, disconnecting the supply when any current is flowing from the live wire to earth, possibly causing an electric shock, rather than returning along the neutral wire. The device operates using the MAGNETIC FIELD produced by a SOLENOID with two coils, through which the live and neutral currents flow in opposite directions, creating a magnetic field that is proportional to the difference in these currents. This device can allow large currents to flow uninterrupted through the circuit, but will break the circuit if only a small current is lost to earth.

earthquake A sudden vibration of the Earth caused by the movement of TECTONIC PLATES, particularly along SUBDUCTION ZONES. *See also* RICHTER SCALE, SEISMIC WAVE, SEISMOLOGY.

eccentricity A number describing the shape of any CONIC SECTION, in particular a measure of the amount by which an ELLIPSE differs from a circle (which has an eccentricity of 1). The equation for an ellipse can be written in POLAR CO-ORDINATES as

$$r = r_0(1 + e\cos\theta)$$

where e is the eccentricity.

echo A reflected sound wave.

eclipse An event in which the Sun, Moon and Earth line up. In a solar eclipse, the Moon

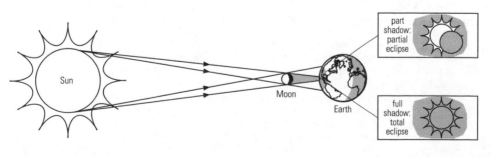

Solar eclipse.

blocks out the Sun as seen from part of the Earth's surface. In a lunar eclipse the Moon moves into the Earth's shadow and appears much darker than normal.

eclipsing binary *See* BINARY STAR.

ecliptic The path traced out by the Sun against the background of stars as a result of the Earth's orbit around the Sun. Seen from the Earth, the Sun appears to move around the ecliptic once every year.

ecliptic plane The plane containing the Sun and the Earth's orbit. The orbits of all the planets lie roughly in the ecliptic plane.

eddy current An electric current induced in a conductor moving through a magnetic field, or which is in an area of changing magnetic field, which circulates in that conductor without doing any useful work. Eddy currents are a potential source of energy loss in many electromagnetic machines. To minimize the effects of eddy currents, iron cores are often made in a LAMINATED construction, from thin sheets of iron coated with an insulating varnish. In this way the eddy currents are broken up into smaller loops without much reduction in the total volume of magnetic material.

efficiency A measure of the performance of a machine. The efficiency of a machine is the fraction of the energy applied to the machine that appears in the desired form at its output.

Efficiency = useful work done/
total energy input

Efficiency is sometimes multiplied by 100 to give a percentage efficiency.

effort The FORCE applied to a MACHINE to move a LOAD.

effusion The flow of a gas through a small opening, comparable in size to the spacing between the molecules in the gas, so that the gas cannot be considered to flow through the opening as a fluid, but rather passes through one molecule at a time. The rate of effusion is proportional to the ROOT MEAN SQUARE speed of the molecules, and thus at a given temperature is inversely proportional to the square root of the RELATIVE MOLECULAR MASS.

elastic (*adj.*) Describing a substance or a process by which the substance is deformed by an applied force, but returns to its original shape once the force is removed. *See also* HOOKE'S LAW, PLASTIC.

elastic collision A collision in which the total

KINETIC ENERGY of the colliding bodies before and after the collision are the same. Collisions involving objects that bounce off each other are partially elastic, whilst collisions between atoms and between sub-nuclear particles are totally elastic unless some energy is used to excite the atom or create a new particle of higher mass. *Compare* HYPERELASTIC COLLISION, INELASTIC COLLISION. *See also* COEFFICIENT OF RESTITUTION.

elastic energy, *strain energy* The energy of an object that has been elastically deformed in some way, such as a stretched spring. The elastic energy E of a spring stretched through a distance x by a force F is

$$E = \tfrac{1}{2}Fx$$

elastic limit The maximum TENSILE STRAIN that can be applied to an ELASTIC material beyond which the material does not return to its original length when the stretching force is removed. *See also* FATIGUE, HOOKE'S LAW.

elastic modulus Any one of the three measures of the stiffness of an ELASTIC material: YOUNG'S MODULUS, the MODULUS OF RIGIDITY or the BULK MODULUS.

electret 1. A plastic material containing POLAR MOLECULES that are aligned in such a way that the material is electrically POLARIZED and carries charges on its surface, even in the absence of an electric field.

    2. A microphone based on such a material. A charged plastic DIAPHRAGM is moved by a sound wave and induces charges on a fixed metal ELECTRODE nearby.

electrical conductor *See* CONDUCTOR, ELECTRICAL.

electrical energy Energy carried by the flow of charge around an electric circuit. For a total charge Q moving through a POTENTIAL DIFFERENCE V the energy gained or lost is QV. On a domestic or industrial scale, electrical energy is also sometimes measured by multiplying the power in kilowatts by the time in hours over which that power is used.

electrical potential energy The energy of a charge in an electric field. The change in electrical potential energy E of a charge q moving through an electrical POTENTIAL DIFFERENCE V is

$$E = qV$$

See also ELECTRIC POTENTIAL.

electric arc A highly luminous spark between two ELECTRODES, which themselves become vaporized by the heat of the arc. Arcs may be used as a source of heat for welding, or as a highly intense source of light. Carbon electrodes are often used, driven forward mechanically as the surface is burnt away by the arc. Arc lights have generally been replaced by more reliable light sources.

electric charge *See* CHARGE.

electric field The region of influence around any electric CHARGE that causes any other charge to experience a force proportional to its charge. Numerically, the strength of an electric field at a point is the force per unit charge experienced by a vanishingly small charge placed at that point. The SI UNIT of electric field is NC^{-1} (NEWTON/COULOMB) or Vm^{-1} (VOLT/METRE).

The electric field produced around a point charge, or outside a spherically symmetric distribution of charge is proportional to the size of the charge and inversely proportional to the distance from the centre of charge. The force F on a charge q in an electric field E is

$$F = Eq$$

The field E at a distance r from a point charge q is

$$E = q/4\pi\varepsilon_0 r^2$$

where ε_0 is the PERMITTIVITY of free space.

Between two parallel metal plates with opposite charges (as in a CAPACITOR for example) the electric field will be constant. The electric field between two parallel plates between which there is a distance d and a POTENTIAL DIFFERENCE V is

$$E = V/d$$

See also ELECTRIC POTENTIAL.

electric field lines Lines drawn on a diagram that show the direction of the force on a positively charged particle placed at that point in an ELECTRIC FIELD. These field lines are drawn from positive charges to negative charges. The closer the lines the greater the ELECTRIC FIELD STRENGTH. Electric field lines always leave a conducting surface at right angles to that surface.

electric field strength A term for ELECTRIC FIELD, emphasizing the numerical value of the field rather than its general direction or pattern. *See* ELECTRIC POTENTIAL.

electric flux A measure of the total amount of ELECTRIC FIELD passing through an area. If the field is E, and it passes through area A at an angle θ, then the electric flux Φ is

$$\Phi = EA\cos\theta$$

electric flux density An obsolete term for ELECTRIC FIELD strength.

electricity The general term describing all effects caused by electric CHARGE, whether at rest (electrostatic; *see* STATIC ELECTRICITY) or in motion (electric CURRENT).

electric motor Any device for converting electrical energy into rotational motion. Generally, motors use the interaction between electric currents and magnetic fields to produce a force (*see* MOTOR EFFECT). *See also* D.C. MOTOR.

electric potential A measure of the WORK needed to move an electric charge to a particular point, or the energy that would be produced if the charge moved away from that point. The electric potential at a given point is defined as the work done per unit charge in bringing a vanishingly small positive charge to that point from infinity. The SI UNIT of electric potential is the VOLT. The electric potential V at a distance r from a point charge q is

$$V = q/4\pi\varepsilon_0 r$$

where ε_0 is the PERMITTIVITY of free space.

The electric field strength at any point is equal to minus the rate of change of electric potential with distance at that point. Between a pair of parallel plates, the electric field strength is equal to the difference in electric potential between the plates divided by the separation between them. The relationship between an electric field E and electric potential V is

$$E = -dV/dx$$

See also ELECTRICAL POTENTIAL ENERGY.

electric power The rate at which energy is carried by an electric circuit. The power is equal to the current multiplied by the voltage. In a circuit containing resistance R, with a voltage V and current I, the power P is given by

$$P = IV = I^2R = V^2/R$$

In a circuit in which alternating current is flowing, the electric power is given by $IV\cos\Phi$, where I and V are the ROOT MEAN SQUARE values of the voltage and current, and Φ is the

PHASE difference between the current and voltage, which is zero in a circuit containing only RESISTANCE and no REACTANCE. The factor $\cos\Phi$ is known as the power factor.

electric shock The reaction of the human body to an electric current flowing through it. The reaction to electric shock depends on the size of current flowing through the body and the duration of current flow. Much of the body's resistance comes from the skin and this is much lower when the skin is wet, which accounts for the particular dangers of handling electrical apparatus with wet hands. Currents of a few milliamperes or so mimic nerve signals and cause tingling sensations and involuntary muscle contraction. Above about 50 mA, currents through any part of the body become painful, and those through the chest can interfere with breathing and heart operation. Large currents may also cause burning, particularly at the points where the current enters and leaves the body. *See also* EARTH, ELECTROCUTION.

electrochemical cell *See* CELL.

electrochemical equivalent The mass of a specified material released or dissolved when one COULOMB of charge flows in an ELECTROLYSIS experiment. The electrochemical equivalent is the mass of the atom in grams divided by the charge of the ion in coulombs (gC^{-1}).

electrocution An ELECTRIC SHOCK causing death, usually by the current interfering with the nerves of the heart. Lower currents, if sustained for long enough, may cause death by interfering with the motion of the chest, preventing breathing.

electrode 1. A conductor through which an electric current passes in or out of an ELECTROLYTE (as in an electrolytic cell; *see* CELL), a gas (as in a discharge tube) or a vacuum (as in a THERMIONIC VALVE).

2. In a SEMICONDUCTOR, an EMITTER or COLLECTOR of electrons or HOLES.

electrode potential The POTENTIAL DIFFERENCE between an IONIC solution and a metal ELECTRODE. The difference between the electrode potential of the two electrodes in a CELL is what gives rise to voltage of that cell. *See also* HALF-CELL.

electrodynamics The study of the ELECTRIC FIELDS and MAGNETIC FIELDS produced around moving charges.

electrolyse (*vb.*) To bring about chemical decomposition by ELECTROLYSIS.

electrolysis The chemical change effected by the passage of an electric current through an IONIC liquid (*see* ELECTROLYTE). In electrolysis, ions are attracted to the oppositely charged ELECTRODE (*see* ANODE, CATHODE, ANION, CATION). When the ions reach the electrode they may gain electrons at the cathode or lose them at the anode and form neutral atoms or molecules. Alternatively, they may react with the electrode, which is then IONIZED and dissolves in the liquid. This latter reaction generally takes place at the anode, as the electrodes tend to be metals that readily give up electrons to form positive ions. Which of the possible reactions takes place depends on the relative reactivities of the substances involved. The more reactive substance will generally form ions or remain in its ionic form. When water acidified with sulphuric acid (to ionize the water so it will conduct) is electrolysed using platinum electrodes, hydrogen gas is formed at the cathode and oxygen gas and water are liberated at the anode, since hydrogen is more reactive than platinum and the hydroxide ion (OH^-) gives up electrons more readily than the sulphate ion (SO_4^{2-}).

Electrolysis is used to extract some metals from their ores; for example, the extraction of aluminium from bauxite. Electrolysis is also used in ELECTROPLATING, where the object to be coated is the negative electrode in a solution of a salt of the coating metal.

See also ELECTROCHEMICAL EQUIVALENT, FARADAY'S LAWS OF ELECTROLYSIS, VOLTAMMETER.

electrolyte Any conducting liquid, through which electric charge flows by movement of IONS. Electrolytes are molten IONIC compounds or solutions of ionic salts or of compounds that ionize in solution. *See also* MOLAR CONDUCTIVITY.

electrolytic capacitor A CAPACITOR in which the insulating layer is formed on one of the plates by ELECTROLYSIS. The result is an insulating layer that is very thin, giving large CAPACITANCES. A disadvantage is that the layer may not be perfectly insulating, so small LEAKAGE CURRENTS may flow. It is also important to use electrolytic capacitors with the voltage in the correct direction. If it is reversed, the leakage current may result in electrolytic processes destroying the insulating layer.

electrolytic cell *See* CELL.

electromagnet A MAGNET that relies on INDUCED MAGNETISM in the iron core of a SOLENOID. Compared to PERMANENT MAGNETS, electromagnets have the disadvantage that they need a power supply. Their advantages are that the magnetism can be turned on and off, and that stronger magnetic fields can be produced. At very high fields, the iron core adds little to the magnetism produced by the current, so high field electromagnets often have no core. Cooling presents a problem for high field electromagnets, due to the very high currents required. In some magnets, superconducting materials are used. These need to be kept cold, but heat does not need to be removed at a high rate provided there is good thermal insulation.

electromagnetic force The force responsible for interactions between electrically charged particles. It is one of the FOUR FORCES OF NATURE. The electromagnetic force is explained in terms of the exchange of PHOTONS. It is responsible for all the non-gravitational interactions between atoms (rather than atomic nuclei) and thus causes friction, contact forces between solid objects, and all forms of chemical bonding. The full quantum mechanical theory of electromagnetism is called QUANTUM ELECTRODYNAMICS (Q.E.D.) and is an important example of the successful links that have been made between QUANTUM MECHANICS and the SPECIAL THEORY OF RELATIVITY in the second half of the 20th century. More recently, the electromagnetic force has been identified as one element of the ELECTROWEAK FORCE in the STANDARD MODEL of particle physics.

electromagnetic induction The process by which an ELECTROMOTIVE FORCE (e.m.f.) is produced in a circuit when the MAGNETIC FLUX LINKAGE in a circuit changes. The flux linkage in a circuit may change in one of two ways: either part of the circuit moves, as in a DYNAMO, effectively producing a change in the area of the circuit; or the field may change, if the magnet producing it is moved or the current in an ELECTROMAGNET changes, as in a TRANSFORMER for example.

For a wire of length l moving through a magnetic field B at a speed v, with the wire, its motion, and the magnetic field direction all at right angles, the induced e.m.f., E, will be

$$E = Blv$$

See also FARADAY'S LAW OF ELECTROMAGNETIC INDUCTION, LENZ'S LAW, MUTUAL INDUCTANCE, SELF-INDUCTANCE.

electromagnetic radiation Energy resulting from the acceleration of electric charge, that propagates through space in the form of ELECTROMAGNETIC WAVES. Alternatively, electromagnetic radiation can be thought of as a stream of PHOTONS travelling at the speed of light (c) each with energy hc/λ, where h is PLANCK'S CONSTANT and λ is the wavelength of the associated wave. *See also* BLACK-BODY RADIATION, ELECTROMAGNETIC SPECTRUM, RADIATION.

electromagnetic spectrum The spread of different frequencies and wavelengths of ELECTROMAGNETIC WAVES with particular reference to the similarities and differences between different parts of this range. In order of increasing frequency, it comprises RADIO WAVES, INFRARED radiation, visible LIGHT, ULTRAVIOLET radiation, X-RAYS and GAMMA RADIATION.

electromagnetic wave WAVES of energy composed of oscillating electric and magnetic fields, IN PHASE but at right angles to one another, propagating through space as a TRANSVERSE WAVE. Since the associated fields are capable of existing in empty space, electromagnetic waves can travel through a vacuum. The detailed properties of the wave depend upon its frequency and wavelength: wavelengths from several kilometres down to 10^{-15} m have been observed. This range is called the ELECTROMAGNETIC SPECTRUM.

Electromagnetic waves travel through a vacuum at a constant speed (the SPEED OF LIGHT) of 3×10^8 ms^{-1}. In other materials they travel more slowly than this, depending on the RELATIVE PERMITTIVITY and RELATIVE PERMEABILITY of the medium.

electromagnetism The branch of physics concerned with the MAGNETISM produced by electric CURRENTS and with the production of, an interaction between, ELECTRIC and MAGNETIC FIELDS.

electrometer An instrument for measuring the ELECTRIC POTENTIAL of electrostatic charges (*see* STATIC ELECTRICITY), or any VOLTMETER with a very high resistance. Historically, such instruments measured the electrostatic force between a charged electrode and one at EARTH potential, and were not very sensitive. Modern instruments use OPERATIONAL AMPLIFIERS with FIELD EFFECT TRANSISTORS in the input stages to combine high sensitivity with very high resistance.

electromotive force (e.m.f.) The amount of electrical energy given to each COULOMB of charge that is driven around a circuit by the source of e.m.f. The SI UNIT of electromotive force is the VOLT. An e.m.f. may be provided by an electrolytic CELL (which converts chemical energy into electrical energy) or an electrical GENERATOR, which relies on the relative motion of a wire though a magnetic field (*see* ELECTRO-MAGNETIC INDUCTION).

electron The lightest charged LEPTON. Electrons are negatively charged ELEMENTARY PARTICLES that occur in all atoms. Atoms consist of a central NUCLEUS surrounded by orbiting electrons. The number and distribution of the electrons in an atom (the electron structure) is responsible for the chemical properties of the atom. An electron that has become detached from an atom is known as a free electron.

The mass of an electron is 9.1×10^{-31} kg and it has a charge of 1.6×10^{-19} C. This charge is the fundamental unit of negative charge, and the electron is the basic particle of electricity. In metals, an electric current consists of a movement of free electrons.

See also MILLIKAN'S OIL DROP EXPERIMENT, NEUTRON, POSITRON, ORBITAL, PROTON.

electron degeneracy pressure The pressure arising from the influence of the PAULI EXCLUSION PRINCIPLE on electrons. This means that extra ENERGY is required to compress a material containing FREE ELECTRONS, such as the PLASMA in a star, regardless of temperature. *See* FERMI DISTRIBUTION. *See also* WHITE DWARF.

electron diffraction The spreading of a beam of electrons as they pass through a narrow aperture or crystal lattice, analogous to the DIFFRACTION of visible light. This effect can only be explained in terms of the WAVE NATURE OF PARTICLES. Electron diffraction is used to measure the lengths and angles of chemical bonds, and in the study of solid surfaces. *See also* NEUTRON DIFFRACTION, X-RAY DIFFRACTION.

electron gun A device for generating a beam of electrons, as in a CATHODE RAY TUBE. An electron gun consists of an electrically heated metal CATHODE together with a number of cylindrical ANODES in an evacuated tube. Electrons are released from the cathode in a process known as THERMIONIC EMISSION, and are accelerated and focused by the anodes. A CONTROL GRID may also be incorporated to alter the intensity of the beam.

electron-hole pair An electron and a HOLE produced at a point in a SEMICONDUCTOR, by the action of light, thermal vibrations or IONIZING RADIATION. The electron and the hole then move in opposite directions under the influence of any electric field. However, an electron may meet another hole and recombine, thus there will be a dynamic EQUILIBRIUM controlling the number of CHARGE CARRIERS.

electronics The science of circuits that contain ACTIVE DEVICES; that is, devices that behave in a more complex way than simple RESISTORS, INDUCTORS and CAPACITORS.

electron lens A set of ELECTRODES, or an ELECTROMAGNET, designed to focus a beam of electrons in the same way that a conventional lens focuses light. Electron lenses are used in ELECTRON MICROSCOPES, in which the beam of electrons are focused by an arrangement of coils that produce a variable magnetic field, and in CATHODE RAY TUBES.

electron microscope An instrument, developed in 1933, that magnifies objects using a beam of electrons. The electrons are accelerated by a high voltage through the object (held in a vacuum) and focused by powerful ELECTRO-

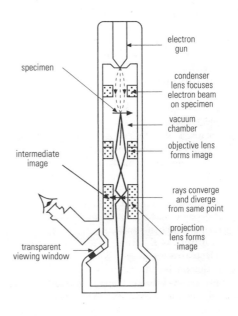

The electron microscope.

MAGNETS (*see* ELECTRON LENS), instead of optical lenses, onto a fluorescent screen for viewing. A camera may be built in to record what is seen on the screen. A beam of electrons of a DE BROGLIE WAVELENGTH of typically 0.005 nm is used, compared to the wavelength of about 500 nm of the light rays used in a light microscope. Therefore the RESOLVING POWER of the electron microscope is vastly improved to about 1 nm, and objects can be magnified more than 500,000 times.

The sample to be viewed is fixed and embedded in araldite, which enables ultra-thin sections to be cut. Living specimens cannot be viewed. There are several types of electron microscope.

electron number A QUANTUM NUMBER believed to be conserved in all interactions. The electron and its NEUTRINO are assigned an electron number of one, their ANTIPARTICLES −1, while all other particles have an electron number of 0.

electron-volt (eV) A unit of energy used in atomic and nuclear physics. One eV is the energy gained by an electron when it moves through a POTENTIAL DIFFERENCE of one volt. It is equivalent to 1.6×10^{-19} JOULES.

electroplating An ELECTROLYSIS technique in which metal is deposited on the CATHODE. For example, when copper sulphate is electrolysed with a copper ANODE, the anode dissolves and copper is deposited on the cathode. Metal objects are often electroplated with silver.

electroscope Any device designed to measure ELECTRIC POTENTIAL. *See also* GOLD-LEAF ELECTROSCOPE.

electrostatic (*adj.*) Related to, producing or caused by STATIC ELECTRICITY.

electrostatic precipitator A device fitted to chimneys, for instance in power stations, to remove small particles from smoke. The smoke is passed between two grids that are oppositely charged; the smoke particles become charged and are attracted to the oppositely charged grid. These grids are then brushed or shaken to collect the dust.

electrostatics (*adj.*) The study of electric charges at rest. *See* STATIC ELECTRICITY.

electroweak force The single interaction now known to account for the effects of the WEAK NUCLEAR FORCE and the ELECTROMAGNETIC FORCE. This example of unification of forces has been an important part of the progress made in theoretical physics since the 1960s. The electroweak unification was suggested in 1967 by three physicists: Sheldon L. Glashow and Steven Weinberg of the United States and Abdus Salam of Pakistan. The W BOSON and Z BOSON predicted by the electroweak theory were discovered at CERN in 1983. *See also* GRAND UNIFIED THEORY, STANDARD MODEL.

element A substance that cannot be broken down into more fundamental constituents by normal chemical means, being made of ATOMS all of the same type (that is, having the same ATOMIC NUMBER). About 110 elements are known, though there are claims for a few more. Of these, 91 are found in nature.

elementary particle, *fundamental particle* Any one of the indivisible particles from which all MATTER is believed to be made up, plus the GAUGE BOSONS that are responsible for the forces between the particles of matter. For every particle of matter there is a similar ANTIPARTICLE, with the same mass, and HALF-LIFE (if unstable), but opposite values of all QUANTUM NUMBERS, such as CHARGE, which remain the same in given types of interaction. Elementary particles are divided into two categories: LEPTONS, which do not feel the STRONG NUCLEAR FORCE, and QUARKS, which do. *See also* BARYON, FOUR FORCES OF NATURE, HADRON, HYPERON, SUPERSTRING THEORY, TACHYON, THEORY OF EVERYTHING.

elevation of boiling point The increase in the boiling point of a solvent caused by the presence of dissolved ions or molecules. This is a COLLIGATIVE PROPERTY, independent of the nature of the particles present, and proportional to their molar concentration, with a constant of proportionality called the ebullioscopic constant. Measurement of the elevation of boiling point caused by dissolving a known mass of substance in a solvent can be used to estimate its RELATIVE MOLECULAR MASS.

ellipse A closed curve that can be formed as a CONIC SECTION, or as the set of points such that the total distance of every point from two fixed points, called the foci, is constant. An ellipse has the appearance of a flattened circle, with the degree of flattening measured by the ECCENTRICITY of the ellipse. The ORBITS of satellites are ellipses or circles. Most PLANETS have orbits that are almost circular, but COMETS travel in highly elliptical orbits.

e.m.f. *See* ELECTROMOTIVE FORCE.

emission spectrum A SPECTRUM produced as excited atoms or molecules return to lower ENERGY LEVELS, giving off ELECTROMAGNETIC RADIATION of specific wavelengths, with all other wavelengths being absent.

emissivity The ability of a surface to give off THERMAL RADIATION when heated. A BLACK BODY, which radiates perfectly, has an emissivity of 1. A non-emitting surface has an emissivity of 0.

emitter The ELECTRODE in a JUNCTION TRANSISTOR from which CHARGE CARRIERS are injected into the BASE before arriving at the COLLECTOR.

empirical (*adj.*) Describing a result that is known purely from experiment, without any theoretical understanding or explanation.

endoscope A device for viewing the inside of otherwise inaccessible structures, such as the inside of the human body, by sending light along a bundle of FIBRE OPTICS.

endothermic (*adj.*) Describing a process, particularly a chemical reaction, in which ENERGY is taken in from the surroundings, usually resulting in a lowering of temperature.

energy A measure of the ability of an object or a system to do WORK. Energy can be broadly divided into two classes: POTENTIAL ENERGY, which arises from an objects state or position (in a gravitational field, for example); and KINETIC ENERGY, which is a function of motion. The kinetic and potential energy that molecules possess as a result of their random thermal motion is sometimes called INTERNAL ENERGY, to distinguish it from the kinetic energy resulting from the motion of a solid body or potential energy that an object possesses as a result of some overall state (such as being in gravitational field). The study of internal energy is called THERMODYNAMICS.

In the SPECIAL THEORY OF RELATIVITY, it is recognized that there is a link between energy and MASS; the greater the energy of a body the more massive it will be, regardless of the form of this extra energy.

The amount of work done, or the energy that gives the capability to do work, is measured in a unit called the JOULE.

The importance of the idea of energy comes from the LAW OF CONSERVATION OF ENERGY, which states that the total energy of any closed system is conserved.

As a result of the increasing demands for energy there has been a constant search for new sources. In the developed world, the chief energy source is the burning of FOSSIL FUELS, but increasing attention is being given to renewable energy sources (*see* RENEWABLE RESOURCE). NUCLEAR FISSION is a major energy source, but there are fears about the safety of the radioactive materials involved, whilst NUCLEAR FUSION appears to be a long way from producing energy on a commercial scale. *See also* CHEMICAL ENERGY, HEAT, MECHANICAL ENERGY.

energy band A range of energies possessed by the electrons in a solid. *See* BAND THEORY.

energy barrier The amount of energy that may need to be provided before a physical or chemical process can take place, even though the process as a whole may release energy. *See* ACTIVATION PROCESS.

energy level Any one of the permitted energy states in which an atom or molecule may exist. Under the rules of QUANTUM MECHANICS, a system can only have certain fixed energies, and each different atom or molecule has a series of possible energy levels. When the system moves from one energy level to another, the energy is absorbed or emitted, often as a PHOTON of light or other ELECTROMAGNETIC RADIATION. The distinct energies of the allowed levels mean that only photons of certain discrete energies are involved. This gives rise to absorption and emission spectra characteristic of the atoms or molecules involved. *See also* ATOMIC ABSORPTION SPECTROSCOPY, ATOMIC EMISSION SPECTROSCOPY, BAND THEORY, SPECTROSCOPY, WAVE NATURE OF PARTICLES.

engine Any machine that converts chemical energy, from the burning of a fuel, to mechanical work via heat energy. Engines are very widely used, as a great deal of energy is available from fuels. However, the SECOND LAW OF THERMODYNAMICS means that a large amount of energy must be carried away in the hot gases produced by burning the fuel. The result is low efficiency, typically 20 to 25 per cent. *See also* CARNOT CYCLE, DIESEL ENGINE, HEAT ENGINE, INTERNAL COMBUSTION ENGINE, PETROL ENGINE.

enlarged (*adj.*) Increased in size, particularly in the context of images formed by CONVERGING LENSES and mirrors.

enriched uranium Uranium in which the concentration of the FISSILE ISOTOPE, uranium–235, has been increased. Natural uranium is a mixture of isotopes but mostly it contains uranium–238, which is not fissile. To produce a fuel for a reactor the uranium must be enriched. This is done by using DIFFUSION or CENTRIFUGE processes, which separate out the two different masses of isotope. Since the mass difference is small, the enrichment process must be repeated many times to produce a useful fuel. *See also* DEPLETED URANIUM, FISSION POWER.

enthalpy A STATE FUNCTION in THERMODYNAMICS, commonly used in the study of chemical reactions. It is the INTERNAL ENERGY of the molecules of a material plus the pressure at which the material is held multiplied by the volume occupied. It is a useful quantity because it is constant in any process that takes place at a constant pressure with no external supply of energy. For a substance where the molecules have energy U and occupy a volume V at a pressure p, the enthalpy is H, where

$$H = U + pV$$

The internal energy of the molecules alone is not constant if there is any change in volume during the process. If a gas is produced in a chemical reaction, for example, work will have to be done at the expense of the internal energy of the molecules in pushing back the surrounding atmosphere to make room for this gas. Enthalpy is also constant during the FREE EXPANSION of a gas.

entropy A measure of the degree of disorder in a thermodynamic system. Typically, this concept is applied to a system containing many molecules. The most probable state in which those molecules are found is the state that can be achieved in the largest number of ways. Entropy can be defined by the equation

$$S = k \ln W$$

where S is the entropy, k is the BOLTZMANN CONSTANT and W is the number of ways in which the molecules in the system can be arranged to produce the specified state.

For any system containing a large number of particles, the most probable state becomes overwhelmingly likely to be the state in which the system is found. This leads to the statement that any irreversible change (not able to happen in reverse) is one which produces an increase in the total entropy of the system. Since such systems tend to move from ordered (low entropy) to disordered (high entropy)

states, the entropy of a system can be used as a measure of the extent to which the energy in a system is available for conversion to WORK.

See also FREE ENERGY, SECOND LAW OF THERMODYNAMICS.

epicentre The point on the surface of the Earth directly above the focus of an EARTHQUAKE. It is the point at which the most damage occurs.

epicycle In the PTOLEMAIC MODEL of the SOLAR SYSTEM, an imaginary circle that carried a PLANET around its circumference whilst the centre of the epicycle rotated around the Earth. One of the weaknesses of the Ptolemaic model was that it failed to provide an explanation for the existence of epicycles, which were introduced to explain the pattern of motion of the planets, particularly their RETROGRADE motion at certain times during the year.

EPROM (erasable programmable read-only memory) ROM that can be erased (usually by exposure to ultraviolet light) so new information can be stored.

equation Any mathematical statement containing an equals sign (=). An equation represents the equality of two expressions of variables and/or constants. They are generally used to determine one of the quantities in the equation in terms of the others. Equations are often used in this way in physics.

equation of motion Any equation that enables the position of an object to be calculated at some future time by knowing its initial position and velocity and how the forces on that object vary with position and velocity. An important class of motion is that where the acceleration is constant. For an object with a constant acceleration a, starting speed u and final speed v, covering distance s in time t:

$$v = u + at$$
$$s = ut + at^2/2$$
$$v^2 = u^2 + 2as$$

equation of state Any equation describing the behaviour of a material under different conditions. In particular, the change in volume under differing temperatures and pressure. For example, the equation of state of an IDEAL GAS is

$$pV = NkT$$

where p is the pressure, V the volume, N the number of molecules, k the BOLTZMANN CONSTANT and T the ABSOLUTE TEMPERATURE.

Ergodic Theory see p 61

equator On the Earth, or any other planet, a line drawn around the surface equally distant from the two POLES.

equilateral (*adj.*) Describing a shape, especially a triangle, where all the sides are the same length.

equilibrium The state of an object or system in which all effects – forces, interactions, reactions, etc. – are balanced so that there is no net change.

In mechanics, an object or system is said to be in a static equilibrium when the total force acting is zero, and when the MOMENTS of the forces, taken about any axis, add to zero. In this case, the system will not accelerate (*see* ACCELERATION) – in particular, if the system starts at rest, it will remain at rest (*see* NEWTON'S LAWS OF MOTION). These conditions for equilibrium are particularly important in the branch of mechanics called STATICS, the study of structures or systems at rest.

If all the forces act in a single plane, the conditions for equilibrium are met if the COMPONENTS of the forces taken in any two non-parallel directions add up to zero, and the moments of the forces about any point add up to zero. If only three non-parallel forces act, the condition that the moments add to zero can be stated as a requirement that the lines along which the three forces act must cross at a single point.

A system is in a state of THERMAL EQUILIBRIUM if there is no net exchange of heat within it or between it and its surroundings. Two objects at the same temperature and in thermal contact will exchange energy when their molecules collide with one another, but the net flow of energy will be zero. This is an example of a dynamic equilibrium, where the processes balance each other out, keeping a system unchanged.

See also STABLE EQUILIBRIUM, THERMODYNAMIC EQUILIBRIUM, UNSTABLE EQUILIBRIUM.

equinox One of two dates in the year when night and day are the same length. They occur in spring and autumn, when the Sun is directly above the equator.

equipartition of energy The principle that states that thermal energy will be distributed equally between the available DEGREES OF FREEDOM.

equipotential A line joining points that are at the same ELECTRIC POTENTIAL. Equipotentials are always at right angles to ELECTRIC FIELD LINES. A conductor will always be at a single potential throughout its volume.

equivalence of mass and energy The concept that when an object gains energy, in whatever form, it gains mass. This is summed up in the famous equation

$$E = mc^2$$

where E is the energy, m the mass and c the SPEED OF LIGHT. The mass changes predicted are normally too small to be detected, but can be measured in atomic nuclei (*see* NUCLEAR BINDING ENERGY). *See also* SPECIAL THEORY OF RELATIVITY.

erasable programmable read-only memory *See* EPROM.

erg A unit of energy in the C.G.S. SYSTEM, the work done when a force of one DYNE moves through a distance of one centimetre. One erg is equivalent to 10^{-7} JOULE.

erosion The wearing down of rock by the action of wind or water.

escape velocity The speed an object must have to be launched into space and completely escape the gravitational field of the planet from which it was launched. The escape velocity is such that the object has sufficient KINETIC ENERGY to lose to compensate for the increase in GRAVITATIONAL POTENTIAL ENERGY in moving away from the surface of the planet. The escape velocity is independent of the mass being launched. For the Earth it is about 11 kms^{-1}. The escape velocity from the surface of a body of mass M and radius R is $(2GM/R)^{-1/2}$, where G is the GRAVITATIONAL CONSTANT.

Euclidean space A flat space in which the normal laws of geometry apply. For example, the angles of a triangle add up to 180º. This will not be true for a triangle drawn on a curved surface. Non-Euclidean spaces in four dimensions, three of space and one of time, are used to describe physical space in the GENERAL THEORY OF RELATIVITY.

eureka vessel *See* DENSITY CAN.

eutectic mixture A SOLID SOLUTION of two or more substances, with the lowest freezing point for any possible mixture of the components. The freezing point of a eutectic mixture is called the eutectic point. On a graph of temperature against composition for a mixture of two substances, which may be either solid or liquid, there are generally four regions. Below the eutectic point, the mixture is entirely solid. This temperature may be lower than the melting point of either substance alone. At higher temperatures, one substance may occur as a solid – generally the one present in higher proportions – together with a liquid that may be regarded as a SATURATED solution of the substance present as a solid. At higher temperatures, both substances will be liquid.

The FREEZING MIXTURE formed by adding sodium chloride to water is an example of a eutectic mixture. A lower liquid temperature can be achieved than would be achieved from either material alone. Not all mixtures form eutectics.

eutectic point The freezing point of of a EUTECTIC MIXTURE.

evaporation The process by which a liquid turns into a gas at a temperature below the boiling point as some of the faster moving molecules escape from the surface of the liquid. Unlike boiling, evaporation takes place only on the liquid surface. The rate of evaporation can be increased by heating, by blowing air over the liquid surface so molecules cannot re-enter the liquid, or by increasing the surface area. A liquid that evaporates easily is called volatile. *See also* COOLING BY EVAPORATION.

event horizon For a BLACK HOLE, the surface on which the ESCAPE VELOCITY is equal to the speed of light. Once an object falling into a black hole reaches the event horizon it is trapped inside the black hole.

exchange In a telephone system, the switching centres that channel the signals to their intended recipients.

excited state A QUANTUM state, in an atom for example, with an energy higher than the lowest allowed energy. *See also* ENERGY LEVEL, GROUND STATE, WAVE NATURE OF PARTICLES.

exciter A DYNAMO used to provide the power supply for the rotating ELECTROMAGNET of a large ALTERNATOR.

excluded volume In a gas, the volume taken up by the molecules themselves. *See* IDEAL GAS.

exhaust stroke The stage in the operation of a PETROL ENGINE at which burnt gases are forced out of the engine.

exitance The amount of ELECTROMAGNETIC RADIATION per unit area emitted by a hot surface. *See also* BLACK-BODY RADIATION.

exosphere The outermost layer of the ATMOSPHERE, beginning at about 400 km.

exothermic (*adj.*) Describing a process, particularly a chemical reaction, in which ENERGY is given out to the surroundings, usually resulting in an increase in temperature.

expansion 1. (*physics*) An increase in size, particularly as a result of an increase in temperature. *See also* THERMAL EXPANSION, FREE EXPANSION.

2. (*mathematics*) The rewriting of a mathematical expression without brackets, or as a sum of a larger number of simpler expressions. For example, $a^2 + 2ab + b^2$ is the expansion of $(a + b)^2$.

expansivity, *coefficient of expansion* A measure of the increase in length (linear expansivity), area, or volume (bulk expansivity) of a material when it is heated. The linear expansivity of a material is the increase in length for a 1°C temperature rise, divided by the original length. If the length, area and volume of an object at 0°C are l_0, A_0, and V_0, then at temperature θ°C they will be l, A and V where

$$l = l_0(1+\alpha\theta)$$
$$A = A_0(1+\beta\theta)$$
$$V = V_0(1+\gamma\theta)$$

with α, β, and γ being the linear, area (or superficial) and volume coefficients of expansivity respectively. Provided the coefficients are small, $\beta = \alpha^2$ and $\gamma = \alpha^3$. *See also* THERMAL EXPANSION.

experiment A set of measurements or observations, often performed on equipment designed specifically for this purpose, and designed to suggest or to test a theory. If the theory satisfies a sufficient range of experimental tests, it may then be used to predict what will happen in similar situations. If an experiment provides results that genuinely contradict the theory, the theory must be revised.

In the design of any experiment, it is important to ensure that only those factors being studied can change, and that all other factors remain constant throughout the experiment.

exponent, *index* A number that expresses how many times another number must be multiplied by itself. For example, in the expression x^n, n is the exponent of x.

exponential 1. *See* E.

2. (*adj.*) In mathematics, frequently used

to describe any rapid growth or decay, but more correctly describing the function e^x where e = 2.713 (the BASE of NATURAL LOGARITHMS). Exponential functions have an important common ratio property – each time the independent variable increases by a certain amount, the function changes by a fixed factor.

The exponential function $y = e^x$ can be defined in various ways, including as the solution to the equation $dy/dx = x$ which has $y = 1$ when $x = 0$.

In an exponential growth of population, the population may double after 10 years, then double again in the following 10 years and so on. In radioactivity, the level of radioactivity falls by a constant factor in equal time intervals. *See also* HALF-LIFE, HALF-THICKNESS, TIME CONSTANT.

extraordinary ray *See* BIREFRINGENCE.

extrinsic semiconductor A SEMICONDUCTOR in which current flows as a result of CHARGE CARRIERS produced by the introduction of impurities into the semiconductor material.

eye The sense organ responding to light. The human eye is roughly spherical and attached to the skull by muscles which allow the eyeball to rotate. Light passes through the external layer of the eye and the CORNEA, and then enters the eye through an aperture called the pupil at the centre of the coloured diaphragm, the IRIS. Most of the REFRACTION of the light entering the eye is carried out by the cornea, which is curved and acts as a fixed lens. Behind the pupil is the lens, which carries out the final focusing of the light entering the eye. The lens can alter its shape due to the action of surrounding muscles. These allow the lens to change shape (fatter for near objects, thinner for distant ones). This allows objects at different distances to be focused, an ability called 'accommodation'. In front of the lens is a chamber containing a clear liquid called aqueous humour, and behind the lens is a larger chamber containing clear, jelly-like vitreous humour, which helps to maintain the shape of the eyeball.

The lens focuses the light entering the eye onto a layer called the RETINA. Light-sensitive cells here (rods and cones) convert the light they receive into nerve impulses that pass along the optic nerve to the brain.

eyepiece In any optical instrument, such as a microscope or telescope, the lens placed closest to the eye. Generally, this acts as a MAGNIFYING GLASS, producing an enlarged image of the image formed in the instrument.

Ergodic Theory

studies of dynamical systems with an INVARIANT measure (measure that is preserved by some function see Krylov–Bogolyubov Theorem) & related problems

initial development [in solving ? problem [was] use of] STATISTICAL physics (statistical mechanics provides a framework for relating microscopic properties of individual atoms & molecules to macroscopic or bulk properties that can be OBSERVED [easily] eg explaining thermodynamics

It works well in classical systems when a number of degrees of freedom . I

Therefore n° of variables is so large exact solutions are not possible

Can describe Non Linear dynamics, Chaos theory, thermal physics, fluid dynamics / particles at a high Knudsen n°s or plasma physics

F

face-centred cubic *See* CUBIC CLOSE PACKED.

facsimile *See* FAX.

factorial (!) A function of a positive integer found by multiplying together all the integers up to and including the given integer, for example:

$$4! = 1 \times 2 \times 3 \times 4$$

0! is given the value 1.

Fahrenheit A TEMPERATURE SCALE obsolete in science but still in everyday use in the UK and US. The fixed points are taken as the freezing point of water, 32°F, and the boiling point of water, 212°F, both at a pressure of one ATMOSPHERE.

Fajan and Soddy's Group Displacement Law An EMPIRICAL law summarizing the changes in ATOMIC NUMBER that take place during ALPHA DECAY and BETA DECAY. In an alpha decay, the new element produced is two places to the left of the decaying element in the PERIODIC TABLE. In a beta decay, the element produced is one place to the right of the original element.

farad (F) The SI UNIT of CAPACITANCE. One farad is the capacitance of a CAPACITOR that stores one COULOMB of charge for each VOLT applied. In practice this is a very large capacitance, and most capacitors have a capacitance in the order of microfarads (μF).

faraday A unit of CHARGE particularly used in the study of ELECTROLYSIS. One faraday is the charge on one MOLE of electrons, i.e. about 96,487 COULOMB.

Faraday cage A container of metal or wire mesh designed to protect equipment inside the cage from the influence of external ELECTRIC FIELDS or ELECTROMAGNETIC RADIATION. Induced charges in the metal produce a field which cancels out the effect of any outside charges.

Faraday screen A metal enclosure used to protect sensitive electrical circuits from the influence of outside electric fields. INDUCED CHARGES within the screen arrange themselves to cancel out the electric field produced by any external charges.

Faraday's constant (*F*) The amount of CHARGE in one FARADAY: 96,487 COULOMBS.

Faraday's law of electromagnetic induction The induced ELECTROMAGNETIC FORCE in a circuit is equal to the rate of change of MAGNETIC FLUX LINKAGE in that circuit. If the rate of change of flux linkage through a circuit is $d\Phi_n/dt$, then the induced e.m.f. in the circuit will be

$$E = -d\Phi_n/dt$$

Faraday's laws of electrolysis Two laws governing the mass of substance dissolved or liberated from a solution in an ELECTROLYSIS. (i) The mass of the substance is proportional to the charge flowing. (ii) This mass is equal to the charge in FARADAYS multiplied by the number of electron charges (positive or negative) carried by the ion concerned multiplied by the RELATIVE ATOMIC MASS of the ion.

fast breeder reactor A type of NUCLEAR REACTOR in which a blanket of uranium–238 is placed around the reactor core. Uranium–238 absorbs NEUTRONS that would otherwise escape and after two BETA DECAYS it becomes plutonium–239, a FISSILE material. The plutonium may be extracted by chemical processing and then used to refuel the reactor.

fatigue The failure or weakening of a material under STRAINS that are applied and removed repeatedly. Even though such strains may not reach the ELASTIC LIMIT of the material, small cracks will form and grow in some materials under these circumstances.

fault A sudden discontinuity between layers of rock. Major faults are formed where TECTONIC PLATES meet and relative movement of rocks on each side of the fault leads to EARTHQUAKES.

fax (abbreviation for *facsimile*) A telecommunications system for the transmission of written or printed text. Documents are scanned and converted into an electric signal, which is sent via the telephone network to a receiving machine where an exact replica of the document is created.

feedback Taking the output of an electronic system (usually an AMPLIFIER) and feeding part of that output back to the input. *See* NEGATIVE FEEDBACK, POSITIVE FEEDBACK.

Fermat's principle, *principle of least time* Any ray of light will take the path through an optical system that takes the least time. Also, all the rays leaving one point and arriving at another (in forming an image, for example) will have taken equal times to pass through the system. The laws of REFLECTION and REFRACTION can be derived from this principle, which is also used to calculate the path of rays through complex optical systems.

fermi (fm) A unit of length, used in nuclear physics. One fermi is equal to 10^{-15} m.

Fermi distribution A mathematical description of the energies of FERMIONS, such as electrons in metals or in certain stars. Although the forces between electrons are often small enough for them to be thought of as forming a gas, the effects of QUANTUM MECHANICS, particularly the PAULI EXCLUSION PRINCIPLE, mean that many of the particles have higher energies than would otherwise be the case. Unless the temperature is very high, it is a good approximation to say that all the ENERGY LEVELS from zero up to a fixed maximum, called the FERMI ENERGY, are filled whilst the rest are empty. This distribution of energies is often only slightly affected by temperature, as the energy of any thermal vibrations is small compared to the Fermi energy, thus only a few particles very close to the Fermi energy can be promoted into higher energy levels. In stars, the Fermi energy increases as a star becomes smaller and this can provide a pressure, called the ELECTRON DEGENERACY PRESSURE, which prevents the star from collapsing completely under its own gravitation.

Fermi energy The energy up to which ENERGY LEVELS are filled in the FERMI DISTRIBUTION. The Fermi energy is the energy at which the probability of an energy level being occupied is ½.

fermion Any particle, such as an electron, with half-integer SPIN; that is spin that is an odd number times $h/4\pi$, where h is PLANCK'S CONSTANT. *See also* FERMI DISTRIBUTION, PAULI EXCLUSION PRINCIPLE.

ferrimagnetic (*adj.*) Describing the alignment of individual magnetic atoms or ions in a solid where adjacent particles have their magnetism aligned in opposite directions, but the net effect is not zero due to the presence of two or more different types of magnetic particle with differing MAGNETIC MOMENTS. The simplest example is the mixed iron oxide ferrite, Fe_3O_4, which contains Fe^{2+} and Fe^{3+} ions with their magnetic moments oppositely aligned.

ferrite Any of a number of mixed oxides of the form $MO.Fe_2O_3$, where M is typically a FERROMAGNETIC element such as iron, cobalt or nickel. Ferrites have ceramic structures and are electrical insulators, but they are also ferromagnetic. This means they can be used as cores in SOLENOIDS at high frequencies without causing EDDY CURRENT losses.

ferroelectric (*adj.*) Describing a DIELECTRIC that retains its electrical POLARIZATION when the polarizing electric field is removed. Many ferroelectric materials are also PIEZOELECTRIC.

ferromagnetic (*adj.*) Describing a strongly magnetic material, such as iron. In such materials the imbalance in ANGULAR MOMENTUM of the electrons in each atom causes the atom to behave like a tiny magnet. Neighbouring atoms line up to form DOMAINS, regions where all the atoms are pointing in the same direction. *See also* CURIE POINT, HARD, PERMANENT MAGNET.

fibre optic, *optical fibre* A solid glass or plastic fibre, typically thinner than the thickness of a human hair, that can transmit light. The light does not leave the fibre when it hits the edge, but passes along the fibre by the process of TOTAL INTERNAL REFLECTION. Light passing along such a fibre can be modulated to convey information (*see* MODULATION).

Many telecommunications systems use optical fibres in place of conventional electrical cables, particularly along trunk routes linking one major city with another. The relatively high frequency of light means that a higher BANDWIDTH is available than with an electrical signal. Higher rates of data transmission are therefore possible in fibres, which also take up less space and are cheaper than electric cables.

Fibre optic.

Bundles of fibre optics can be used to convey an image from one place to another along a flexible cable, such as in an ENDOSCOPE. *See also* GRADED INDEX, MONOMODE, STEP INDEX.

field A region of influence around a mass, charge or current (including internal currents in magnetic atoms), where another mass, charge or current will experience a force. *See* ELECTRIC FIELD, GRAVITATIONAL FIELD, MAGNETIC FIELD.

field coil In a motor or other electromagnetic machine, coils that provide a magnetic field.

field effect transistor (FET) An electronic device similar in its uses to a JUNCTION TRANSISTOR, but constructed in a different way. A field effect transistor has a CHANNEL of semiconducting material (*see* SEMICONDUCTOR) along which a current flows between terminals called the SOURCE and the DRAIN. A third ELECTRODE, called the GATE, is formed either on an insulating layer (in an IGFET – insulated gate FET) or from a REVERSE BIASED PN JUNCTION DIODE (JFET – junction FET). As the voltage on the gate is increased (made more negative in the case of an n-channel FET), its electric field narrows the region in which the CHARGE CARRIERS can move along the channel, thus reducing the current in the channel.

The chief advantage of the field effect transistor over the junction transistor is its very high input resistance (the input voltage divided by the input current). At high powers, FET's are also less prone to the problems of THERMAL RUNAWAY.

filament A fine metal wire, such as in a light bulb, that produces visible light when heated by the passage of an electric current. Tungsten is usually used for such filaments owing to its high melting point. Even at temperatures close to the melting point of tungsten, most of the energy produced is radiated from the filament in the form of infrared radiation rather than visible light, making the efficiency of a filament lamp very low compared to a FLUORESCENT TUBE.

film badge A device that records the total level of IONIZING RADIATION to which it has been exposed. It contains a piece of photographic film covered by a number of 'filters' of different materials and thicknesses. Ionizing radiation has a similar effect on a photographic film to light. When processed, the badge gives information about the amount and type of radiation to which it has been exposed. Film badges are often worn by those who work with ionizing radiation to monitor the level of radiation they are receiving.

filter 1. A fine porous material through which a liquid can pass, but solid particles cannot. Filters are widely used to remove solid particles from liquids.

2. *light filter* A piece of coloured glass or plastic that is transparent to certain wavelengths of light but absorbs others. Thus a red filter, for example, allows long–wavelength (red) visible light to pass, but absorbs other wavelengths.

finite (*adj.*) Describing a quantity or number that is neither large without limit, or small without limit. *Compare* INFINITY, INFINITESIMAL.

first law of thermodynamics When heat is supplied to an isolated system, the amount of heat energy, ΔQ, equals the increase in INTERNAL ENERGY, ΔU, plus the mechanical WORK, ΔW, done by the system:

$$\Delta Q = \Delta U + \Delta W$$

For a gas at constant pressure p, the work is done by a change in the volume of the gas, ΔV, so

$$\Delta Q = \Delta U + p\Delta V$$

This statement is equivalent to the LAW OF CONSERVATION OF ENERGY.

fissile (*adj.*) Describing an ISOTOPE, such as uranium–235 or plutonium–239, that exhibits INDUCED FISSION.

fission power The use of NUCLEAR FISSION to produce energy on a commercial scale. *See* NUCLEAR REACTOR.

fixed point A temperature that can easily be maintained, so it can be used as the basis of a TEMPERATURE SCALE. Fixed points usually involve two or more phases of a particular material being in THERMAL EQUILIBRIUM, often at a specified pressure.

flavour In particle physics, the property that distinguishes one type of QUARK from another. The six flavours of quark, which make up the STANDARD MODEL of particle physics, are up, down, strange, charm, top and bottom.

Fleming's left-hand rule A rule to find the direction of the force on a current in a magnetic field. If the fingers of the left hand are held at right angles to one another with the first finger pointing in the direction of the magnetic field and the second finger pointing in the direction of the CONVENTIONAL CURRENT, the thumb will point in the direction of the force.

Fleming's right-hand rule A rule to find the direction in which a current will flow when a voltage is induced in a wire as a result of its motion in a magnetic field. If the thumb and first two fingers of the right hand are held at right angles to one another, with the thumb in the direction of motion of the wire, and the first finger in the direction of the magnetic field, the induced voltage will be such as to drive CONVENTIONAL CURRENT through the wire in the direction of the second finger.

flip flop *See* CLOCKED BISTABLE.

flotation The tendency of an object to float to the surface of a fluid in which it is immersed. This is caused by a force called UPTHRUST, which is due to the pressure of the lower parts of an immersed object being greater than those on the upper parts. The upthrust is equal to the weight of the fluid displaced (pushed out of the way) by the immersed object. If the average density of the immersed object is greater than the density of the fluid, the upthrust is greater than the weight, and the object will float.

Ships float even if they are made of steel, which is much denser than water. This is because the part of the ship below the water – much of which is air inside the ship – weighs the same as the water it displaces. When a ship is fully loaded, it will float lower in the water because of the need to displace a greater amount of water to balance the greater weight of the ship.

The fact that an object appears lighter when immersed in a fluid is also used as a way of measuring density. The difference between the weight an object appears to have in air and when immersed in water is equal to the weight of water displaced by the object. From this, its volume, and thus its density can be found (*see* ARCHIMEDES' PRINCIPLE).

fluid Any substance that can flow: a LIQUID or a GAS.

fluid mechanics The science of fluids (GASES and LIQUIDS), their motion and interaction with solid matter. The BERNOULLI EFFECT and the onset of turbulence are both of critical importance in studying fluid flow. *See also* BERNOULLI'S THEOREM, REYNOLD'S NUMBER, TURBULENT.

fluorescence An effect in which ultraviolet light is absorbed and then re-emitted immediately as visible light. This is used in whitening agents that are added to paper and some washing powders to produce a brighter appearance. *Compare* PHOSPHORESCENCE, where the effect takes place over a longer period of time.

fluorescent (*adj.*) Describing a material that produces visible light by the process of FLUORESCENCE.

fluorescent tube A light source comprising a long glass tube containing mercury vapour at low pressure and an ELECTRODE at each end. To start the tube, filaments at each end heat the mercury to increase the VAPOUR PRESSURE, but these are disconnected once the tube is lit. The GAS DISCHARGE in the mercury vapour produces ULTRAVIOLET light, which is converted to visible light by a FLUORESCENT coating on the inside of the tube. Fluorescent tubes are a more efficient source of light than FILAMENT lamps, but the tube itself and the associated circuitry are more expensive, although this is offset by the longer life of the tube.

flyback The period during a TIMEBASE or RASTER when the electron beam is being returned to its starting position. *See also* CATHODE RAY OSCILLOSCOPE.

flywheel A heavy wheel used to store KINETIC ENERGY, between the POWER STROKES of a PETROL ENGINE for example.

FM *See* FREQUENCY MODULATION.

focal length The distance between a lens or curved mirror and the plane in which parallel rays of light are brought to a FOCUS. In the case of a DIVERGING LENS or CONVEX mirror, the rays appear to emerge from a VIRTUAL FOCUS, and the focal length is often quoted as a negative number.

focal plane A plane associated with a lens or curved mirror in which rays of light that are parallel to one another before striking the lens or mirror are brought to a FOCUS.

focus 1. (*n.*) A point at which rays of light come together.

2. (*vb.*) To adjust the arrangement of an optical system so that rays of light come together at a chosen point – on the film in a camera for example.

fog Water droplets in the air at ground level, reducing the visibility below 1,000 m. *See also* ADVECTION FOG, MIST, RADIATION FOG, WEATHER SYSTEMS.

forbidden band *See* BAND THEORY.

force Any agency that tends to change the state of rest or motion of a body; that is, one that tends to cause a body to accelerate (*see* ACCELERATION). Force is defined as being proportional to the rate of change of MOMENTUM of a body. It is a VECTOR quantity, and the SI UNIT of force is the NEWTON.

For a body of mass m, travelling at velocity v, the momentum is mv. The force is given by $d(mv)/dt$. If mass remains constant, then the force is equal to $mdv/dt = ma$, where a is acceleration.

See also CONTACT FORCE, FOUR FORCES OF NATURE, NEWTON'S LAWS OF MOTION, STRESS, STRAIN.

forced oscillation The motion of a system that would exhibit SIMPLE HARMONIC MOTION were it displaced from its EQUILIBRIUM position and released, but which is driven by some external force (called the driving force), which is itself oscillating. The frequency at which the system would oscillate were it simply displaced and left to oscillate is called the natural frequency of the system and the frequency at which the driving force oscillates is called the driving frequency.

If the driving frequency is large compared to the natural frequency, then the driven system will oscillate very little and almost exactly OUT OF PHASE with the driving force. If the driving force is well below the natural frequency, the response will again be small, but now IN PHASE with the driving force. When the driving frequency is close to the natural frequency, there will be a much greater response, reaching a maximum when the two frequencies are equal. *See also* RESONANCE.

Fortin barometer A version of the MERCURY BAROMETER equipped with an adjustment to allow the level of the open mercury surface to be set to a reference mark and a VERNIER scale for precise measurement of the height of the mercury column.

forward biased (*adj.*) Describing a junction between a P-TYPE SEMICONDUCTOR and a N-TYPE SEMICONDUCTOR to which a voltage is applied enabling CHARGE CARRIERS to carry a current across the junction – that is, with the p-type material positive and the n-type negative. *See also* PN JUNCTION DIODE, REVERSE BIASED.

fossil fuels Petroleum oil, coal and natural gas. Petroleum and natural gas are produced by the decay of marine life. Coal forms as a result of similar geological processes compressing decayed forests. There is increasing pressure to reduce the consumption of fossil fuels as they are NON-RENEWABLE. The burning of such fuels also releases carbon dioxide into the atmosphere, which is believed to contribute to the GREENHOUSE EFFECT.

four forces of nature The four fundamental forces by which all matter is believed to interact. They are the gravitational force (*see* GRAVITY), the ELECTROMAGNETIC FORCE, the STRONG NUCLEAR FORCE and the WEAK NUCLEAR FORCE. Other forces, such as FRICTION are just large-scale effects of these four forces acting on a microscopic level.

At a fundamental level, all matter is believed to be made up of a limited number of ELEMENTARY PARTICLES, which cannot be divided into smaller objects. These particles interact with one another by the exchange of other particles called GAUGE BOSONS. Different elementary particles are each subject to the interactions carried by one or more of the gauge bosons, so 'feel' one or more of the four forces.

See also ELECTROWEAK FORCE, GRAND UNIFIED THEORY, STANDARD MODEL, THEORY OF EVERYTHING.

four-stroke cycle The sequence of events in most PETROL ENGINES, which requires the PISTON to make two movements in each direction to complete the cycle. *See also* TWO-STROKE CYCLE.

frame of reference A set of directions in space and time together with a reference point called an origin. The position of any other point is described by how far away it is from the origin in each of the directions. The comparison between how events are seen by observers who use frames of reference in motion relative to one another is at the heart of the theory of RELATIVITY. *See also* INERTIAL REFERENCE FRAME.

Fraunhofer diffraction A form of DIFFRACTION in which the light rays that interfere are parallel on reaching and on leaving the APERTURE. They must then either have been FOCUSED by a lens or else the light source and the screen on which the diffraction is observed must be sufficiently far away for the rays to be effectively parallel. *Compare* FRESNEL DIFFRACTION.

free electron An electron in a METAL or SEMICONDUCTOR that is not bound to any single atom but is free to move, carrying its charge and energy through the material. The large numbers of free electrons in metals make them good conductors of electricity (charge flow) and heat (energy flow).

free energy A measure of the energy released or absorbed during a reversible process. The Gibbs free energy change, ΔG, in a reaction under constant temperature and pressure is defined as:

$$\Delta G = \Delta H - T\Delta S$$

where ΔH is the ENTHALPY change, ΔS is the change in ENTROPY and T is the ABSOLUTE TEMPERATURE.

The Helmholtz free energy change, ΔF, is defined by:

$$\Delta F = \Delta U - T\Delta S$$

where ΔU is the change in INTERNAL ENERGY. The Helmholtz free energy is a measure of the maximum work that may be done by a reversible process at constant temperature.

free expansion Any process by which a gas expands without doing any WORK against the surroundings, such as is the case where a gas expands freely into a vacuum. Most gases cool in these circumstances, as work is done against attractive INTERMOLECULAR FORCES. Thus cooling in a free expansion is a method by which gases can be cooled sufficiently to turn them into liquids.

At higher pressures and temperatures, the repulsive part of the intermolecular forces is more effective and gases become warmer in a free expansion. With hydrogen and helium this happens at room temperature and pressure. Before liquefaction, such gases must be pre-cooled to a temperature called the INVERSION TEMPERATURE, below which the gas cools in a free expansion.

free fall The motion of an object on which gravity is the only force acting. Because the gravitational force acting on an object increases with its mass, whilst the acceleration produced by that force decreases with mass, the acceleration produced on an object by a gravitational field does not depend on the mass of the falling object. On Earth, for example, all objects in free fall have an acceleration of roughly 9.8 ms^{-2}. A feather will fall more slowly than a coin, but this is due to AIR RESISTANCE – place them both in a vacuum, so they are genuinely falling

freely, and they will accelerate together. This concept was first demonstrated in an experiment performed in 1604 by Galileo Galilei (1564–1642), when he dropped two balls of different mass from the Leaning Tower of Pisa.

The motion of an object in free fall is not affected by any horizontal motion – a ball launched horizontally will hit the ground at the same time as one that is simply dropped, provided the ground is flat. If the horizontal speed is large enough, the curvature of the Earth will have to be taken into account, thus an object will fly further than expected before reaching the ground. If the horizontal speed is very large, it is possible for an object to be constantly falling toward the centre of the Earth without ever hitting it. In this case the object is in orbit, but it is also still in free fall.

An astronaut inside an orbiting spacecraft feels weightless for just the same reason as the occupant of a lift with a broken cable. In each case the container and its occupant are falling freely with the same acceleration, so there is no CONTACT FORCE between them. The only difference is that the lift will eventually hit the ground, whilst the spacecraft will remain in orbit until something is done to bring it back to Earth.

freezing mixture A mixture of two or more components designed to produce a solution with a temperature below 0°C. The most common example is a mixture of common salt (sodium chloride), ice and water. The energy needed to dissolve the salt and that needed to melt the ice both come from the INTERNAL ENERGY of the molecules and the material cools. The ice continues to melt as a result of the EUTECTIC MIXTURE formed between the ice and salt, which is initially above its EUTECTIC POINT. The mixture will continue to cool until all the ice has melted, all the salt dissolved or else the eutectic temperature is reached.

frequency 1. The number of oscillations completed in one second. The frequency of waves or of an oscillating system is measured in HERTZ (Hz).

2. The number of times a particular event occurs or the number of times a variable is found to have a particular value or to lie within a particular range.

frequency distribution *See* DISTRIBUTION.

frequency modulation (FM) A type of MODULATION system in which the frequency of a

CARRIER WAVE is varied, to convey information such as a speech or music signal. Frequency modulation requires a wider BANDWIDTH than AMPLITUDE MODULATION, but is less prone to interference so is increasingly used for broadcast radio systems. *See also* FREQUENCY SHIFT KEYING, PHASE MODULATION, PULSE-CODE MODULATION.

frequency shift keying A BINARY form of FREQUENCY MODULATION, widely used in the transmission of DIGITAL data along radio links and telephone lines.

Fresnel diffraction A type of DIFFRACTION in which the waves that interfere to produce a diffraction pattern are not parallel when they come together. *Compare* FRAUNHOFER DIFFRACTION.

Fresnel's biprism A variation on YOUNG'S DOUBLE SLIT EXPERIMENT in which the double slit is replaced with a flat glass PRISM (called a biprism), which refracts light (*see* REFRACTION) to form two images of a single slit. Light from these two images then interferes (*see* INTERFERENCE).

friction The force produced when one object slides over another (DYNAMIC FRICTION) or which prevents one object sliding over another (STATIC FRICTION). Friction always acts to prevent or reduce the speed of the motion, and so tends to reduce the KINETIC ENERGY of the system, converting it into heat.

Friction can be explained in terms of the forces between the atoms of the two surfaces. Whilst these surfaces may appear smooth they are often quite rough on an atomic scale. Very smooth surfaces, such as a pair of very flat glass sheets, often produce surprisingly large frictional forces. This is contrary to the usual idea that friction is less for smooth surfaces, which follows from lumps on one rough surface becoming jammed in hollows in the other.

The details of friction are very difficult to model mathematically; a simple and surprisingly effective model for dynamic friction takes the frictional force as being proportional to the NORMAL REACTION, and as independent of the area of the sliding surfaces and of the speed at which they slide over one another. *See also* COEFFICIENT OF FRICTION, LIMITING FRICTION.

front In meteorology, the boundary between two AIR MASSES. *See also* COLD FRONT, WARM FRONT.

front-silvered (*adj.*) Describing a mirror with the reflecting coating on the front of a glass support.

frost Ice formed as air becomes SATURATED when it cools at night, with the temperature of the air being below the freezing point of water. *See also* WEATHER SYSTEMS.

fuel Any material that burns to provide a source of heat or energy.

fulcrum A fixed PIVOT about which a LEVER rotates.

function If a change in x produces a change in y, then x is said to be a function of y, expressed as

$$y = f(x)$$

In this case, y is called the dependent variable and x the independent variable. The independent variable, can take on any value within a specified range, called the domain. For example, the expression $y = x + 3$ is a function that gives a value which is always greater than the independent variable by 3.

fundamental The lowest FREQUENCY of STANDING WAVE that can be supported by a system.

fundamental constant Any quantity believed to have the same value throughout all space and time and which does not depend on the value of any other such quantity. Examples of fundamental constants include the SPEED OF LIGHT in a vacuum and the charge and mass of an electron. *See also* BOLTZMANN CONSTANT, GRAVITATIONAL CONSTANT, PERMEABILITY, PERMITTIVITY, PLANCK'S CONSTANT, RYDBERG CONSTANT.

fundamental particle *See* ELEMENTARY PARTICLE.

fundamental unit *See* BASE UNIT.

fuse A safety device that uses the heating effect of an electric current. A fuse is made from a thin piece of wire designed to get hot and melt once a certain current is exceeded. This protects the rest of a circuit from the damage that might result from overheating elsewhere caused by excessively large currents, produced by a SHORT-CIRCUIT for example.

fusion power The proposed extraction of energy on a commercial scale from NUCLEAR FUSION. Most schemes are based on the D-T REACTION, which is the fusion of deuterium and tritium nuclei.

G

gain The amount by which a signal leaving an AMPLIFIER is greater than the input signal. Gain is sometimes measured in DECIBELS.

galactic halo The roughly spherical region around a galaxy in which gas clouds and GLOBULAR CLUSTERS are found.

galaxy A large volume of space containing millions of STARS held together by gravity. Galaxies occur in various shapes, but the most common is a flat disc with stars arranged in spiral arms. The MILKY WAY, a starry band across the sky, is our own galaxy, seen edge on. Galaxies are surrounded by smaller, roughly spherical groupings of stars called GLOBULAR CLUSTERS. *See also* QUASAR.

Galilean relativity The RELATIVITY PRINCIPLE as applied to NEWTON'S LAWS OF MOTION, in which the motion of an object in one INERTIAL REFERENCE FRAME could be computed from a knowledge of its motion in another inertial frame by the addition of the RELATIVE VELOCITY of the two frames. *See also* GALILEAN TRANSFORMATION.

Galilean satellite Any of the four largest planetary satellites of JUPITER, discovered by Galileo shortly after the invention of the telescope. They are IO, Europa, Ganymede and Callisto.

Galilean transformation Addition of the RELATIVE VELOCITY of two INERTIAL REFERENCE FRAMES to convert the velocity of an object in one frame to its velocity in the other. *See also* GALILEAN RELATIVITY.

galvanometer A sensitive instrument for detecting and measuring small electric currents. *See* MOVING-COIL GALVANOMETER. *See also* AMMETER.

gamma camera An array of GAMMA RADIATION detectors used to detect the motion of a gamma ray source around a patient's body. A typical source may be a gamma ray-emitting gas, inhaled to check for blockages within the lungs, for example.

gamma radiation, *gamma ray* High energy, short wavelength ELECTROMAGNETIC RADIATION, emitted as an atomic nucleus rearranges itself into a lower energy state after an ALPHA DECAY or BETA DECAY. Most alpha decays and many beta decays produce gamma rays. Gamma radiation typically has a wavelength between 10^{-10} and 10^{-4} m and an energy between 10^{-15} and 10^{-12} J (10 keV to 10 MeV).

Gamma radiation is only weakly IONIZING, so has a very long range in air, effectively falling off in an INVERSE SQUARE LAW as the radiation spreads out. A dense material such as lead will provide some reduction in the intensity of gamma rays, provided a thickness of several centimetres is used. The intensity of gamma radiation falls off exponentially (*see* EXPONENTIAL) in an absorbing material, at a rate that depends on the nature of the absorbing material and on the distance from the gamma ray source. The thickness of a given material which is needed to reduce the level of radiation by a factor of one half is called the HALF-THICKNESS.

gamma ray *See* GAMMA RADIATION.

gas The state of matter in which a substance will expand to fill its container. In gases, the molecules are much more widely spaced than they are in solids, so the forces between them are much weaker; thus the density of a gas is much less than for a solid.

At high temperatures and pressures, the distinction between liquid and gas can disappear. The temperature above which this happens is called the critical temperature. Oxygen is an example of a substance that is above its critical temperature at room temperature and cannot be turned into a liquid simply by compressing it – such gases are called PERMANENT GASES.

See also BOYLE'S LAW, CHARLES' LAW, DALTON'S LAW OF PARTIAL PRESSURES, IDEAL GAS, IDEAL GAS EQUATION, KINETIC THEORY, PRESSURE LAW, TRANSPORT COEFFICIENT.

gas constant *See* MOLAR GAS CONSTANT.

gas discharge The flow of electric current through a gas, often at reduced pressure. The electric field must be strong enough to accelerate ions rapidly enough for them to create further IONIZATION when they collide with gas molecules; thus gas discharges generally occur only at relatively high voltages. As the ions recombine with electrons, light is given out, with a colour characteristic of the gas used. *See also* GLOW DISCHARGE.

gas giants Large planets, composed chiefly of hydrogen and other gases. In order of distance from the Sun, they are JUPITER (the largest), SATURN, URANUS and NEPTUNE. The gas giants show active weather systems and cloud patterns, including Jupiter's GREAT RED SPOT, which is believed to be a long-lived storm system. All the gas giants are known to have RING SYSTEMS.

gas laws The three laws, BOYLE'S LAW, CHARLES' LAW and the PRESSURE LAW, that between them describe the properties of IDEAL GASES. They contain the same information as is contained in the IDEAL GAS EQUATION, but relate only to a fixed mass of gas, i.e. a fixed number of molecules.

gas turbine A JET ENGINE in which some of the energy of the hot gases leaving the engine is used to drive a TURBINE attached to a shaft that drives a compressor forcing fresh air into the engine. The shaft may also be used to extract energy from the engine for other purposes – to drive a propeller on a ship or aircraft (where such engines are called TURBOPROP engines), or to generate electricity.

Modern aircraft often use gas turbine engines, and modern airliners are fitted with HIGH-BYPASS ENGINES, or turbofans. In these engines, much of the air from the compressor leaves the engine without having fuel burnt in it. This provides increased fuel economy, whilst the relatively cool air surrounds the air heated by combustion and reduces the noise

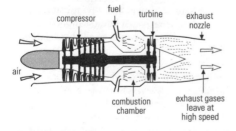

Gas turbine.

level of the engine. In military aircraft, fuel consumption and noise are of less concern than power and additional fuel may be burnt in the engine after the main turbine in a system known as an afterburner or REHEAT SYSTEM.

gate 1. In electronics, the ELECTRODE in a FIELD EFFECT TRANSISTOR at which an applied ELECTRIC FIELD controls the current flowing in the CHANNEL from the SOURCE to the DRAIN. The electrode with a similar function in a THYRISTOR or TRIAC is also called the gate.

2. *See* LOGIC GATE.

gauge boson A particle that is exchanged between two particles, carrying energy and MOMENTUM. This is the quantum mechanical (*see* QUANTUM MECHANICS) interpretation of a force acting between the two particles. The ELECTROMAGNETIC FORCE acts by the exchange of PHOTONS, the WEAK NUCLEAR FORCE by the exchange of W BOSONS and Z BOSONS and the STRONG NUCLEAR FORCE by the exchange of GLUONS. The name GRAVITON is used for the hypothetical exchange particle in a quantum theory of gravitation (*see* GRAVITY), but as yet there is no completely satisfactory theory of quantum gravity, nor any experimental evidence for the graviton.

gauge theory Any theory of interactions between ELEMENTARY PARTICLES that exhibits properties associated with certain forms of SYMMETRY. The details are complex, but in essence the interactions between particles of MATTER are explained in terms of the exchange of particles called GAUGE BOSONS. All currently accepted theories of the quantum mechanical interactions of matter are gauge theories.

gauss Unit of MAGNETIC FIELD strength in the C.G.S. SYSTEM. One gauss is equal to 10^{-4} TESLA.

Gaussian curve *See* NORMAL DISTRIBUTION CURVE.

Gauss' law The total ELECTRIC FLUX through any closed surface is equal to the total charge enclosed by the surface divided by the PERMITTIVITY of free space.

gears A pair of rotating wheels, forming a MACHINE based on the LEVER principle. The wheels are provided with teeth to make them mesh (rotate together). One wheel, the pinion, is smaller than the other and rotates more quickly. The teeth act as a pivot between the two rotating wheels. The rotational speeds of the wheels are inversely proportional to the number of teeth on the wheel.

gedankenexperiment (German = thought experiment) An imaginary experiment, carried out with idealized apparatus, designed to illustrate a particular concept. Many complex concepts in QUANTUM MECHANICS are explained by *gedankenexperiments. See also* HEISENBERG'S MICROSCOPE, SCHRÖDINGER'S CAT.

Geiger counter A RADIATION DETECTOR containing a GEIGER-MÜLLER TUBE and its associated electronics.

Geiger-Müller tube A device for detecting IONIZING RADIATION using the flow of current in a sample of low pressure gas. *See* RADIATION DETECTORS.

general theory of relativity A theory proposed by Albert Einstein (1879–1955) in 1915, which extended the ideas of the SPECIAL THEORY OF RELATIVITY to cover observers in NON-INERTIAL FRAMES OF REFERENCE; that is, observers who would see one another as accelerating.

One of the predictions of the general theory is the deflection of light by gravity. This was shown to be true during a solar eclipse in 1919, when it was possible to observe the apparent shift in the position of stars that appeared in the same part of the sky as the Sun. The effect of gravity on light also led to the prediction of BLACK HOLES.

Another result that pointed to the truth of the theory was the explanation of the PRECESSION of the PERIHELION of Mercury. The orbit of the planet Mercury is a flattened circle, or ELLIPSE. This ellipse does not remain fixed in space as predicted by NEWTONIAN MECHANICS, but instead rotates very gradually. Much of this rotation could be explained in part by the gravitational influence of other planets, but the general theory of relativity explained the rest of the motion.

The mathematics behind the general theory of relativity is based on the idea that the gravity associated with a massive object is not a force in the conventional sense, but rather is a distortion of the space through which the object moves. In the general theory of relativity, the Universe is described in four dimensions of space and time. This SPACE-TIME continuum obeys non-Euclidean geometry (*see* EUCLIDEAN SPACE). *See also* CURVATURE OF SPACE, GRAVITATIONAL WAVE, PRINCIPLE OF EQUIVALENCE.

generator A machine for converting mechanical energy into electrical energy. The term is also used to refer to the combination of a generator with a small INTERNAL COMBUSTION ENGINE to provide a source of electricity away from a mains supply. A generator may be a DYNAMO or an ALTERNATOR, but modern machines are usually alternators.

geodesic line The path of shortest time between two points, and the non-Euclidean equivalent of a straight line. *See* CURVATURE OF SPACE, EUCLIDEAN SPACE.

geology The study of the solid part of the Earth.

geometric mean *See* MEAN.

geometric progression A SERIES of numbers in which each number is a constant multiple of the previous number in the series. Thus if the nth member of the series is a_n, the $(n+1)$th member will be

$$a_{n+1} = ka_n$$

where k is a constant called the common ratio. For a geometric progression of N terms, the sum S_N of the series will be

$$S_N = (1 - a_n)/(1 - a)$$

geometry The study of the properties of points, lines and planes, and of curves, shapes and solids. The study of two-dimensional shapes on flat surfaces is called plane geometry, and this, along with solid geometry (in three dimensions) make up pure geometry. In analytical, or co-ordinate geometry, problems are solved using algebraic methods. *See also* EUCLIDEAN SPACE.

geophysics The branch of science that applies the principles of physics and mathematics to the study of the Earth, its interior and CLIMATE. *See also* SEISMOLOGY.

geostationary orbit A circular ORBIT over the equator, moving in the same direction as the Earth's rotation and with the same period, i.e. 24 hours. As seen from the rotating Earth, an artificial satellite in such an orbit always appears in the same place in the sky. An altitude of 36,000 km corresponds to such an orbit.

Geostationary orbits are very important for communications satellites and also for DIRECT BROADCAST SATELLITE systems, which broadcast television signals straight into the homes of the viewers, as the receiving AERIALS do not then need to track the motion of the satellite. Geostationary orbits are also used by some weather satellites, as they provide a constant view of one half of the Earth.

See also CIRCUMPOLAR ORBIT.

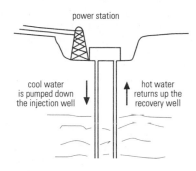

power station

cool water
is pumped down
the injection well

hot water
returns up the
recovery well

Geothermal energy.

geothermal energy Heat extracted from the Earth's CORE by pumping water through hot rocks. The hot water can either be used directly, for heating, or to generate electricity. In some countries, such as Iceland, this is an important source of energy. Experimental schemes have been tried elsewhere, but the technical problems inherent in pumping water in and out of deep rocks are too great at present for such schemes to be viable in most locations.

getter A device for removing small amounts of gas from vacuum vessels in order to improve the vacuum. A getter consists of a wire made of a reactive metal, such as magnesium. The vessel is evacuated and sealed and the getter vaporized by passing a current through it. The metal atoms react with any air molecules remaining in the vessel to form a non-volatile compound.

Gibbs free energy *See* FREE ENERGY.

giga- (G) Prefix used before a unit to indicate that the size of the unit is multiplied by 10^9. For instance, one gigawatt (GW), is one billion watts.

glass Any of a number of transparent brittle materials. Common glass is made by melting together sand (mostly silicon dioxide) with lime (calcium oxide) and soda (mostly sodium carbonate). AMORPHOUS materials are sometimes referred to as glasses, as glass is an example of an amorphous material. In some ways glasses are more like liquids than solids: they have no melting point, but simply flow more readily as they are heated. *See also* DISORDERED SOLID.

global warming An increase in the overall temperature of the Earth, believed to be caused by the increasing level of GREENHOUSE GASES, particularly carbon dioxide, in the atmosphere. *See* GREENHOUSE EFFECT.

globular cluster A roughly spherical grouping of a few hundred stars found in orbit around a GALAXY. Globular clusters are of particular interest to astronomers as they provide a group of stars all at much the same distance from the Earth. *See also* GALACTIC HALO.

glow discharge A GAS DISCHARGE in a low pressure gas, producing a steady luminous glow.

gluon The GAUGE BOSON that is exchanged in the STRONG NUCLEAR FORCE, named as the glue that holds QUARKS together in a HADRON. *See also* QUANTUM CHROMODYNAMICS.

gold-leaf electroscope An ELECTROSCOPE in which the electrical repulsion between a conductor made of a metal plate and a thin sheet of gold causes the gold to move away from the plate against the force of gravity.

graded index (*adj.*) Describing a FIBRE OPTIC in which the core is made of a material with a REFRACTIVE INDEX that decreases steadily from the centre to the edge. This means that the light rays that travel almost parallel to the axis of the fibre, and which therefore have a shorter path, are slowed down relative to those at larger angles that are reflected off the outer layer more often. The change in refractive index is designed to compensate for these changes in path length, resulting in less spreading of individual light pulses over a long length of fibre, and enabling data to be sent at higher rates. *See also* MULTIMODE, STEP INDEX.

Graham's law of diffusion The rate of DIFFUSION of a gas is inversely proportional to the square root of the RELATIVE MOLECULAR MASS. Thus light gases, such as hydrogen, diffuse more quickly than heavier ones, such as carbon dioxide.

grain boundary In a POLYCRYSTALLINE material, the boundary between one orientation of the crystal lattice and another.

gram A unit of mass, nowadays defined as one thousandth of a KILOGRAM.

grand unified theory (GUT) A hypothetical unification of the ELECTROWEAK FORCE and STRONG NUCLEAR FORCE, and possibly GRAVITY too. Such theories make few predictions that can be tested at energies available from current or conceivable PARTICLE ACCELERATORS. The predictions they do make are also sometimes at odds with experimental observation, suggesting, for example, that the proton may be unstable and decay into lighter particles. Despite intensive searches for PROTON DECAY, no evidence has been found. *See also* STANDARD MODEL, THEORY OF EVERYTHING.

graph A way of showing how one variable behaves as a function of another. Points are plotted on a two-dimensional surface according to the values of the two variables. The variables are usually referred to as *x*, measured along a horizontal AXIS, and *y*, measured up a vertical axis. The independent variable is generally plotted along the *x*-axis, with the dependent variable up the *y*-axis. If statistical or experimental data is plotted, a series of measurements will build up a SCATTER DIAGRAM. A mathematical function may be represented by a line or curve on the graph.

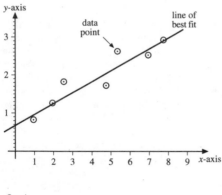

Graph.

graticule A scale, similar to a ruler, mounted on a transparent support and placed in the EYE-PIECE of a TELESCOPE or MICROSCOPE, or on the screen of a CATHODE RAY OSCILLOSCOPE, to enable the size of features to be measured.

gravitation *See* GRAVITY.

gravitational anomaly A small change in the GRAVITATIONAL FIELD STRENGTH on the surface of the Earth caused by the uneven distribution of mass in the locality.

gravitational collapse The coming together of a number of objects, or the increase in density of a gas cloud, as a result of the gravitational attraction between different parts of the

system. Gravitational collapse is the process by which STARS are formed (*see* PROTOSTAR) and may also lead to the Universe ending in a BIG CRUNCH.

gravitational constant (*G*) The constant that measures the overall strength of the gravitational force, the force between two 1 kg masses 1 m apart. It is equal to 6.67×10^{-11} Nm^{2}kg^{-2}. *See also* CAVENDISH'S EXPERIMENT, GRAVITY, NEWTON'S LAW OF GRAVITATION.

gravitational field The region of influence around any mass that will cause any other mass within that region to experience a force proportional to its mass. *See also* GRAVITATIONAL POTENTIAL.

gravitational field strength The force per unit mass experienced by an object placed at the point where the GRAVITATIONAL FIELD strength is being measured. The SI UNIT of gravitational field strength is the Nkg^{-1} (newton per kilogram) or ms^{-2} (metre per second squared).

Since GRAVITATIONAL MASS and INERTIAL MASS appear to be the same for all objects, an object released in a gravitational field (on the surface of the Earth for example) will always fall with the same acceleration (the ACCELERATION DUE TO GRAVITY) regardless of its mass, provided no other forces act. This acceleration will be equal to the gravitational field strength. The gravitational field strength on the surface of the Earth is about 9.8 Nkg^{-1}, but it varies slightly from place to place due partly to the non-spherical shape of the Earth and partly to variations in the density of nearby rocks. These local variations, called gravitational anomalies, have been used as a way of studying the geology of the underlying rocks.

At a distance *r* from the centre of a body of mass *M* and radius *R*, if *r* is greater than *R*, the gravitational field is

$$g = GM/r^2$$

gravitational lens A massive object, such as a QUASAR or BLACK HOLE, that acts to bend light arriving on Earth from some more distant object, resulting in the formation of a double image or a distorted image of the more distant object. Originally a hypothetical prediction of the GENERAL THEORY OF RELATIVITY, a number of examples of gravitational lensing have now been observed experimentally.

gravitational mass MASS measured as an object's response to a GRAVITATIONAL FIELD, as opposed

to INERTIAL MASS. *See also* PRINCIPLE OF EQUIVA-
LENCE.

gravitational potential The gravitational potential at a point is the WORK needed to bring a
1 kg mass from infinity to that point. Because
gravity is always an attractive force, systems
tend to collapse under gravity rather than
expand, thus gravitational potentials are
always negative. The difference in gravitational
potential between two points is a measure of
the energy needed per unit mass to move an
object from one point to the other.

At a distance *r* from the centre of a body of
mass *M* and radius *R*, if *r* is greater than *R*, the
gravitational potential is

$$V_g = -GM/r$$

and the relationship between GRAVITATIONAL
FIELD *g* and gravitational potential V_g is

$$g = -dV_g/dx$$

gravitational potential energy The energy of a
mass in a GRAVITATIONAL FIELD. For an object of
mass *m* moving through a height *h* in a gravitational field *g*, the change in gravitational
potential energy is *mgh*.

gravitational wave A wave of distortion of
SPACE-TIME, carrying energy at the speed of
light and predicted by the GENERAL THEORY OF
RELATIVITY. Evidence for the existence of gravitational waves has recently been obtained from
the changes in motion of a BINARY STAR, one
element of which is a PULSAR. As the double star
system loses energy by gravitational waves, its
period should change. This can be detected in
such a system by observing changes in the pattern of pulses from the pulsar, which are not
quite regular, but influenced by the DOPPLER
EFFECT as a result of the orbital motion.

graviton The hypothetical GAUGE BOSON responsible for GRAVITY in a QUANTUM THEORY of gravity. No such theory has yet been satisfactorily
developed.

gravity, *gravitation* The force of attraction
between all objects dependent on their MASS.

Gravity is a weak force, so it is normally
only noticeable when at least one of the masses
is very large, such as in the case of a planet or
star. Whilst the GENERAL THEORY OF RELATIVITY
provides a good explanation of gravity at large
scales, it is widely believed that at small scales
and very high energies there must be a theory
that links gravity and QUANTUM MECHANICS

and explains gravity in terms of the exchange
of a particle called the GRAVITON.

At the scales currently studied by particle
physics, and for laboratory-sized objects,
gravity is generally too weak a force to be
important. It only becomes important at large
scales because it is a long range force with no
cancellation between the effect of positive and
negative charges as there is with the ELECTRO-
MAGNETIC FORCE. The gravitational force acts
on every particle in an object but can be considered as a single force acting at a point called
the CENTRE OF MASS or centre of gravity.

See also FREE FALL, GRAND UNIFIED THEORY,
GRAVITATIONAL FIELD, GRAVITATIONAL FIELD
STRENGTH, GRAVITATIONAL POTENTIAL, GRAVITA-
TIONAL WAVE, NEWTON'S LAW OF GRAVITATION.

gray (Gy) The SI UNIT of absorbed IONIZING
RADIATION. One gray is equal to an energy of
one JOULE absorbed from the radiation. *See
also* DOSE.

Great Red Spot A long-lived cloud feature in the
atmosphere of JUPITER, believed to be some
form of storm system.

greenhouse effect A supposed increase in the
average surface temperature of the Earth as a
result of changes in the composition of its
atmosphere. Since the temperature of the Earth
is much lower than that of the Sun, whilst the
electromagnetic radiation absorbed from the
Sun mostly lies in or near the visible part of the
ELECTROMAGNETIC SPECTRUM, the radiation
given off by the Earth is in the infrared part of
the electromagnetic spectrum. Some of this is
absorbed by gases in the atmosphere, particularly carbon dioxide and water vapour, thus the
concentration of such gases (called GREEN-
HOUSE GASES) has a marked effect on the surface
temperature of the Earth. This effect is called
the greenhouse effect since the Earth's atmosphere behaves in a way similar to the glass in a
greenhouse, which also allows in visible radiation but absorbs infrared.

Fears have been expressed over the
increase in carbon dioxide in the atmosphere
caused by the burning of increasing quantities
of FOSSIL FUELS over the last century, and
attempts are being made to reduce carbon
dioxide emission by increasing use of renewable energy sources (*see* RENEWABLE RESOURCE).
The systematic destruction of vast areas of for-
est has also contributed to the increase of carbon dioxide levels. It is believed that the

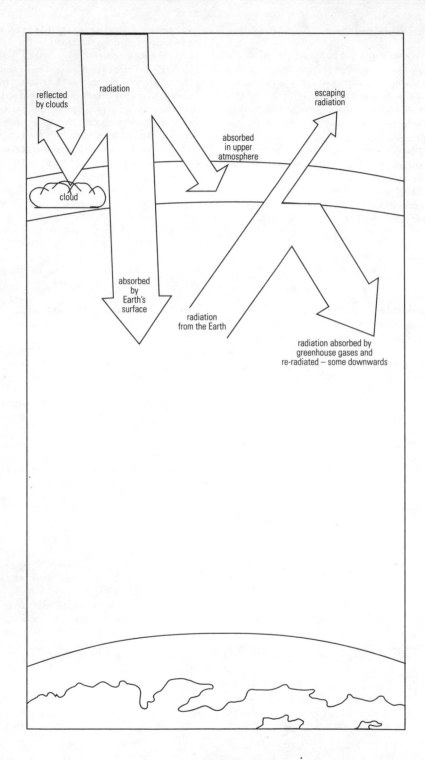

reflected
by clouds

radiation

escaping
radiation

absorbed
in upper
atmosphere

cloud

absorbed
by
Earth's
surface

radiation
from the Earth

radiation absorbed by
greenhouse gases and
re-radiated – some downwards

The greenhouse effect.

increasing levels of carbon dioxide may increase the average temperature of the planet, a phenomenon called GLOBAL WARMING, which may also lead to other changes in climate. Other greenhouse gases include methane (a by-product from agriculture), water vapour (as a by-product from industry) and chlorofluorocarbons (from refrigerators, aerosol sprays and polystyrene).

The problem with continued global warming is that it will cause the polar ice caps to melt, resulting in a rise in sea levels and consequent flooding of low-lying land, which could include whole countries and many world capital cities. A change in the climate would also affect crop growth. It is not clear exactly what or how rapid the consequences of the greenhouse effect will be, because many effects will interact with one another.

greenhouse gas Any gas, such as carbon dioxide or methane, that contributes to the GREEN-HOUSE EFFECT in the Earth's atmosphere by absorbing infrared radiation.

ground (*US*) *See* EARTH.

ground state The lowest energy state of a system, such as an atom, from which it can be excited to higher energy states. *See* ENERGY LEVEL. *See also* EXCITED STATE.

group A column of elements in the PERIODIC TABLE having similar chemical properties resulting from a similar arrangement of ELECTRONS in their outer SHELL. An example is group 1, the alkali metals, which all form positive ions by losing their single outer electron, and react with water to form hydrogen gas plus a metal hydroxide.

group displacement law *See* FAJAN AND SODDY'S GROUP DISPLACEMENT LAW.

gyroscope A rapidly rotating metal wheel, that will tend to retain the same direction in space regardless of any motion of its support. Such devices are used to measure and control the direction of aircraft and spacecraft. *See also* LAW OF CONSERVATION OF ANGULAR MOMENTUM, PRECESSION.

H

Hadley cell A closed loop of CONVECTION within the Earth's atmosphere. *See* CLIMATE.

hadron Any of a class of subatomic particles that are influenced by the STRONG NUCLEAR FORCE. Hadrons can be subdivided into BARYONS (which include PROTONS and NEUTRONS) and MESONS. It is now known that all hadrons are composed of particles called QUARKS held together by the strong nuclear force. Hadrons other than protons and neutrons may be produced in experiments using a high energy beam from a PARTICLE ACCELERATOR.

hail Frozen water droplets that fall from CUMULONIMBUS clouds, often in association with a THUNDERSTORM. The CONVECTION CURRENTS in the clouds cause raindrops to move up and down the cloud, freezing, collecting a layer of water, freezing again and so on, until they become too massive to be supported by the rising air in the cloud. *See also* WEATHER SYSTEMS.

half-cell A single ELECTRODE immersed in an ELECTROLYTE. In use, two half-cells are connected together by a salt bridge – a piece of absorbent paper soaked in an electrolyte, usually potassium chloride. The voltage between the two electrodes is equal to the difference in their ELECTRODE POTENTIALS. By changing just one of the electrodes whilst leaving the other one unchanged, the different electrode potentials can be compared.

half-life The time taken for one half of the radioactive nuclei originally present in a sample to decay (*see* RADIOACTIVITY). The half-life τ is related to the DECAY CONSTANT λ by the formula

$$\tau = \ln 2/\lambda$$

half-thickness The thickness of a given material needed to reduce the intensity of GAMMA RADIATION by one half. The half-thickness of a material depends on the energy SPECTRUM of the gamma radiation concerned.

Hall effect The production of a POTENTIAL DIFFERENCE between the edges of a conductor or semiconductor carrying a current in a magnetic field. The MAGNETIC FORCE on the CHARGE CARRIERS produces a build-up of charge at one side of the conductor. The potential difference between the two edges of a conductor in a magnetic field, measured at right angles both to the field and to the direction of current flow, is called the HALL VOLTAGE. The electric field resulting from this build-up of charge limits the size of the Hall voltage. In metals it is very small, even for strong fields. Semiconductors have a far lower number of free charge carriers per metre cubed and so produce larger Hall voltages.

Halley's comet The brightest and best known of the periodic COMETS. It has a period of 76 years and last reached PERIHELION in 1986.

Hall probe A device used to measure magnetic fields, using the HALL EFFECT in a small piece of semiconductor.

Hall voltage Voltage produced by the HALL EFFECT. The Hall voltage V_H produced between the edges of a conductor of thickness t carrying a current I in a magnetic field B and having n free CHARGE CARRIERS per metre cubed, each of charge e is given by

$$V_H = BI/net$$

hard (*adj.*) **1.** Describing any material that is not easily scratched.

2. Describing FERROMAGNETIC materials in which the DOMAIN walls (boundaries between one domain and the next) are held in place (by carbon atoms in steel for example), so the material retains its magnetism when the magnetizing field is removed. Magnetically hard alloys are used in the manufacture of PERMANENT MAGNETS and for materials such as magnetic tapes and computer disks, which use magnetic fields to store information.

hardness Resistance to scratching.

harmonic A whole number multiple of a given FREQUENCY, such as the FUNDAMENTAL frequency, in a system that supports STANDING WAVES. The frequency that is twice the fundamental is called the second harmonic, that which is three times the fundamental is called the third harmonic etc.

Hawking radiation Electromagnetic radiation arising from the creation of ELECTRON-POSITRON pairs in the intense gravitational field of a BLACK HOLE with only one member of the pair falling into the black hole.

H-bomb *See* HYDROGEN BOMB.

heat The energy that is transferred from one body or system to another as a result of a difference in temperature. Heat was originally believed to be some kind of fluid, called caloric, which could be squeezed out of materials or released when they were burned, but careful experiments by James Joule (1818–1889), using a falling weight to drive a paddle wheel which churned up water and increased its temperature, led to the recognition that heat is a form of energy. In SI UNITS, heat, like all other forms of energy, is measured in JOULES. *See also* CONVECTION, HEAT CAPACITY, LATENT HEAT, LAW OF CONSERVATION OF ENERGY, THERMAL CONDUCTION, THERMAL RADIATION, THERMODYNAMICS.

heat capacity The amount of heat needed to change the temperature of an object by one degree CELSIUS (or one KELVIN, which is the same size). If the temperature is increasing, this much heat energy will have been taken in by the object; if it is decreasing, the energy will have been given out. *See also* CONSTANT FLOW METHOD, METHOD OF MIXTURES, MOLAR HEAT CAPACITY, RATIO OF SPECIFIC HEATS, SPECIFIC HEAT CAPACITY.

heat engine Any machine for converting HEAT energy to mechanical WORK. *See* CARNOT CYCLE, KELVIN STATEMENT OF THE SECOND LAW OF THERMODYNAMICS.

heat exchanger A device for transferring heat energy from one fluid to another, without contact between the two fluids. In many applications, a hot liquid, often water, needs to be cooled, giving up its heat to the surrounding air. The liquid flows through pipes to which thin metal plates are attached. Air is then forced over the plates carrying away heat. *See also* COUNTERCURRENT SYSTEM.

heat pump A machine for removing heat from one system and depositing it in a second, hotter system. Energy must be supplied to make the heat flow in this direction (*see* SECOND LAW OF THERMODYNAMICS). Refrigerators and air-conditioning systems are examples of heat pumps. Heat pumps have also been proposed as a more energy efficient means of heating buildings in cold weather, but, because of the increased complexity compared to direct heating, they are not widely used.

heat radiation *See* THERMAL RADIATION.

heat reservoir A hypothetical object of infinite HEAT CAPACITY, which heat can enter or leave without producing a change in temperature.

heat treatment The heating and cooling of a material, usually a metal or alloy, under controlled conditions to produce changes in the mechanical properties of the material which remain after it has returned to room temperature. *See* ANNEALING, QUENCHING.

heavy water Water in which both hydrogen atoms have been replaced with the DEUTERIUM ISOTOPE. Heavy water is used as a MODERATOR in some forms of nuclear reactor.

Heisenberg's microscope A thought experiment designed to illustrate the effect OF HEISENBERG'S UNCERTAINTY PRINCIPLE on experimental observations. A particle is viewed through a microscope but the image is affected by DIFFRACTION at the lens of the microscope, producing an uncertainty in position. To reduce this, shorter wavelength light could be used, but this will result in a greater uncertainty in the MOMENTUM transferred to the particle when this light is scattered into the microscope, since the only way the exact path of the PHOTON could be known is by using a microscope with a very small opening, leading to greater diffraction so again a larger uncertainty in position.

Heisenberg's uncertainty principle A consequence of the WAVE NATURE OF PARTICLES. Loosely stated, this means that the more accurately we know the position of a particle the less sure we can be of the associated wavelength and hence the MOMENTUM, and vice versa. More precisely, if the position is known with an uncertainty Δx and the momentum with an uncertainty Δp then

$$\Delta x \Delta p \geq h/\pi$$

where h is PLANCK'S CONSTANT. *See also* HEISENBERG'S MICROSCOPE, VIRTUAL PARTICLE.

helix A spiral in three dimensions. A helix can be described by the PARAMETRIC EQUATION:

$$x = r\cos\omega t, y = r\sin\omega t, z = kt$$

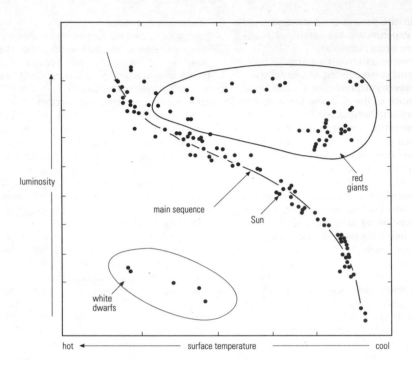

Hertzsprung–Russell diagram.

The pitch of a helix is the distance moved along the axis of the helix in the course of one complete rotation. For the above equation the pitch is equal to $2\pi k/\omega$.

Helmholtz coils A pair of flat parallel coils separated by a distance equal to their radius. Helmholtz coils give a magnetic field that changes very little in the space between the coils.

henry (H) The SI UNIT of SELF-INDUCTANCE and MUTUAL INDUCTANCE. An inductance of one henry will result in an induced ELECTROMOTIVE FORCE (e.m.f.) of one VOLT when the current producing the induced e.m.f. is changing at the rate of one AMPERE per second.

hertz (Hz) The SI UNIT of FREQUENCY. One hertz is a frequency of one oscillation per second.

Hertzsprung–Russell diagram A plot of the LUMINOSITY of various stars against their SPECTRAL CLASS (which is effectively a measure of temperature). Most stars lie in a diagonal band from top left (hot bright stars) to bottom right (cool dim stars) called the main sequence. *See also* MAIN SEQUENCE STAR.

heterogeneous (*adj.*) Describing a substance whose properties vary from one place to another. *Compare* HOMOGENEOUS.

hexagonal close packed (*adj.*) Describing a crystalline structure in which each layer of atoms is CLOSE PACKED, with each atom surrounded by 6 others in that layer. The second layer of atoms lies above the gaps in the first, whilst the atoms in the third layer lie above gaps in the second and directly above the atoms in the first. If the layers of atoms are labelled *A* and *B*, the structure can be described as *ABAB*... Many metals occur as hexagonal close packed structures, including magnesium and zinc, though the CUBIC CLOSE PACKED structure is also common.

Higgs boson A hypothetical particle in the STANDARD MODEL of particle physics. The Higgs boson is responsible for giving mass to all the charged LEPTONS and QUARKS and is regarded as one of the more unsatisfactory aspects of the standard model. Its mass is about 300 times that of the proton, too heavy to have been found in the present generation of particle physics experiments, but within the reach of those now being planned.

high In DIGITAL electronics, a voltage level or other signal corresponding to a BINARY 1.

high-bypass engine, *turbofan* A GAS TURBINE in which much of the air leaves the back of the engine without being used to burn fuel.

high-temperature superconductor A material that remains a SUPERCONDUCTOR at temperatures that can be reached using liquid nitrogen (77 K/–196°C) rather than liquid helium (4 K/–269°C). The ultimate goal is to produce a material which is superconducting at room temperature. The high-temperature superconductors discovered so far are ceramic materials and are chemically rather unstable. Their poor mechanical properties have made them less commercially valuable than was originally hoped.

Hofmann voltammeter A VOLTAMMETER designed to collect gases given off in ELECTROLYSIS.

hole A vacant space in the electron structure of a SEMICONDUCTOR. A neighbouring electron may move to fill this hole, which then appears to move through the semiconductor in the opposite direction to the electron flow. Holes can be thought of as behaving like positive CHARGE CARRIERS.

hologram A three-dimensional image recorded on a flat piece of photographic film. In the simplest method of producing a hologram, a beam of COHERENT light, usually from a laser, is split into two by a semi-reflecting mirror. One beam (the signal beam) is diffracted by the object being recorded onto a piece of photographic film or plate. The other beam (the reference beam) falls directly onto the film, where it interferes with the signal beam. The INTERFERENCE pattern thus produced forms the hologram.

Once the film has been processed, the hologram can be viewed by illuminating it with a coherent beam of light (usually of the same wavelength as the original beam). As the viewer changes his angle of view, he sees the object from a different perspective. Each small part of the hologram effectively contains a whole image of the object, but only from a single viewpoint.

As with any other interference effect, distances comparable to the wavelength of light are critical. This has lead to interference

holograms, used to study vibrations of solid objects. Because holograms are hard to make, they are also used as security labels in devices such as credit cards.

holography The process of making HOLOGRAMS.

homogeneous (*adj.*) Describing a substance that is the same throughout, such as a material made from a single compound. *Compare* HETEROGENEOUS.

Hooke's Law For certain materials, up to a point called the ELASTIC LIMIT, the amount by which the length of the material increases in response to an applied force is proportional to the size of that force. If the force is removed, the material will return to its original size and shape.

For example, if a spring is stretched by 2 cm by a force of 1 N, it will stretch by 4 cm with a force of 2 N. For a force F, the extension x of a spring will be such that

$$F = kx$$

where k is a constant, called the spring constant.

horsepower (hp) A unit of power, obsolete in science but still widely used for measuring the power output of INTERNAL COMBUSTION ENGINES. One hp is equal to 746 WATTS. *See also* BRAKE HORSEPOWER.

horseshoe magnet A MAGNET in the form of a horseshoe, a bar with its ends bent round so the ends are next to one another, with a POLE at each end of the bar.

hot-air balloon A cloth container filled with hot air from which a basket is suspended to carry passengers and fuel to heat the air. A hot-air balloon rises because the density of the warm air in the balloon is less than that of the surrounding cooler air. The balloon stops rising when the average density of the warm air plus the passengers and their basket is equal to the density of the surrounding air. *See also* FLOTATION.

Hubble constant The ratio between the speed of a galaxy and its distance from the Earth (*see* HUBBLE'S LAW). Experimental values vary due to the difficulty in estimating the distance of a galaxy, but generally range from 50 to 150 kms^{-1} per megaparsec (*see* PARSEC).

Hubble's law Distant galaxies appear to be receding with a speed proportional to their distance from the Earth. Evidence for this comes from a measurement of the RED-SHIFT in the SPECTRAL LINES of stars in distant galaxies.

Difficulties in measuring the distances of these distant galaxies lead to considerable uncertainties in the measurement of the HUBBLE CONSTANT, the constant of proportionality linking a galaxy's distance with the speed of its motion. Hubble's law provides important evidence for the BIG BANG theory. *See also* COSMOLOGY, DOPPLER EFFECT.

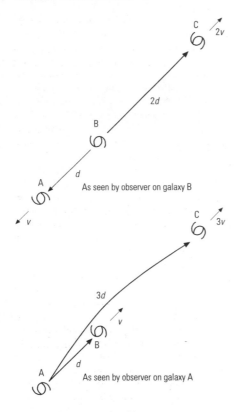

Hubble's law.

Hubble Space Telescope A large astronomical TELESCOPE placed in orbit around the Earth in 1990. It operates in the visible, ultraviolet and near-infrared regions of the spectrum.

humidity The amount of water contained by an AIR MASS. This is often expressed as a RELATIVE HUMIDITY – the proportion of water vapour in the air compared to the maximum amount of water vapour the air can hold as a gas (at which point the air is said to be saturated). Humidity may also be expressed as a DEW POINT – the temperature to which the air would have to be cooled to become saturated. The humidity of the air depends on the amount of water it has passed over, and its temperature, since water evaporates more readily at high temperatures. Instruments for measuring humidity are called HYGROMETERS. *See also* WET AND DRY BULB HYGROMETER.

Hund's rule of maximum multiplicity A consequence of the repulsion between electrons in a given ORBITAL. The rule states that the electrons in p-, d- or f-orbitals tend to arrange themselves with one electron in each available orbital before a second electron enters any of the orbitals. Thus iron, for example, which has six 3d electrons, will have one D-ORBITAL with two electrons and four with a single electron.

hurricane A tropical CYCLONE characterized by strong surface winds in excess of $120\,\mathrm{kmh^{-1}}$. Winds spiral around a low pressure centre in a clockwise direction in the northern hemisphere and anticlockwise in the southern hemisphere. Hurricanes are called typhoons in the North Pacific.

Huygens' construction A system for predicting the position of a WAVEFRONT from that of the previous wavefront. Each point on a wavefront is imagined to be a source of circular waves (called SECONDARY WAVELETS), and the combined effect of these wavelets gives the position of the next wavefront. In DIFFRACTION, for example the constructive and destructive interference (*see* INTERFERENCE) between different parts of the wavefront account for the observed effects.

hybrid orbital A superposition of the WAVEFUNCTIONS of a number of ORBITALS with the same energy to form a composite orbital. This is the way orbitals sometimes behave in the formation of COVALENT BONDS. Thus carbon, containing two electrons in a 2s orbital and two in 2p orbitals, forms four sp^3 hybrid orbitals, each containing one electron. These orbitals have the shape of a lobe directed towards the corners of a TETRAHEDRON centred on the carbon atom, and it is these orbitals that produce the well-known TETRAHEDRAL structure of carbon's covalent bonds.

hydraulic (*adj.*) Describing any device in which a liquid, such as oil, is used to transmit FORCES from one place to another. By making the PISTONS that produce and receive the pressure different sizes, the pressure can be used to exert

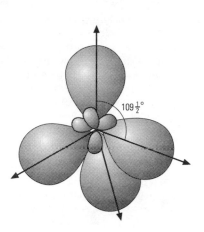

$109\tfrac{1}{2}°$

Hybrid orbital.

forces of different sizes, producing a MACHINE. Such systems are used in the brake systems of cars and to control many large machines, from aircraft to excavators. *See also* PNEUMATIC.

hydraulics The study of fluids in motion or at rest. It is the practical application of HYDROSTATICS and HYDRODYNAMICS.

hydrodynamics The study of the motion of fluids, particularly liquids.

hydroelectric power The generation of electricity from falling water, normally obtained by damming a river and allowing the water from the resulting lake to fall through a TURBINE. Such schemes are very effective in the production of electricity in mountainous regions, such as Scotland and Switzerland, but in developing countries there have been concerns about the environmental impact of flooding large areas of land.

hydrogen bomb, *H-bomb* A weapon that derives its energy from NUCLEAR FUSION of hydrogen nuclei.

hydrogen bond A weak bond formed between molecules that contain a hydrogen atom linked by a COVALENT BOND to an atom of a reactive element. Such covalent bonds are POLAR and the hydrogen bond is an extreme form of the attraction between opposite ends of the DIPOLES formed in neighbouring molecules. Hydrogen bonding results in higher than expected melting and boiling points and , is particularly important in water, accounting for many of its unusual properties, such as the

fact that it expands on cooling below 4°C, and expands further on freezing.

hydrogen electrode A platinum ELECTRODE, coated with finely divided platinum (platinum black) over which hydrogen is passed at a pressure of one ATMOSPHERE, in contact with hydrogen ions at a concentration of one mole per decimetre cubed. *See also* STANDARD ELECTRODE POTENTIAL.

hydrogen spectrum The EMISSION SPECTRUM of hydrogen. As an atom of hydrogen has only a single electron its spectrum is fairly simple. It comprises a number of series of SPECTRAL LINES – the Lyman series in the ultraviolet; the BALMER SERIES in visible light; and the Paschen, Ritz and Brackett-Pfund series in infrared.

Each series corresponds to the electron moving to a particular ENERGY LEVEL from one of the higher levels. In the Lyman series, the electron transitions are all between the GROUND STATE and higher levels, in the Balmer series they all connect with the energy level above the ground state, in the Paschen series with the next higher energy level and so on. This pattern was first described empirically in the RYDBERG EQUATION, and the successful explanation of this pattern was an early triumph for QUANTUM MECHANICS.

See also BOHR THEORY, RUTHERFORD–BOHR ATOM.

hydrometer A device for measuring the DENSITY of a fluid by means of a weighted float with a long stem. The level to which the float sinks in the liquid is a measure of the liquid's density. Hydrometers are often used to determine the concentrations of certain solutions, such as sugar and ethanol in water when brewing beer, or sulphuric acid in a LEAD-ACID CELL.

hydrostatic pressure PRESSURE produced by FLUIDS. Hydrostatic pressure acts in all directions through a fluid. A fluid will transmit any pressure applied at one point to all other points in the fluid. One important source of hydrostatic pressure is GRAVITY, which pulls down on a fluid producing a pressure due to the weight of the fluid. An example of this is ATMOSPHERIC PRESSURE, produced by gravity acting on the Earth's atmosphere. For hydrostatic pressure produced by gravity:

pressure = depth of fluid × density
× strength of gravity

hydrostatics The study of the effects of fluids at rest, particularly the effects of HYDROSTATIC PRESSURE on an immersed object.

hygrometer An instrument for measuring HUMIDITY. One common form is the WET AND DRY BULB HYGROMETER.

hyperbola A CONIC SECTION in which the plane intersects the CONE at an angle to the axis of cone that is smaller than that made by the cone itself. This results in an open curve, that is asymptotic (*see* ASYMPTOTE) to two straight lines. If these lines are at right angles, the curve is called a rectangular hyperbola, and can be represented by the function $y = 1/x$.

hyperelastic collision Not a collision in the normal sense, but a process in which the KINETIC ENERGY after the event is greater than before. This means that some energy must have been converted to kinetic energy from some other form, for example from chemical energy in an explosion. *See also* COEFFICIENT OF RESTITUTION.

hyperon Any BARYON more massive than the NEUTRON. All hyperons have very short HALF-LIVES and decay into NUCLEONS.

hypotenuse The longest side in a right-angled triangle.

hypothesis A statement that has not been proved but which is used as a starting point for a logical or mathematical argument. If this argument leads to an obviously false conclusion, the hypothesis or the argument must be incorrect, a procedure called *reductio ad absurdum*, literally 'reduction to the absurd'.

hysteresis A loss of ENERGY as heat in a material as it is taken around some closed cycle, returning to its original conditions. In particular, hysteresis is the energy lost when a FERROMAGNETIC material is magnetized and then demagnetized (*see* HYSTERESIS LOOP), or when certain materials, such as rubber, are deformed and then allowed to return to their original shape.

hysteresis loop The shape of a graph of the level of INDUCED MAGNETISM in a FERROMAGNETIC material against the applied magnetic field over successive cycles of magnetization first in one direction then the other. The area of the loop is proportional to the energy needed to move the material through one magnetic cycle.

I

iceberg A floating mass of ice, of which only about one-fifth is above the water.

icosahedron A POLYHEDRON with 20 plane faces.

ideal gas A hypothetical gas that obeys the GAS LAWS perfectly at all temperatures and pressures. For this to happen, the forces between the molecules of the gas must be negligible except during collisions, and the volume of the gas molecules must be negligible compared to the volume occupied by the gas itself. REAL GASES behave in approximately this way provided the temperature is not too low, so the KINETIC ENERGY of the molecules is large compared to the POTENTIAL ENERGY of the INTER-MOLECULAR FORCES, and provided that the density is not too high so that the volume of the molecules (the EXCLUDED VOLUME) is not too large a fraction of the total volume of the gas. *See also* BOYLE'S LAW, CHARLES' LAW, IDEAL GAS EQUATION, JOULE'S LAW, KINETIC THEORY, PRESSURE LAW.

ideal gas equation, *universal gas equation* The EQUATION OF STATE for an IDEAL GAS. One particular consequence of the ideal gas equation is that the volume taken up by a gas at a given temperature and pressure depends only on the number of molecules present and not on the nature of those molecules. In particular, at STANDARD TEMPERATURE AND PRESSURE (0°C and one ATMOSPHERE), one MOLE of molecules of any gas will take up a volume of 22.4 dm³.

For a gas containing n moles of molecules, at a pressure p in a volume V:

$$pV = nRT$$

where T is the ABSOLUTE TEMPERATURE and R is the MOLAR GAS CONSTANT, $R = 8.31\ \text{J}\,\text{K}^{-1}\text{mol}^{-1}$, or if the gas contains N molecules:

$$pV = NkT$$

where k is the BOLTZMANN CONSTANT, $k = 1.38 \times 10^{-23}\ \text{J}\,\text{mol}^{-1}$.

ideal gas temperature scale A TEMPERATURE SCALE that defines ABSOLUTE TEMPERATURE as being proportional to the pressure exerted by a fixed mass of IDEAL GAS held in a constant volume. The size of the temperature unit, the KELVIN, is fixed by defining the TRIPLE POINT of water to have a temperature of 273.15 K. The ideal gas scale is used in the CONSTANT VOLUME GAS THERMOMETER.

IGFET (insulated gate field effect transistor) *See* FIELD EFFECT TRANSISTOR.

illuminance The amount of light per unit area falling on a surface. The SI UNIT of illuminance is the LUX.

image A pattern of light rays coming from an object and passing through an optical system such that the light rays from each point on the object meet at, or appear to have come from, a single point on the image. *See also* REAL IMAGE, VIRTUAL IMAGE.

imaginary number A quantity used to represent the square ROOT of a negative number. All such numbers can be represented as a REAL NUMBER multiplied by the square root of −1, which is given the symbol i. A number that is made up of the sum of a real and an imaginary number is called a COMPLEX NUMBER. Imaginary numbers obey all the usual rules of algebra, with the additional rule

$$i^2 = -1$$

impedance The total RESISTANCE of a circuit to the passage of electric current. In a circuit carrying ALTERNATING CURRENT, it is equal to the peak value of the voltage divided by the peak current, or equivalently the ROOT MEAN SQUARE (r.m.s.) voltage divided by the r.m.s. current. The SI UNIT of impedance is the OHM. If the circuit has a total resistance R in series with a REACTANCE X, then the impedance Z is given by

$$Z^2 = R^2 + X^2$$

impulse The effect that causes a change in MOMENTUM: the force applied to an object multiplied by the time for which that force acts.

inactive electrode An ELECTRODE used in ELECTROLYSIS made from a material such as platinum or graphite that does not play any chemical role in the electrolysis.

incandescence The radiation of visible light from an object as a result of its high temperature. *See also* BLACK-BODY RADIATION.

incident ray An incoming ray of light.

independent variable *See* VARIABLE.

index *See* EXPONENT.

induced charge Electric charge produced on an object as a result of the flow of electrons under the influence of a nearby charged object. The ELECTRIC FIELD produced will cause equal and opposite charges on opposite sides of an originally uncharged object. *See also* CHARGING BY INDUCTION.

induced fission NUCLEAR FISSION that occurs shortly after the nucleus has been struck by a neutron.

induced magnetism Temporary magnetism produced by the DOMAINS of a FERROMAGNETIC material being temporarily aligned by an external magnetic field.

inductance The property by which a change in the current flowing in an electric circuit produces an ELECTROMOTIVE FORCE either in the same circuit (SELF-INDUCTANCE) or in a neighbouring circuit with which it is magnetically linked (MUTUAL INDUCTANCE). *See also* ELECTROMAGNETIC INDUCTION, HENRY.

induction coil A device based on the TRANSFORMER principle used to produce high voltage pulses from a DIRECT CURRENT supply. Induction coils are used to generate the high voltages needed to produce sparks in PETROL ENGINE ignition systems. A steady current flows through a PRIMARY COIL containing a relatively small number of turns of thick wire wound around an iron core. A SECONDARY COIL containing many turns of thinner wire is wound on top of the primary. The current in the primary is broken by a device called a CONTACT BREAKER – a switch operated by the rotation of the engine. The very rapid change in MAGNETIC FLUX LINKAGE in the secondary when the primary current stops flowing produces a large voltage, which is used to produce the spark needed to ignite the petrol and air mixture in the engine.

induction motor A MOTOR that operates on ALTERNATING CURRENT supplies only. Induction motors use an arrangement of FIELD COILS to produce a rotating magnetic field pattern. The ARMATURE has no coils or connections to the supply, but EDDY CURRENTS set up within it

produce forces tending to make the armature rotate with the rotating field pattern.

induction stroke In a PETROL ENGINE, the movement of the engine that draws fuel and air into the CYLINDER.

inductor A coil or any other circuit component with an INDUCTANCE.

inelastic collision A collision in which the total KINETIC ENERGY after the collision is less than that before the collision. An example of this is the case of two objects that coalesce (stick together), the kinetic energy lost being converted to heat energy. *Compare* ELASTIC COLLISION, HYPERELASTIC COLLISION. *See also* COEFFICIENT OF RESTITUTION.

inequality, *inequation* A mathematical statement that compares two numbers or algebraic expressions, describing one as being greater than ($>$) or less than ($<$) the other, or as 'greater than or equal to' or 'less than or equal to' ($\geq$, $\leq$). The rules for handling inequalities are essentially the same as those for other EQUATIONS, but whenever an inequality is multiplied or divided by a negative number, or inverted, the sign of the inequality must be reversed.

inequation *See* INEQUALITY.

inertia The resistance of an object to having its motion changed. According to Newton's second law, this depends on an object's mass. This idea of mass is sometimes called INERTIAL MASS to distinguish it from GRAVITATIONAL MASS. *See also* NEWTON'S LAWS OF MOTION.

inertial confinement A technique for obtaining the extreme temperatures needed for NUCLEAR FUSION by heating material so rapidly that it does not have time to expand before fusion begins. This is the method used in NUCLEAR WEAPONS but does not seem likely to be of much use for the peaceful use of nuclear fusion, where a continuous release of energy is required.

inertial mass MASS as a measure of resistance to change in motion. *Compare* GRAVITATIONAL MASS. *See also* NEWTON'S LAWS OF MOTION.

inertial navigation system (INS) A system used to enable aircraft and guided missiles to calculate their position by knowing their starting point and the duration of all accelerations in any direction. Such systems have the advantage that they do not rely on signals transmitted from satellites or ground-based stations.

inertial reference frame One of a series of co-ordinate systems, none of which seems to be accelerating when viewed from any other. In particular, co-ordinate systems that seem to be moving at a constant speed relative to the centre of mass of the observed Universe. *See also* FRAME OF REFERENCE.

infinity A quantity that is so large that the largest imaginable number is not large enough to describe it. Infinity is represented in mathematics by the symbol ∞.

infinitesimal (*adj.*) A vanishingly small quantity, larger than zero but smaller than the smallest non-zero number.

inflation In COSMOLOGY, a period of extremely rapid growth in the size of the Universe. *See* BIG BANG.

infrared ELECTROMAGNETIC WAVES with wavelengths from about 1 mm to 7×10^{-7} m. Infrared radiation is emitted by all hot objects and can be detected by the heating effect it produces when absorbed by a blackened surface (*see* BOLOMETER, THERMOPILE). Infrared with a wavelength close to the visible part of the ELECTROMAGNETIC SPECTRUM (called near infrared) can also be detected by modified versions of photographic film and electronic devices used to detect visible light. The fact that warm objects produce more infrared than cold objects has led to the development of many military applications based on infrared cameras for night-time surveillance, and heat-seeking missiles.

infrared spectroscopy The study of the INFRARED ABSORPTION SPECTRUM of molecules. This is a useful technique for determining molecular structure, as many COVALENT BONDS have resonant frequencies corresponding to the stretching or bending of the bonds which lie in the infrared region of the electromagnetic spectrum. Infrared radiation from a hot source is passed through the sample and then via a DIFFRACTION GRATING to a detector. The extent to which the radiation is absorbed is then represented as a plot of absorption against WAVENUMBER.

infrasound Sound of a frequency too low to be detected by the human ear.

in phase Describing two OSCILLATIONS that are exactly in step with one another.

insulated gate field effect transistor *See* FIELD EFFECT TRANSISTOR.

insulator A material through which current or heat cannot flow. Except for graphite, all non-metals are electrical insulators in their solid form. Organic liquids are also electrical insulators, as are all gases unless ionized. *See also* BAND THEORY, CONDUCTOR, SEMICONDUCTOR.

integer A whole number, positive or negative or zero. The numbers −1, 0 and 273 are all integers.

integral 1. In mathematics, a function that, when differentiated (*see* CALCULUS), gives a particular function. For example, if y and z are functions of x and $dy/dx = z$, then y is the integral of z with respect to x, represented as

$$y = \int z\, dx$$

On a graph, an integral is represented by the area between the curve that represents the function and the x-axis. An integral is generated by adding up the value of the function over a range of values of x.

Integrals may be either indefinite, with no specified range, or definite, taken between two specified limits. The indefinite integral of the function x^n is

$$x^{n+1}/(n+1) + c$$

where c is a constant, called the constant of integration, which can take on any value.

2. (*adj.*) Relating to an INTEGER.

integrated circuit, *silicon chip* A miniature electronic circuit made of a complex array of ACTIVE DEVICES produced on a single silicon wafer and designed to perform an electronic function. Most integrated circuits form elements in DIGITAL electronic circuits.

The development of digital integrated circuits has proceeded rapidly since the TRANSISTOR was invented in 1947. Techniques for producing large wafers of pure silicon and for DOPING very small regions of this base material to produce complex patterns of active devices have led to a steady increase in the number of elements that can be manufactured in a single integrated circuit. Further reduction in size is currently limited by the accuracy of the doping techniques and problems in removing the heat generated within an integrated circuit when it is operating. Ultimately it is hoped to be able to produce devices with elements that are limited in size only by the finite size of the atoms from which they are made. Reductions in the size of integrated circuits not only mean that more computing power can be contained in a single integrated circuit, but have also led to circuits that operate more rapidly, as the speed of operation is limited by the time taken for CHARGE CARRIERS to diffuse through an active device. These advances have led to a huge reduction in size and cost and a huge increase in the power of computers.

integration In mathematics, the process of finding an INTEGRAL. This may be done analytically, following a set of algebraic procedures, or numerically, adding up the value of a function for a range of values of the independent variable. Numerical integration is often performed using a computer.

integrator A circuit that produces an output proportional to the integral of the input voltage over time. They are usually based on an OPERATIONAL AMPLIFIER using a CAPACITOR in the FEEDBACK circuit.

intensity The power per unit area carried by a wave. The intensity of a wave is proportional to the square of its AMPLITUDE. The SI UNIT of intensity is the Wm^{-2} (watt per metre squared), but light waves may have their intensity quoted in LUX, while DECIBELS are often used for sound waves.

interaction In general, any process in which one object exerts a force on another, or where an object changes its nature in some way. In particle physics, four distinct forms of interaction are known, sometimes called the FOUR FORCES OF NATURE. These are the STRONG NUCLEAR FORCE, the WEAK NUCLEAR FORCE, ELECTROMAGNETIC FORCE, and gravitation (see GRAVITY). See also ELECTROWEAK FORCE, GRAND UNIFIED THEORY, STANDARD MODEL.

intercept The point at which a line or curve crosses a specified AXIS on a GRAPH.

interface An electronic system for gathering non-electronic information and feeding it into an electronic system, such as a computer, or for connecting two electronic systems together in such a way that information can be passed from one to another.

interference 1. (*optics*) The effect of two or more WAVES arriving at the same point at the same time. Interference is described as constructive if it leads to a greater AMPLITUDE and destructive if the amplitude is reduced. In light, interference is visible only if the interfering sources are COHERENT. If an incoherent light source is used, interference may be observed by splitting a single light beam in two

and allowing it to recombine on a screen or photographic plate. The beam may be split either by diffracting it though a narrow slit (interference by DIVISION OF WAVEFRONT), or by separating one beam into two by partial reflection (interference by DIVISION OF AMPLITUDE).

The effects of interference are studied using the PRINCIPLE OF SUPERPOSITION, which states that the effects of the interfering waves can be calculated by simply adding together (superposing) the effects that the individual waves would have on their own.

See also BEATS, DIFFRACTION, INTERFEROMETRY, PATH DIFFERENCE, THIN-FILM INTERFERENCE.

2. (*telecommunications*) An unwanted signal, either man-made or naturally occurring, which arrives at the receiving end of the system along with the desired signal. See also NOISE.

interferometer A device for splitting a beam of light into two, usually by partial reflection, and then recombining those beams after they have travelled along different paths. Observation of the INTERFERENCE patterns produced enables the difference in path length to be measured very precisely. See also INTERFEROMETRY.

interferometry A technique for combining light or radio waves received at two different points, taking due note of their relative PHASES. Interferometry is used for precise measurement of distances, and is accurate to a fraction of a wavelength of light.

Radio interferometry, combining the signals obtained by two or more separate RADIO TELESCOPES some distance away from one another is a valuable technique in astronomy. Interferometry involving radio telescopes in different continents has also provided direct evidence of CONTINENTAL DRIFT.

intermolecular forces The forces that act between one molecule and another in a substance. If these forces are strong enough compared to the energy of any thermal vibrations – in other words, if the substance is cold enough – the intermolecular forces will hold the substance together as a solid or liquid. At higher temperatures, or with weaker forces, the material will behave as a gas. In an IDEAL GAS there are no intermolecular forces.

All intermolecular forces are ELECTROSTATIC in origin and are the result of forces between the electrons and the nuclei of the molecules involved, governed by the rules of quantum mechanics. In all cases, there is a correlation between the strength of the intermolecular forces and the melting point and also the amount of energy needed to melt the material (the LATENT HEAT). The way in which the intermolecular forces change with distance is linked to the stiffness of the resulting material. However most materials are much weaker than might be expected from the strengths of the individual intermolecular forces, owing to the existence of defects in the lattice structure, which also account for PLASTIC behaviour. Within the ELASTIC region, the extent to which the intermolecular forces vary linearly with separation between the molecules is related to the extent to which the material obeys HOOKE'S LAW.

See also HYDROGEN BOND, IONIC SOLID, MACROMOLECULE, METAL, VAN DER WAALS' FORCE.

internal combustion engine The most important engine of the 20th century, fuelled by petrol or diesel fuel. See DIESEL ENGINE, PETROL ENGINE. See also TURBOCHARGING.

internal energy The energy that atoms or molecules in a substance possess as a result of forces between themselves and the KINETIC ENERGY of their random thermal motion, as opposed to any bulk motion of the object, or external forces acting on the object. When an object is heated, provided no external WORK is done, the increase in the internal energy of the object is equal to the amount of heat energy supplied. In an IDEAL GAS, the internal energy is purely due to the kinetic energy of the molecules. See also FIRST LAW OF THERMODYNAMICS, JOULE'S LAW.

internal resistance The RESISTANCE a source of electrical energy appears to possess. When a battery or any other source of electrical energy is connected to a LOAD (some device that draws current from the supply of electricity), the voltage across the terminals of the supply will fall. The greater the load (i.e. the lower its resistance), the greater the fall in voltage. For simplicity in calculations, the supply can be treated as an ideal supply (one which can provide an unlimited current with no drop in voltage) in series with an internal resistance, across which there is a voltage (called the lost volts), which increases with the current drawn from the supply. In applications where high currents are required, it is important to keep the internal resistance as low as possible. For example, the LEAD-ACID CELLS used in car batteries may be required to provide currents of several hundred amperes to start the car whilst the voltage across the terminals remains close to the level it would be with no load. See also MAXIMUM POWER THEOREM.

International Practical Temperature Scale (IPTS) A TEMPERATURE SCALE designed to conform as close as possible to thermodynamic temperature. The unit of temperature is the KELVIN (K). The latest version of the scale, devised in 1990, has 16 FIXED POINTS. The IPTS supersedes all other temperature scales for scientific purposes.

intersection The point at which two lines or curves cross.

interstellar (*adj.*) Between stars. The term is particularly used to describe large distances and also the low density gas and dust that fills the spaces between stars in a galaxy.

interstitial (*adj.*) Describing an atom located in the interstices, or spaces, between atoms in a regular crystal lattice. For example, in steel, carbon atoms occupy the interstices between the much larger iron atoms.

intrinsic semiconductor A SEMICONDUCTOR with no DOPING and an equal number of electrons and HOLES.

inverse square law The behaviour of any quantity that radiates from a point source or the surface of a sphere with its strength falling in such a way that it is reduced by a factor of four when the distance from the source is doubled. ELECTRIC FIELDS and GRAVITATIONAL FIELDS, and the intensity of ELECTROMAGNETIC RADIATION all exhibit inverse square law behaviour.

inversion temperature The temperature below which a gas will cool in a FREE EXPANSION.

inverted (*adj.*) Upside down, particularly in reference to optical images formed by CONVERGING LENSES and mirrors.

inverter 1. *NOT gate* A LOGIC GATE with a single input and a single output that is the opposite of the input.

2. A device for producing an ALTERNATING CURRENT from a DIRECT CURRENT supply. The output is often converted to a higher voltage using a TRANSFORMER.

inverting amplifier An AMPLIFIER with a negative GAIN; that is, a positive input produces a negative output.

inverting input One of the inputs of a DIFFEREN-TIAL AMPLIFIER. Signals applied to this input are amplified with a negative GAIN.

Io One of the satellites of JUPITER. Io is remarkable as the only place beyond the Earth where active volcanoes are observed. Io is heated by its interaction with Jupiter's strong magnetic field.

ion An ATOM or MOLECULE that is not electrically NEUTRAL, having gained or lost ELECTRONS. An atom that has lost one or more electrons (cation) is positively charged, whereas one that has gained electrons (anion) is negatively charged. *See also* IONIZATION, IONIZATION ENERGY, IONIC BOND.

ionic (*adj.*) Describing a material, particularly a solid or a solution, that contains IONS or is held together by IONIC BONDS.

ionic bond A bond between two types of ATOM in which one or more ELECTRONS are transferred from one atom to another, creating positive and negative IONS. The attractive forces between these ions then hold the material together as a solid. These solids are hard, brittle materials with high melting points. The solids do not conduct electricity, but they do conduct when molten as the ions are then free to move around, carrying charge. IONIC SOLIDS are also usually soluble in water, forming conducting solutions.

ionic solid A crystalline solid in which adjacent atoms gain and lose one or more electrons to form a regular lattice of IONS. Each ion is attracted to its nearest neighbours with the opposite charge. There is also a repulsive element in the interatomic force at short distance, caused by the effect of the PAULI EXCLUSION PRINCIPLE, which begins to promote electrons to higher ENERGY LEVELS as the atomic electron clouds of neighbouring atoms start to overlap. As a result of the strong nature of the ionic attractions, ionic materials tend to have melting and boiling points above room temperature and are usually fairly hard materials. *See also* COVALENT CRYSTAL.

ionization The process of creating IONS in a substance that previously contained neutral atoms or molecules, such as by a chemical reaction or IONIZING RADIATION. *See also* HEAT OF IONIZATION.

ionization energy, *ionization potential* The minimum energy needed to remove an electron infinitely far away from an atom, often specified either in ELECTRON-VOLTS or per mole of atoms. The first ionization energy relates to the removal of the first electron, the second ionization energy is the additional energy needed to remove a second electron and so on.

ionization potential *See* IONIZATION ENERGY.

ionized (*adj.*) Describing a material that contains IONS – for example as a result of chemical action as in acidified water, or IONIZING RADIATION.

ionizing radiation Any RADIATION that creates IONS in any matter through which it passes. Ionizing radiation may be a stream of high-energy particles (such as ELECTRONS, PROTONS or ALPHA PARTICLES) or short-wavelength ELECTROMAGNETIC RADIATION (ULTRAVIOLET radiation, X-RAYS or GAMMA RADIATION).

For all types of ionizing radiation, the level of radiation can be reduced by passing the radiation through a material containing as many electrons as possible, to provide the greatest number of opportunities for the radiation to lose energy by IONIZATION. Thus lead, which has a large number of electrons per unit volume, so is a dense material, is a much better absorber of ionizing radiation than aluminium.

See also BACKGROUND RADIATION, BECQUEREL, DECAY CONSTANT, RADIATION DETECTORS, RADIOCARBON DATING, TRACER TECHNIQUES.

ionosphere The outer layer of the ATMOSPHERE that is ionized by X-rays from the sun.

iris A coloured diaphragm of circular and radial muscles in the eye that can alter the size of the PUPIL to control the amount of light entering the eye.

iron losses In a TRANSFORMER, those energy losses attributable to the iron parts of the transformer. Iron losses are made up of energy losses due to HYSTERESIS and EDDY CURRENTS.

irradiance The total power of ELECTROMAGNETIC RADIATION of all wavelengths arriving per unit area on a surface, measured in watts per square metre.

irradiation The process of bombarding a surface with RADIATION, particularly IONIZING RADIATION. Especially used in the context of using ionizing radiation to kill bacteria in foodstuffs.

irreversible (*adj.*) Describing a change that cannot happen in reverse, usually because to do so would violate the SECOND LAW OF THERMODYNAMICS.

isentropic (*adj.*) Describing a process that takes place without any change in ENTROPY; that is, a reversible process.

isobar 1. A line joining points of equal ATMOS-PHERIC PRESSURE on a weather map. The CORIO-LIS FORCE means that wind does not blow at right angles to the isobars, but more or less parallel to them.

2. Any one of a series of atomic NUCLEI having the same MASS NUMBER, and hence approximately the same MASS.

isochronous (*adj.*) Describing an OSCILLATION whose period does not depend on its AMPLI-TUDE. *See also* SIMPLE HARMONIC MOTION.

isomorphic (*adj.*) Describing two crystalline materials with the same lattice structure. Thus sodium chloride and potassium bromide are isomorphic, as they each form a crystalline structure in which each ANION is surrounded symmetrically by six CATIONS, resulting in a cubic structure described as CUBIC CLOSE PACKED.

isosceles (*adj.*) Describing a triangle that has two equal sides.

isostasy In geology, the balance between the weight of a TECTONIC PLATE and the upthrust from the magma in which it is floating. Changes in this balance cause continents to rise or sink.

isothermal (*adj.*) Describing a change that takes place at constant temperature.

isotone Any one of a series of atomic NUCLEI having the same number of NEUTRONS.

isotope Any one of a series of atomic nuclei of the same element, each with the same number of PROTONS but different numbers of NEU-TRONS, hence different masses. *See also* NUCLEUS, RELATIVE ATOMIC MASS.

isotropic (*adj.*) Describing a material, usually a crystalline solid, that has the same physical properties, such as thermal or electrical CON-DUCTIVITY, regardless of the direction in which these properties are measured. *Compare* ANISOTROPIC.

iteration The process of solving an equation by a series of repeated calculations. The result from one calculation is used in the next calculation, with successive answers converging to give an increasingly accurate solution.

J

J/ψ (J/psi) A MESON containing the charmed (*see* CHARM) QUARK and antiquark. Discovered independently by two experimental groups in 1974, it was the first evidence for the existence of charm. The J/ψ has a mass about 3.2 times that of the proton and a HALF-LIFE of around 10^{-20} s.

jet engine An engine that burns fuel continuously to provide an outflow of high speed, high temperature gas. Jet engines are widely used to propel aircraft. Most modern jet engines are of the GAS TURBINE type, which are also used to propel ships and to generate electricity. *See also* HIGH-BYPASS ENGINE, PULSE-JET, RAMJET, TURBOPROP.

jet propulsion The use of a fast-moving stream of gas, usually from a JET ENGINE, to exert a forward force on a vehicle, usually an aircraft. By Newton's third law (*see* NEWTON'S LAWS OF MOTION), the rearward force exerted on the gas to expel it from the engine exerts a forward force on the vehicle.

jet stream A particularly fast, narrow current of wind that occurs at an altitude of 10 to 12 km. Wind speeds typically reach between 60 to 125 kmh^{-1}.

JFET (junction field effect transistor) *See* FIELD EFFECT TRANSISTOR.

joule (J) The SI UNIT of WORK and ENERGY. One joule of work is done when a force of one NEWTON moves through a distance of one metre in the direction of the force.

Joule's law For an IDEAL GAS, the INTERNAL ENERGY depends only on temperature, not on pressure or volume. REAL GASES show slight departures from Joule's law, most cooling as they expand, to compensate for work done against INTERMOLECULAR FORCES, though some gases, such as hydrogen, show an increase in temperature.

junction field effect transistor *See* FIELD EFFECT TRANSISTOR.

junction transistor A SEMICONDUCTOR device that exhibits the important property of GAIN – a small current can be used to control a much larger one. The commonest type of junction transistor is the NPN TRANSISTOR. It consists of a thin layer of lightly doped (*see* DOPING) P-TYPE SEMICONDUCTOR, called the BASE, sandwiched between two more heavily doped pieces of N-TYPE SEMICONDUCTOR material, the COLLECTOR and the EMITTER. The base-emitter junction is FORWARD BIASED and a current flows through it, mainly as a flow of electrons from the emitter to the base. Since the base is thin, many of these electrons diffuse through the base and enter the collector, which is held at a more positive potential than the base. Thus the current flowing into the base is small, but controls a much larger collector current. The PNP TRANSISTOR operates in the same way, but with all the polarities reversed. *See also* THERMAL RUNAWAY.

Jupiter The largest planet in the SOLAR SYSTEM. It has a mean diameter of 138,000 km, 11 times that of the Earth, and a mass of 1.9×10^{27} kg, 320 times the Earth's mass. Jupiter is composed mainly of hydrogen, so its average density is relatively low at 1,300 kg m^{-3}. Jupiter is the fifth planet in order of distance from the Sun, with a mean orbital radius of 5.2 AU (230,000,000 km). Jupiter takes 11.9 years to complete one orbit of the sun, yet rotates on its own axis in just 8 hours. Jupiter has no visible solid surface, but a complex atmosphere with active weather patterns is clearly visible. The most famous of these surface features is the GREAT RED SPOT. Numerous satellites are known and new small ones are routinely discovered. The four largest satellites, called the GALILEAN SATELLITES, are clearly visible with a small telescope. Jupiter has a faint RING SYSTEM, similar to that of SATURN but less developed.

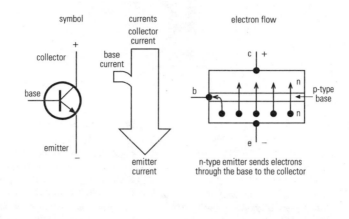

symbol
currents
electron flow

collector current

base current

collector

base

emitter

+

−

collector current

emitter current

n-type emitter sends electrons
through the base to the collector

c | +

n

b

p-type
base

n

e | −

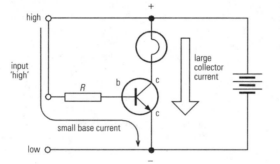

+

high

input
'high'

R

b c

c

large
collector
current

small base current

low

−

Junction transistor.

K

kaon *See* MESON.

kelvin (K) The SI UNIT of TEMPERATURE. The size of the kelvin is the same as the degree CELSIUS. The TRIPLE POINT of water is fixed at exactly 273.16 K: the kelvin is defined as the fraction 1/273.16 of the temperature above ABSOLUTE ZERO of the triple point of water. *See also* ABSOLUTE TEMPERATURE, TEMPERATURE SCALE.

Kelvin statement of the second law of thermodynamics No system can convert heat energy entirely to mechanical WORK – a certain amount of heat must be given out to the cooler surroundings. For example, the exhaust gas of an INTERNAL COMBUSTION ENGINE contains waste heat energy. If a HEAT ENGINE takes in heat at an ABSOLUTE TEMPERATURE T_1 and gives out waste heat at a lower temperature T_2, the maximum EFFICIENCY permissible by the SECOND LAW OF THERMODYNAMICS, even if there are no avoidable losses, such those due to friction, is $(1 - T_2/T_1)$.

Kepler's laws A set of three laws originally based on empirical observation of the SOLAR SYSTEM but now recognized as a consequence of the laws of mechanics and GRAVITY.

Kepler's first law sets out the arrangement of the Solar System, saying that all planets move in elliptical ORBITS, which have the Sun as a common FOCUS.

Kepler's second law is a consequence of the LAW OF CONSERVATION OF ANGULAR MOMENTUM, and the fact that the force of gravity acts along a line between the Sun and the planet. It states that the line joining the planet to the Sun sweeps out equal areas in equal intervals of time. Thus a planet moves fastest in its orbit when it is closest to the Sun.

Kepler's third law states that the square of the time taken to complete an orbit (called the period of the orbit) is proportional to the cube of the average radius of the orbit. Thus a planet close to the Sun, such as Mercury, orbits more quickly (88 days), than one further away (such as Jupiter, which has a period of

11.9 years). This law is a consequence of the fact that the force of gravity obeys an INVERSE SQUARE LAW; that is, the strength of gravity falls to one quarter of its original strength at double the distance from the Sun.

kilo- (k) A prefix before a unit indicating that the size of the unit is to be multiplied by 10^3, for example kilowatt (kW) is equivalent to one thousand watts.

kilogram The SI UNIT of mass. One kilogram is defined as the mass of the International prototype kilogram, a cylinder of platinum-iridium alloy kept at Sèvres, near Paris, France.

kinematics The study of objects moving with specified ACCELERATIONS, as opposed to DYNAMICS, which also studies the forces that produce those accelerations.

kinetic energy The energy of movement. The kinetic energy, E, of a mass m moving at a speed v is

$$E = \tfrac{1}{2}mv^2$$

kinetic theory The part of physics that explains the physical properties of matter in terms of the motion of its component atoms and molecules. The temperature of a body is dependant on the INTERNAL ENERGY, and therefore the velocity, of its molecules. If heat is supplied to a substance the velocity of the particles increases and the temperature rises.

The very different properties of solids, liquids and gases are explained in terms of the different arrangements of molecules. The solid and liquid forms of a material have similar densities, far higher than the density of a gas, because the molecules in solids and liquids are closely packed. This also explains why solids and liquids are far harder to compress than gases. In a solid, the molecules are arranged in fixed positions in a lattice, whilst in liquids and gases they are free to move around at random: this explains why solids retain their own shape, whilst liquids and gases flow freely.

Kinetic theory explains the pressure of a gas in terms of the impacts of its molecules on to the walls of its container. If the temperature rises, the number of impacts per second increases, and the pressure increases. By making various assumptions about the molecules of an IDEAL GAS (such as negligible forces between molecules, the molecules themselves take up a negligible volume, etc.), kinetic theory gives rise to the GAS LAWS.

Kinetic theory also explains such physical phenomenon as THERMAL EXPANSION, CHANGES OF STATE and change of RESISTANCE in terms of the thermal motion of the particles.

See also BROWNIAN MOTION, STATISTICAL MECHANICS.

Kirchhoff's laws A set of two laws that can be used to calculate the currents and voltages in a complex circuit. Kirchhoff's first law states that at any junction in a circuit the total current entering the junction is equal to the total current leaving the junction. If currents entering the junction are taken as positive, whilst those leaving are negative, the sum of all the currents at any junction will be zero. This law is in effect a LAW OF CONSERVATION OF CHARGE: the charge per second entering the junction must equal the charge per second leaving the junction if charge is not to be created or destroyed. At any junction:

$$\Sigma I_{in} = \Sigma I_{out}$$

Kirchhoff's second law states that around any closed loop in a circuit, the sum of all the ELECTROMOTIVE FORCES (e.m.f.'s) in the loop must be the same as the sum of all the POTENTIAL DIFFERENCES (p.d.'s) across RESISTANCES in the loop. This is a consequence of the LAW OF CONSERVATION OF ENERGY, since the e.m.f.'s represent electrical energy gained per unit charge whilst the p.d.'s represent electrical energy lost per unit charge. Around any loop:

$$\Sigma E = \Sigma IR$$

for a loop containing e.m.f.'s E, currents I and resistances R.

klystron A device for the production of MICROWAVES. Electrons in a beam produced by THERMIONIC EMISSION from a hot filament have their speeds varied by a second ELECTRODE (the buncher). The faster electrons catch up with the slower ones and so bunches are formed. These produce a varying current at a collecting electrode (the catcher) and this is fed back to the buncher. The resulting oscillations have a frequency which depends on the voltage used to accelerate the electrons and the separation between the buncher and catcher. The buncher and catcher form cavities resonant (see RESONANT CAVITY) at the frequency of the electromagnetic radiation generated.

In a variation on this design, called the reflex magnetron, only a single cavity is used, with the electron beam being repelled back to the cavity by a negative electrode. See also MAGNETRON.

knocking A phenomenon occurring in PETROL ENGINES that reduces the power output. Knocking sounds result from explosions of unburned fuel-air mixture before it is ignited. The extent of knocking depends on the fuel composition. Lead is used in petrol to reduce knocking but is being phased out due to its association with mental retardation in children.

Knudsen regime The situation in a gas at a very low pressure, where the MEAN FREE PATH of a molecule becomes large compared to the size of the apparatus, and the TRANSPORT COEFFICIENTS no longer correctly describe the behaviour of the gas. Similarly, a gas will no longer support sound waves once the pressure is low enough for the mean free path of the gas molecules to be comparable to the wavelength of the sound.

Knudsen Number (dimensionless)

$$K_N = \frac{\lambda}{L}$$

where;
λ mean free path
L representative scale for an ideal gas

$$K_N = \frac{k_B T}{\sqrt{2} \pi \ell^2 pL}$$

where
K_b BOLTZMANN constant
T theoretical temp (°K?)
σ particles hard shell ϕ [?]
p total pressure

L

lagging Thermal insulation, particularly around pipes or hotwater tanks to prevent loss of heat or damage by the water freezing.

lambda particle (Λ) A neutral BARYON. The lambda particle is the lightest baryon to contain the STRANGE QUARK, with a mass about 1.2 times that of the proton and a relatively long HALF-LIFE of 2.6×10^{-10} s. *See also* SIGMA PARTICLE.

laminar flow In a fluid, smooth flow in which the motion of one part of the fluid is very similar to that in other nearby regions and does not change over time. *See also* DRAG.

laminated (*adj.*) Describing a construction made from many thin layers. *See also* EDDY CURRENT.

Large Electron Positron *See* LEP.

laser Acronym for *l*ight *a*mplification by STIMULATED EMISSION of RADIATION. A device for producing an intense parallel beam of COHERENT light.

A laser comprises a RESONANT CAVITY filled with some 'lasing' medium, with a mirror at either end, one of which is semi-reflecting. The atoms in the lasing medium are 'pumped' to an EXCITED STATE by an external source of energy, such as flashlight, an electric current or another laser. This produces a POPULATION INVERSION, with more of the atoms in an excited state than in the GROUND STATE. An electromagnetic STANDING WAVE set up in the cavity then causes stimulated emission in which atoms move to a lower energy state, giving out coherent radiation. The radiation emitted by the atoms is reflected back and forth between the two mirrors, stimulating more emissions and so amplifying the amount of radiation. The radiation emerges from the semi-reflecting mirror as a powerful beam of light.

The lasing medium can be solid, liquid or gas, but infrared carbon dioxide lasers, red helium neon lasers and green argon ion lasers are amongst the most important technologically. Infrared semiconductor lasers are used to read COMPACT DISCS. Lasers are used as an alternative to scalpels in surgery, where they have the advantage that, since they cut tissue by heating it, there is less bleeding. *See also* MASER.

laser-guided weapon A missile that finds its target by homing in on a laser beam aimed at the target by an air or ground-based observer.

latent heat The heat energy taken in or given out when a material changes its physical state at a constant temperature. Energy is taken in when a solid turns to a liquid, or a liquid to a gas, as energy is needed to overcome the forces holding the molecules together in a solid or liquid. When a gas condenses, or a liquid turns into a solid, energy is released as chemical bonds form holding the molecules together.

Latent heat can be measured in a CALORIMETER either by using a measured amount of electrical energy to bring about the change of state, or by a variation of the METHOD OF MIXTURES, in which a solid is placed in a liquid and melts. Since the substance will not start or end up at its melting point, the latter method also requires knowledge of the HEAT CAPACITY of the material.

See also SPECIFIC LATENT HEAT OF FUSION, SPECIFIC LATENT HEAT OF VAPORIZATION.

latitude A measure of position on the surface of the Earth, or any other planet. Latitude is measured in degrees north or south of the equator. *See also* LONGITUDE.

lattice An array of points arranged in space in some symmetrical pattern, such as the positions occupied by atoms in a crystal. *See also* ISOMORPHIC.

lattice energy The energy required to separate the molecules or ions of a crystal so they are infinitely far apart from one another. Often quoted for one mole of the material.

law of conservation of angular momentum If there are no MOMENTS acting on an object, its ANGULAR MOMENTUM will be constant. This law leads to Kepler's second law (*see* KEPLER'S LAWS). It also explains, for example, why the rotational speed of a rotating skater increases if she pulls her arms in. As the MOMENT OF INERTIA is reduced with the mass being concentrated closer to the axis, the ANGULAR VELOCITY must increase to keep the angular momentum constant.

law of conservation of charge In any electric circuit or interaction between ELEMENTARY PARTICLES, the total charge remains unchanged. *See also* KIRCHHOFF'S LAWS.

law of conservation of energy In any closed system, where no energy can enter or leave, the total amount of energy remains constant. Thus energy can never be created or destroyed, though energy may be transferred from one form into another.

When energy is converted from one form to another, there is always a certain amount of heat energy produced as a by-product of the conversion process (*see* SECOND LAW OF THERMODYNAMICS). For example, burning fuel to produce mechanical energy in a motor car, or electricity in a power station, also results in the production of hot gases. This means that, as energy is converted into more useful forms, more and more of the energy is lost as low-grade heat.

According to the SPECIAL THEORY OF RELATIVITY, energy and MASS are equivalent (*see* EQUIVALENCE OF MASS AND ENERGY), thus any change in energy is accompanied by an equivalent change in mass. Thus both mass and energy are conserved. In practice, the mass changes resulting from energy changes are far smaller than the REST MASS of the materials involved, except for processes involving the atomic NUCLEUS (*see* MASS DEFECT) and some processes of ELEMENTARY PARTICLE PHYSICS.

Thus, mass and energy are usually regarded as being conserved separately (in chemical reactions, for example).
See also KIRCHHOFF'S LAWS.

law of conservation of momentum For any system in which no external forces act, the total MOMENTUM of the system is constant. In a system of particles interacting with each other, such as two particles colliding with one another, Newton's third law means that IMPULSES will occur in equal and opposite pairs. This means that the total change in momentum will be zero. This law is particularly useful in the study of collisions, which are divided into three classes: INELASTIC, ELASTIC and HYPERELASTIC.

For two objects of mass m and M, moving with velocities u and U respectively before collision and v and V after collision:

$$mu + MU = mv + MV$$

See also NEWTON'S LAWS OF MOTION.

LCD *See* LIQUID CRYSTAL DISPLAY.

LDR *See* LIGHT-DEPENDENT RESISTOR.

lead-acid cell A rechargeable electrochemical CELL with ELECTRODES of lead and lead sulphate and a sulphuric acid ELECTROLYTE, often used where high currents are needed, such as in motor vehicles.

leakage current A small current flowing through a material that is designed to be an insulator, such as the insulating layer in a CAPACITOR.

LED *See* LIGHT-EMITTING DIODE.

Lee's disc A method of measuring the THERMAL CONDUCTIVITY of a poor conductor. The sample is shaped into a thin disc and sandwiched between two metal plates, each drilled with a

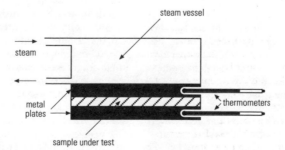

Lee's disc.

hole for a thermometer. The THERMAL RESIS-TANCE of these plates is much less than that of the sample, so the thermometers effectively enable the temperature at each surface of the sample to be measured. One plate is heated whilst the other loses heat by CONVECTION. Once a steady state has been established, the rate of heat flow can be found by removing the heat source and measuring the rate of temperature drop of the other metal plate. Knowing the mass and SPECIFIC HEAT CAPACITY of the lower plate, it is possible to find the rate at which heat must have been flowing through the sample to maintain this plate at a constant temperature.

Lennard–Jones 6-12 potential A form of interatomic potential function used to model the effects of VAN DER WAALS' FORCES. The repulsive core of the forces is modelled as a positive term in the potential which falls off as the twelfth power of the separation between the centres of the two molecules, whilst the attractive force between induced dipoles is a negative contribution which falls off with the sixth power of separation.

lens In optics, a curved piece of glass designed to refract light rays to form an image (*see* REFRACTION). The curved surfaces of a lens are described as CONCAVE if they curve inwards or CONVEX if they curve outwards. A lens with two concave surfaces will be diverging whilst a lens with two convex surfaces will be converging. Single lenses form images that are not perfect, but suffer form a number of defects called ABERRATIONS. Lenses that are used in optical instruments such as cameras or microscopes are almost always COMPOUND LENSES made up of several separate lenses made from different materials and with different curvatures to minimize the effect of these aberrations. *See also* CONVERGING LENS, DIVERGING LENS.

Lenz's law A law concerning the direction of the current produced by any process involving ELECTROMAGNETIC INDUCTION. It states that any such current will be in a direction that opposes the change that created it.

LEP (Large Electron Positron) Currently the world's largest SYNCHROTRON. Located at CERN, it is 27 km in circumference and accelerates electrons and positrons to 50 GeV for COLLIDING BEAM EXPERIMENTS.

lepton Any of a class of ELEMENTARY PARTICLES not affected by the STRONG NUCLEAR FORCE.

Three charged leptons are known, the ELECTRON, the MUON and the TAU LEPTON, in order of increasing mass and decreasing HALF-LIFE (the electron is stable). For each of these particles there is also a NEUTRINO, a light (probably massless) particle carrying no charge, but the same QUANTUM NUMBER as the charged lepton (such as ELECTRON NUMBER, muon number, etc.). This quantum number is conserved in all interactions. Thus, for example, the decay of a muon produces an electron, a muon neutrino and an electron antineutrino, so the total muon number remains at 1 whilst the electron and the electron antineutrino have opposite electron number. *See also* STANDARD MODEL.

lepton number In any particle physics process, the number of LEPTONS taking part in the process less the number of corresponding antiparticles. The lepton number appears to be conserved in any process.

lever A simple MACHINE in which a support rotates about a PIVOT, or FULCRUM. For a lever,

effort force × effort to pivot distance = load force × load to pivot distance

See also EFFORT, LOAD.

lift An aerodynamic force produced by the motion of a solid object through a fluid, such as air or water. The lift force acts at right angles to the direction of motion and gives the object the ability to move upwards through the fluid. For example, lift gives an airplane the ability to climb into the air and holds it there during flight. The amount of lift produced by an object depends on its ANGLE OF ATTACK and its velocity through the fluid. It also depends on the shape of the object; those shapes designed to produce lift are known as AEROFOILS. *See also* COEFFICIENT OF LIFT.

light ELECTROMAGNETIC RADIATION of a wavelength to which the human eye is sensitive (visible light). The eye is sensitive to wavelengths from about 7×10^{-7} m (red) to 4×10^{-7} m (violet). The term is also commonly used for wavelengths outside this narrow range but where the properties being exploited are similar to those of visible light. Longer wavelengths are described as INFRARED and shorter wavelengths as ULTRAVIOLET.

Visible light is produced by BLACK-BODY RADIATION from very hot objects (e.g. stars, light bulb filaments) and also when electrons move from one ENERGY LEVEL to another in an atom. The wavelengths of light emitted depend on the nature of the atom and this is an important tool in chemical analysis, particularly in cases where a direct sample cannot be obtained (for example, in analysing the light given out by stars).

There has been a long debate about the nature of light. Isaac Newton (1642–1727), who first described the REFRACTION of light by a PRISM, and recognized that white light was a mixture of wavelengths, believed light to be a stream of small particles which he called corpuscles. In YOUNG'S DOUBLE SLIT EXPERIMENT however, DESTRUCTIVE INTERFERENCE was observed, showing that light has wave-like properties and enabling its wavelength to be measured. Einstein's explanation of the PHOTO-ELECTRIC EFFECT once again led to an understanding of light in terms of particles called PHOTONS. The modern understanding of light, based on the ideas of WAVE-PARTICLE DUALITY suggests that it has aspects of both wave and particle in its behaviour. The SPEED OF LIGHT in a vacuum, originally thought by some to be infinite, was first measured by Römer in 1674. It is now fixed by definition at $299{,}792{,}458 \text{ ms}^{-1}$, and is the same as the speed of all ELECTROMAGNETIC WAVES, equal to $(\varepsilon_0\mu_0)^{-1/2}$ where ε_0 is the PERMITTIVITY of free space and μ_0 is the PERMEABILITY of free space.

· *See also* COLOUR, LUMINOUS FLUX, LUMINOUS INTENSITY, SPECTRUM.

light-dependent resistor (LDR) A device whose RESISTANCE changes on exposure to light. An example of such a material is cadmium sulphide, which is an insulator in darkness: exposure to light releases FREE ELECTRONS, enabling a current to flow. LDRs are sometimes used in circuits that are required to react to changes in light level, though PHOTODIODES respond more quickly to changes in illumination, and can be made sensitive to a greater range of wavelengths of light.

light-emitting diode (LED) A SEMICONDUCTOR device that converts electrical energy into visible or infrared light. It comprises a DIODE in which the energy loss for CHARGE CARRIERS crossing the DEPLETION LAYER is large enough for them to produce a PHOTON of light. Thus light is produced when the diode is FORWARD BIASED. By making packages of LED's with suitable shaped segments, displays that form letters or numbers can be manufactured.

light filter *See* FILTER.

light gate An electronic timing device that measures the speed of a moving object by finding the time for which a light beam is obstructed by a card of known length attached to the object.

lightning The flash of light seen from a spark within a CUMULONIMBUS cloud, between two such clouds, or between a cloud and the ground. These sparks occur as a result of the build up of electrostatic charges (*see* STATIC ELECTRICITY) from friction between rain and hail in the turbulent core of the cloud. The thunder that accompanies a flash of lightning is produced by the expansion of the heated air.

A spark of lightning generated within a cloud cannot usually been seen directly from the ground, but instead produces a general illumination of the sky. This effect is called sheet lightning. Forked lightning is the direct observation of a spark between one cloud and another or between the cloud and the ground.

See also CHARGING BY FRICTION, WEATHER SYSTEMS.

lightning conductor A metal rod connected between the top of a tall building and the ground that protects the building from damage by LIGHTNING. The rod builds up a large enough charge in the presence of a thundercloud to spray ions into the cloud, neutralizing it and reducing the chance of a lightning strike. If a lighting strike does occur, the lightning conductor will provide a low resistance path to Earth. Without this, a large amount of current would flow through the building, with the heat produced causing a great deal of damage.

light pollution In astronomy, man-made light sources scattering off particles in the atmosphere and making astronomical observations more difficult.

light year A unit of distance used in astronomy, the distance travelled by light in one year, equal to $9.46 \times 10^{12} \text{ km}$.

limit 1. One of the two ends of the range over which a sum or INTEGRAL is evaluated.

2. The value to which some function tends more and more closely as the independent variable moves closer and closer to a particular value, with the function not being defined for that value. For example, the limit of the function $(\sin x)/x$ is 1 as x tends to 0.

3. The value to which a sum or other SERIES approaches more and more closely as more and more terms in the series are taken into account.

limiting friction The maximum amount of STATIC FRICTION that is produced just before two objects start to slide over one another, often slightly greater than the DYNAMIC FRICTION.

linear (*adj.*) Relating to a line, in particular describing a relationship that would be shown by a straight line on a graph, with a given change in one quantity always causing a fixed change in some other quantity.

linear accelerator A PARTICLE ACCELERATOR in which charged particles are accelerated through a series of metal cylinders called DRIFT TUBES. Alternate drift tubes are connected to opposite sides of an ALTERNATING CURRENT supply. The frequency of the supply and the lengths of successive tubes are arranged so that the particles being accelerated, typically electrons, are in the field-free region inside the tubes whilst the polarity of the supply reverses, so they always experience an accelerating field.

linear expansivity *See* EXPANSIVITY.

line of action Of a force, a line that runs in the same direction as the force and passes through the point at which the force acts.

line of best fit A line drawn on a graph on which some data points are also plotted so as to minimize the sum of the squares of the distances of the points from the line.

line of force An imaginary line in an ELECTRIC FIELD or MAGNETIC FIELD that allows the direction of the associated force to be visualized.

line spectrum An EMISSION SPECTRUM or ABSORPTION SPECTRUM in which the wavelengths absorbed or emitted form a number of separate very narrow ranges or lines. Each line in a spectrum represents a closely defined wavelength, corresponding to the PHOTON energy equal to the difference in energy between the ENERGY LEVELS that produced the line.

liquid The state of matter in which a material is able to flow freely and take up the shape of its container, but where there is a distinct boundary between the material and its surroundings. Liquids are materials in which the molecules are closely spaced (like solids they are hard to compress, and they have similar densities to solids), but unlike most solids the molecules are randomly arranged. The forces between

the molecules in a liquid are weak enough to allow liquids to flow, so they will take up the shape of any container in which they are placed, but the forces are strong enough to prevent the molecules from moving off into space, so a liquid has a fixed volume and will not expand to fill a volume.

liquid crystal Any liquid made from molecules that tend to line up under the influence of INTERMOLECULAR FORCES, producing long range ordering of a type more commonly associated with solid crystalline materials. Under the influence of an electric field, this ordering can

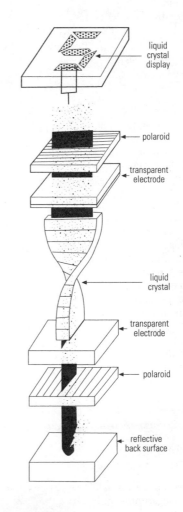

liquid crystal display

polaroid

transparent electrode

liquid crystal

transparent electrode

polaroid

reflective back surface

Liquid crystal display.

extend throughout the whole of the liquid. *See also* LIQUID CRYSTAL DISPLAY.

liquid crystal display (LCD) An electronic display device that uses a LIQUID CRYSTAL material in which all the molecules can be aligned by an electric field (such materials are described as nematic). The molecules of the liquid crystal are optically active, rotating the plane of polarization of light. By placing a piece of POLAROID in front of a suitable thickness of liquid crystal, areas of a display can be made which are clear in the absence of an electric field, but appear dark when an electric field is produced by applying a voltage to suitably shaped ELECTRODES.

liquid drop model A model of the atomic NUCLEUS that attempts to calculate the NUCLEAR BINDING ENERGY and other properties of the nucleus by treating it as a drop of liquid with a fixed DENSITY, SURFACE TENSION, SPECIFIC LATENT HEAT OF VAPORIZATION and so on. This model is most appropriate to nuclei containing large numbers of NUCLEONS. *See also* SHELL MODEL.

litre (l) A unit of volume, correctly called the decimetre cubed (dm^3) in the SI system of units. One litre is equal to $10^{-3} m^3$.

live That part of a mains electricity supply that is at a high ELECTRIC POTENTIAL relative to EARTH.

load 1. In electricity and electronics, any device to which electrical power is supplied.

    2. In mechanics, an object on which WORK is done by some machine, such as a pulley system, or the force acting on that object.

locus A series of points all satisfying a certain equation, and forming a continuous line. For example, a circle can be thought of as the locus of points that are all a certain distance (the RADIUS) away from a single fixed point (the centre of the circle).

logarithm The logarithm of a number to a given BASE is that number which, when the base is raised to the power of the logarithm, will give the original number. Thus if n is the logarithm to base x of y,

$$n = \log_x y$$

then x raised to the power n will give y,

$$y = x^n$$

Logarithms are usually taken to base 10 (common logarithms) or to base e (NATURAL LOGARITHMS).

logic circuit An electronic system handling DIGITAL signals and containing one or more LOGIC GATES.

logic gate The basic building block of any DIGITAL electronic system. A logic gate gives an output that is high or low depending on the state of one or more inputs. The simplest logic gate is the INVERTER or NOT gate, where the output is high if the input is low and vice versa. An AND GATE has an output only if all the inputs are high, whilst an OR GATE has a high output if any of its inputs is high. Particularly important is the NAND GATE (NAND = not and), where the output is high unless all the inputs are high. This gate is simple to manufacture and can be used as a building block for many more complex devices.

lone pair A pair of electrons in a single ORBITAL in the VALENCE SHELL of an atom. Lone pairs do not take part in the usual covalent bonding (*see* COVALENT BOND) of the atom.

longitude A measure of position on the surface of the Earth, or any other planet. Longitude is measured in degrees east or west of the prime meridian; a line running from pole to pole and passing through a specified point. On Earth, the prime meridian is fixed as passing through Greenwich in the UK. *See also* LATITUDE.

longitudinal wave A WAVE in which the motion of the particles that make up the wave is to-and-fro along the direction in which the wave is travelling. Such a wave can be thought of as a series of compressions and rarefactions; areas where the density of the particles is alternately higher and lower than normal. SOUND is an example of this kind of wave motion. *Compare* TRANSVERSE WAVE.

long wave RADIO WAVES with a wavelength longer than about 600 m.

Lorentz contraction The apparent reduction in length of a moving object compared to its length at rest, according to the SPECIAL THEORY OF RELATIVITY. If an observer at rest with respect to an object observes it as having a length l_0, then an observer moving at a speed v relative to the first will observe length l with

$$l = l_0 \sqrt{(1 - v^2/c^2)}$$

where c is the SPEED OF LIGHT.

Lorentz transformation The mathematical transformation needed to move from co-ordinates based on one INERTIAL REFERENCE FRAME to another in the SPECIAL THEORY OF RELATIVITY.

Love wave, *L-wave* In SEISMOLOGY, the low frequency transverse surface wave produced by an EARTHQUAKE. *See* SEISMIC WAVE.

low In DIGITAL electronics, a voltage level or other signal corresponding to a BINARY 0.

lowering of vapour pressure The reduction in the SATURATED VAPOUR PRESSURE of a solvent, depending on the concentration of dissolved material. Provided the dissolved material does not itself have a significant vapour pressure, the reduction in vapour pressure is an approximately COLLIGATIVE PROPERTY, depending far more on the molar concentration of the dissolved material than its chemical nature.

low-grade heat Heat energy that has been used to raise the temperature of a large quantity of matter by a small amount and thus is of little use for further energy transformations. *See also* HEAT ENGINE.

lumen (lm) The SI UNIT of LUMINOUS FLUX. One lumen is equal to the amount of visible LIGHT energy falling on one square metre at a distance of one metre from a point source of LUMINOUS INTENSITY of one CANDELA.

luminance The amount of visible LIGHT given off by a luminous surface, per second per square metre of its area, measured in CANDELA per square metre.

luminescence The emission of visible LIGHT from an object as a result of something other than high temperature, such as FLUORESCENCE or PHOSPHORESCENCE. *See also* THERMOLUMINESCENCE.

luminosity The total amount of visible LIGHT energy produced per second, such as by a star. *See also* APPARENT LUMINOSITY.

luminous flux The total amount of visible LIGHT arriving at a given surface per second. The SI UNIT of luminous flux is the LUMEN.

luminous intensity The amount of visible LIGHT given off per second by a light source. The SI UNIT of luminous intensity is the CANDELA.

lux The SI UNIT of ILLUMINANCE One lux is an illuminance of one LUMEN per square metre, or the illuminance at a distance of one metre from a light source of LUMINOUS INTENSITY of one CANDELA.

L-wave *See* LOVE WAVE.

M

machine A device for transferring energy from one form to another, or enabling energy to be used more effectively.

An important class of machine are those that work by the transfer of MECHANICAL ENERGY only. Such machines, called simple machines, are usually based on moving a small FORCE through a large distance in order to make a larger force move through a smaller distance. The force applied to a machine is called the effort, and this force is converted into one that acts at the other end of the machine on the load. Since the load moves through a shorter distance than the effort, it also moves more slowly. This is expressed by the velocity ratio of the machine – the ratio of distance moved by the effort to distance moved by the load.

The MECHANICAL ADVANTAGE is the ratio of load to effort, and the purpose of simple machines is generally to exert a force on the load that is greater than the effort. If there were no loss of energy, as a result of FRICTION for example, the machine would be completely efficient and the mechanical advantage would be equal to the velocity ratio. There is always some energy loss, however, making the mechanical advantage smaller than the velocity ratio by a factor called the EFFICIENCY of the machine. This is the amount of useful work done by the machine divided by the total amount of energy supplied.

In general, the efficiency of machines varies depending on the type of energy conversion being performed; ELECTRIC MOTORS, for example, can be as much as 95 per cent efficient. Since most waste energy ends up as heat, heaters are usually close to 100 per cent efficient, though the heat may not be produced where it is needed – in a gas boiler for example, much heat will be lost in the hot gases released to the surrounding atmosphere.

See also LEVER, PULLEY.

Mach number The speed of an object, particularly an aircraft, expressed as a ratio to the SPEED OF SOUND in the surrounding air. Thus Mach 2 is equivalent to twice the speed of sound, etc.

MACHO (Massive Compact Halo Object) An invisible massive object in orbit around a galaxy. It is a possible form of missing matter in the Universe. *See* COSMOLOGY.

macromolecule A very large MOLECULE, in particular a crystal structure in which the atoms are held together by COVALENT BONDS, so that the whole crystal is effectively a single giant molecule. Diamond and silicon dioxide are examples of such a structure, which are both tetrahedral in shape.

macroscopic (*adj.*) Visible to the naked eye. In science, a macroscopic state is a description of the behaviour of large-scale features of a system, such as temperature and pressure, as opposed to that of the individual atoms or molecules from which a system in made up.

magic numbers In nuclear physics, a number of PROTONS or NEUTRONS in a NUCLEUS marking the end of a periodic variation of certain properties of the nucleus, such as binding energy per nucleon (*see* NUCLEAR BINDING ENERGY). There are unusually large numbers of stable ISOTOPES of elements with magic ATOMIC NUMBERS. Doubly-magic nuclei with magic numbers of protons and neutrons have very high binding energies. The magic numbers are 2, 8, 20, 28, 50, 82 and 126.

magnet Any object that is surrounded by a MAGNETIC FIELD and attracts or repels other magnets. Magnets also attract unmagnetized pieces of iron, nickel or cobalt and some alloys containing these elements (such materials are described as FERROMAGNETIC). This is the result of the magnet temporarily magnetizing the ferromagnetic material; a phenomenon known as INDUCED MAGNETISM.

Magnets are of two kinds: ELECTRO-MAGNETS and PERMANENT MAGNETS. Electromagnets are SOLENOIDS, usually with an iron core, which rely on an electric current for their magnetism and so can be turned on and off. In a permanent magnet, the magnetism is present continuously, without the need for a current. Permanent magnets can be made in many shapes, but the most common are bar magnets, with a POLE at each end of a bar, horseshoe magnets, a bar bent into an arc of a circle with a pole at each end of the arc, and slab magnets, flat pieces of magnetic material with a pole on each of the two large faces.

magnetic bottle Any pattern of MAGNETIC FIELDS that can be used for the MAGNETIC CONFINE-MENT of a PLASMA.

magnetic circuit A closed path of MAGNETIC FIELD LINES around a CORE and through an air gap. If a small air gap is made in an otherwise closed core, producing a horseshoe ELECTRO-MAGNET, the effect will be to greatly reduce the MAGNETIC FLUX through the core. The air gap has the same effect on the flux as a resistance on the flow of electric current. This effect is called RELUCTANCE. *See also* MAGNETOMOTIVE FORCE.

magnetic compass A small MAGNET, pivoted so it can freely move horizontally, used to find direction. The magnet (called the compass needle) aligns itself with the MAGNETIC FIELD OF THE EARTH, so that it points to the MAGNETIC NORTH. A scale under the needle shows the points of the compass, for use in navigation. *See also* PLOTTING COMPASS.

magnetic confinement A technique for containing a PLASMA by the use of MAGNETIC FIELDS whilst the plasma is heated to the very high temperatures required for NUCLEAR FUSION. *See also* TOKAMAK.

magnetic dipole The simplest form of magnet, having a single north-seeking POLE and a single south-seeking pole. The strength of a magnetic dipole is measured by its MAGNETIC MOMENT. *See also* DIPOLE, MAGNETIC MONOPOLE.

magnetic field The region surrounding a magnet or a current-carrying conductor, in which a moving charge will experience a force or a MAGNETIC DIPOLE will experience a TORQUE. A small permanent magnet, such as a PLOTTING COMPASS, will tend to turn to point in the direction of the field. The SI UNIT of magnet field is the TESLA.

The direction of the force on a current in a magnetic field is given by FLEMING'S LEFT-HAND RULE. The size of the force is proportional to the field strength, to the current and to the length of the current-carrying conductor in

the field. It is also proportional to the sine of the angle between the current and the field: if the field and the current are in the same direction there will be no force; if the two are at right angles, the force will be a maximum. The force F on a wire of length l, carrying a current I at an angle θ to a magnetic field B is given by

$$F = BIl\sin\theta$$

Since a current is comprised of moving charges, the force on an electric current can be seen as a force on the individual moving CHARGE CARRIERS. The force on a moving charge is proportional to the strength of the magnetic field, the size of the charge and the speed at which it is moving. The force F on a charge q moving at a speed v at right angles to a field B is given by

$$F = Bqv$$

An electric current will produce a magnetic field. For a straight wire, the field strength is proportional to the current and inversely proportional to the distance from the wire. The MAGNETIC FIELD LINES form circles around the wire, the direction of which can be found using the RIGHT-HAND GRIP RULE. The magnetic field B at a distance r from a long straight wire carrying a current I is given by

$$B = \mu_o I / 2\pi r$$

where μ_o is the PERMEABILITY of free space.

The quantity B is also known as the MAGNETIC FLUX density, and is the magnetic flux per unit area of a magnetic field at right angles to the MAGNETIC FORCE. The numerical value of a magnetic field can also be given in terms of the magnetic field strength, H, where

$$H = B/\mu$$

where μ is the permeability of the medium. Both B and H are vector quantities.

See also MAGNETIC FIELD OF THE EARTH, SUSCEPTIBILITY.

magnetic field lines, *magnetic flux lines* Lines that show the direction of a MAGNETIC FIELD at each point. This is the direction in which a PLOTTING COMPASS would point if it were placed at that point. The closer the lines the stronger the field at that point.

magnetic field of the Earth The Earth, and many other planets, produce their own MAG-

NETIC FIELDS, believed to be due to electric currents within their molten metallic CORES. This means that a freely suspended BAR MAGNET will tend to point in one direction. The POLE that points roughly north is called the north-seeking pole whilst the pole which points roughly south is the south-seeking pole. *See also* MAGNETIC NORTH.

magnetic field strength (*H*) *See* MAGNETIC FIELD.

magnetic flux (Φ) A measure of the total amount of MAGNETIC FIELD passing through a given area. The magnetic flux is the magnetic field multiplied by the area over which the flux is being found and the COSINE of the angle between the field and the direction perpendicular to the plane of the area. If a coil has an area *A* with a magnetic field *B* passing through this area at an angle θ to the perpendicular to the circuit, and the coil contains *N* turns, the magnetic flux Φ will be

$$\Phi = BA\cos\theta$$

and the MAGNETIC FLUX LINKAGE ϕ_n will be

$$\Phi_n = BAN\cos\theta$$

See also MAGNETOMOTIVE FORCE.

magnetic flux density (*B*) *See* MAGNETIC FIELD.

magnetic flux lines *See* MAGNETIC FIELD LINES.

magnetic flux linkage The MAGNETIC FLUX multiplied by the number of times this flux passes through the circuit.

magnetic force The force produced by a MAGNETIC FIELD on a magnet, an electric current or a moving charge. For a current *I* at an angle α to a magnetic field *B*, the magnetic force is

$$F = BIl\sin\alpha$$

where *l* is the length over which the current flows in the field. For a charge *q* moving at speed *v* at an angle α to the field, the force is

$$F = Bqv\sin\alpha$$

magnetic moment A measure of the strength of a MAGNETIC DIPOLE. The SI UNIT of magnetic moment is the AMPERE metre squared (Am^2), and is equivalent to the strength of field produced at large distances from a loop carrying a current of one ampere and enclosing an area of one square metre. Alternatively, this is the strength of magnetic dipole which will experience a TORQUE of one NEWTON metre (1 Nm) when placed at right angles to a magnetic field of one TESLA. *See also* MAGNETON.

magnetic monopole A hypothetical source of MAGNETIC FIELD, a north-seeking POLE without a south-seeking pole or vice versa. Despite extensive searches, no magnetic monopoles have been found. *See also* MAGNETIC DIPOLE.

magnetic north The direction in which a MAGNETIC COMPASS points. It is not quite the direction to the north pole because the Earth's magnetic poles do not quite coincide with its geographical poles – the points about which it rotates. The position of the Earth's magnetic poles also changes with time and the MAGNETIC FIELD OF THE EARTH has completely changed direction many times over its lifetime. *See also* MAGNETIC VARIATION.

magnetic resonance imaging (MRI) An imaging technique increasingly used in medicine. The patient is placed in a magnetic field, which aligns the SPINS of the protons in hydrogen atoms throughout the patient. A pulsed electromagnetic field destroys this alignment, and the signals produced as the protons realign themselves have frequencies characteristic of the chemical environment of the individual protons. TOMOGRAPHY techniques are used to produce a picture of the body in a series of slices. Unlike X-RAYS, MRI is non-ionizing (*see* IONIZING RADIATION) and can produce clear images of soft tissues. *See also* NUCLEAR MAGNETIC RESONANCE.

magnetic variation The angle between true north and MAGNETIC NORTH. Magnetic variation varies with time and with position on the surface of the Earth.

magnetism The collective term for all the effects resulting from the presence of MAGNETIC FIELDS. Magnetic fields are produced by electric currents and by many ELEMENTARY PARTICLES, including the electron. The magnetic effects of electrons are responsible for magnetic properties of several elements. *See also* DIAMAGNETISM, ELECTROMAGNET, ELECTROMAGNETIC INDUCTION, FERRIMAGNETIC, FERROMAGNETIC, INDUCED MAGNETISM, MAGNET, MAGNETIC FORCE, MAGNETOMOTIVE FORCE, PARAMAGNETISM, PERMANENT MAGNET.

magnetohydrodynamics (MHD) The study of electrically conducting fluids. MHD covers both molten metals and the properties of PLASMAS, such as those in a NUCLEAR FUSION reactor and the interaction of the SOLAR WIND with planetary magnetic fields.

magnetometer A device for measuring a MAGNETIC FIELD, particularly that of the Earth.

Local variations of the Earth's magnetic field are sometimes indicative of disturbance of the underlying soil, so magnetometers are used by archaeologists as a surveying tool.

magnetomotive force The influence that creates the MAGNETIC FLUX in a MAGNETIC CIRCUIT. For a SOLENOID, this is equal to the product of the current in the coil and the number of turns (called the number of AMPERE-TURNS).

magneton A unit of MAGNETIC MOMENT used in atomic and nuclear physics. The Bohr magneton is 9.27×10^{-24} Am^2. The nuclear magneton is 5.05×10^{-27} Am^2.

magnetosphere The region in space that is influenced by the MAGNETIC FIELD OF THE EARTH.

magnetostriction The change in length of a FERROMAGNETIC material with changes in MAGNETIC FIELD. The effect results from the contribution of magnetic forces to the intermolecular forces within the material. This effect is the basis of some TRANSDUCERS used to convert motion into electrical signals and vice versa.

magnetron A device for generating MICROWAVES. It comprises a filament surrounded by several ANODES, each forming a cavity resonant (*see* RESONANCE) at the frequency of microwaves being generated, and placed in a magnetic field. Electrons are emitted by the filament by the process of THERMIONIC EMISSION, accelerated towards the anodes and deflected by the magnetic field so that they move in a spiral path. Interactions between the electrons and the electric fields at the anodes set up oscillations at the RESONANT FREQUENCY of the cavities. The magnetron is often operated in pulses to generate microwaves for RADAR transmitters. *See also* KLYSTRON.

magnification The extent to which an image is larger than the original object, in a microscope or telescope for example.

magnification = image size/object size

In the case of a telescope, the term is used to denote angular magnification:

angular magnification = angle subtended by image/angle subtended by object

magnifying glass A CONVERGING LENS used with the object viewed less than one FOCAL LENGTH away from the lens, producing an enlarged VIRTUAL IMAGE.

magnitude Of a VECTOR, the size of the vector quantity, regardless of direction. If the COMPONENTS of a vector in the x, y and z directions are a, b and c, the magnitude of the vector is $\sqrt{(a^2 + b^2 + c^2)}$.

main sequence star Any star lying on the main sequence of the HERTZSPRUNG–RUSSELL DIAGRAM, the diagonal band running from top left (hot bright stars) to bottom right (cool dim stars).

Stars are formed by the GRAVITATIONAL COLLAPSE of clouds of interstellar gas, which is made up of about 75 per cent hydrogen and 25 per cent helium. As the cloud collapses, provided it has a mass greater than about 0.08 SOLAR MASSes, it will heat up to a temperature at which further collapse is opposed by the release of energy from NUCLEAR FUSION, in which hydrogen is converted to helium. The chain of reactions by which this takes place is called the PP CHAIN. During the time when these reactions are taking place the star remains in more or less the same position on the Hertzsprung–Russell diagram, on the main sequence. The time spent as a main sequence star accounts for about 90 per cent of a star's lifetime, which explains why most of the stars observed are on the main sequence. Once the hydrogen in the core has all been fused the star moves away from the main sequence. *See also* RED GIANT, WHITE DWARF.

malleable (*adj.*) Able to be beaten into a new shape without breaking. Malleability is an important property of many metals.

Maltese cross tube A device used to demonstrate the effects of THERMIONIC EMISSION. It comprises an electrically heated filament in an evacuated glass vessel, which also contains a hollow metal cylinder and a metal cross. The end of the tube opposite the filament is coated with a PHOSPHOR. The metal cylinder and the cross are made positive and the filament negative with a POTENTIAL DIFFERENCE of several thousand volts. The screen glows where it is struck by the electrons from the filament but there is a 'shadow' on the screen showing that electrons travel in straight lines and cannot pass through the metal of the cross.

manganese-alkaline cell A common type of electrochemical CELL, non-rechargeable and more expensive than ZINC-CARBON CELLS, but longer lasting.

manometer An instrument for measuring pressure, comprising a U-shaped glass tube filled with a liquid. The difference in the height of the liquid in the two arms of the tube is proportional to the pressure difference between the two ends of the manometer, one of which is usually left open to ATMOSPHERIC PRESSURE.

mantle The layer of molten rock surrounding the Earth's CORE on which TECTONIC PLATES float, forming the Earth's CRUST.

maritime (*adj.*) Describing an AIR MASS that has travelled mostly over water.

Mars The fourth planet in order from the Sun, with an orbital radius of 1.52 AU (228 million km). It is rather smaller than the Earth, with a diameter of 6,780 km (0.53 times that of the Earth), and a mass of 6.4×10^{23} kg (0.11 times that of the Earth). Mars orbits the Sun every 1.9 years and rotates on its own axis every 24.6 hours.

As a result of its small size, Mars has only a very thin atmosphere. This atmosphere supports some weather effects, such as dust storms and OROGRAPHIC cloud, but has not produced significant erosion of the surface. Whilst the surface of the planet contains no liquid water, there is evidence for fluid flow at some time in the past, though it may have been short lived. In 1996 it was announced that some evidence had been found for the existence of primitive life forms at some stage in the planet's history.

maser Acronym for MICROWAVE *a*mplification by STIMULATED EMISSION of RADIATION. A device similar to a LASER, but operating in the microwave region of the ELECTROMAGNETIC SPECTRUM. Masers are sources of microwaves with very accurate frequencies and as such are used in many high-precision clocks.

mass A measure of the total amount of MATTER in an object, expressed either in terms of the resistance of an object to having its motion changed (INERTIAL MASS) or the effect of a GRAVITATIONAL FIELD on the object (GRAVITATIONAL MASS). The SI UNIT of mass is the KILOGRAM. *See also* REST MASS.

mass defect The difference between the mass of an atomic NUCLEUS and the mass of the NEUTRONS and PROTONS from which it is made. A consequence of the SPECIAL THEORY OF RELATIVITY is that a loss of energy will result in an equivalent loss of mass, so that when nuclei were formed in stars, and energy was released, mass was lost. *See also* NUCLEAR BINDING ENERGY.

mass-energy equation The equation

$$E = mc^2$$

where E is energy, m mass and c the speed of light in a vacuum. This equation, an important consequence of the SPECIAL THEORY OF RELATIVITY, states that an object with more energy will appear more massive. *See also* EQUIVALENCE OF MASS AND ENERGY, REST MASS ENERGY.

Massive Compact Halo Object *See* MACHO.

mass number, *nucleon number* The total number of NUCLEONS (NEUTRONS and PROTONS) in a particular atomic NUCLEUS.

mass spectrometer An instrument for the measurement of the mass of atoms, molecules or fragments of a molecule, and the relative abundance of each mass present. A sample of material is IONIZED by bombardment with an electron beam. If the sample is molecular, the molecule will also be broken into fragments. The charged particles are accelerated in an electric field. The ions then enter a VELOCITY SELECTOR, a region in which electric and magnetic fields are applied at right angles to one another to produce opposing forces on the ions. These forces balance out only for particles moving at one particular speed, which then enter the next region of the device, where there is a magnetic field that deflects particles according to their charges and masses. A detector produces a reading of abundance against mass/charge ratio.

As well as measuring the relative abundance of ISOTOPES, the mass spectrometer is an important tool in determining the structure of organic molecules.

mass spectroscopy The use of a MASS SPECTROMETER to determine the structure of a compound or the relative abundances of various ISOTOPES.

matter The collective term for all ATOMS. Matter is any substance with MASS and which is not ANTIMATTER. In particle physics, matter is defined as all substances with a positive BARYON or LEPTON NUMBER.

maximum The largest of a set of numbers, or the largest value taken by a function. A local maximum is a value of a function that is larger than the value of that function for values of independent variable adjacent to that giving the maximum value, but which may be exceeded when the independent variable takes on other, very different, values.

maximum and minimum thermometer A thermometer for recording the highest and lowest temperatures reached since the thermometer was last reset. Traditionally, such thermometers use the expansion of alcohol to push a mercury thread around a U-shaped CAPILLARY TUBE. Steel markers are pushed along the tube by the mercury, but are left behind when the mercury moves away from them. These markers can be repositioned using a magnet. Modern maximum and minimum thermometers are electronic.

maximum power theorem In electricity, the POWER delivered to a LOAD is a maximum when the RESISTANCE of the load is equal to INTERNAL RESISTANCE of the power supply. If the load resistance is too large only a small current will flow, if the resistance is too small there will be too great a drop in the voltage across the load. In cases where it is important to deliver as much power to the load as possible, its resistance should be matched to the internal resistance of the supply.

maxwell Unit of magnetic flux in the C.G.S. SYSTEM. One maxwell is equal to 10^{-8} WEBER.

Maxwell–Boltzmann distribution A description of the range of energies possessed by molecules in a system in which those molecules are able to exchange energy freely with one another to maintain THERMAL EQUILIBRIUM. In the Maxwell–Boltzmann distribution the number of particles with an energy greater than E is proportional to $e^{-E/kT}$, where T is the ABSOLUTE TEMPERATURE and k is the BOLTZMANN CONSTANT. This quantity is called the BOLTZMANN FACTOR, and if E is the ACTIVATION ENERGY, the rate at which an ACTIVATION PROCESS takes place will depend on this factor. The Maxwell–Boltzmann distribution can also be used to describe range of speeds of molecules in a gas.

The situation is a little more complicated with FERMIONS, particularly at low temperatures, due to the PAULI EXCLUSION PRINCIPLE, which prevents all the particles from occupying the lowest energy states. The FREE ELECTRONS in a metal, for example, have far higher speeds than might be expected.

See also FERMI DISTRIBUTION.

Maxwell's equations A set of four equations describing the interaction of CURRENTS and CHARGES with ELECTRIC and MAGNETIC FIELDS. The first equation is essentially GAUSS' LAW. The second equation describes the creation of magnetic fields by currents and by changing electric fields. The third equation records the absence of MAGNETIC MONOPOLES, and the fourth is essentially FARADAY'S LAW OF ELECTROMAGNETIC INDUCTION. Taken collectively, they led to the discovery of ELECTROMAGNETIC WAVES and the recognition that light is an electromagnetic wave. The SPEED OF LIGHT in a vacuum, c, is related to the PERMITTIVITY of free space, ε_0, and the PERMEABILITY of free space, μ_0, by the equation

$$c^2 = 1/(\varepsilon_0\mu_0)$$

mean 1. *arithmetic mean* The sum of a set of numbers divided by the number of elements in the set. *See also* STANDARD DEVIATION.

2. *geometric mean* The nth ROOT of the product of all the values in a set, where n is the number of values in the set.

mean free path The average distance that a molecule in a fluid, usually a gas, travels before colliding with another molecule. This depends on the number of molecules per metre cubed and on the size of the two molecules, as measured by the COLLISION CROSS-SECTION. In liquids the mean free path is similar to the size of a molecule, and the molecules cannot be thought of as moving freely, as they can in gases.

For a gas with n molecules per metre cubed, each having a collision cross-section of σ, the mean free path is λ, where

$$\lambda = (\sqrt{2}n\sigma)^{-1}$$

See also TRANSPORT COEFFICIENT, KNUDSEN REGIME.

mean free time The average time for which a particle in a fluid, usually a gas, travels between collisions.

mean solar day *See* DAY.

mechanical advantage The factor by which the FORCE applied to a LOAD by a machine is greater than the EFFORT force:

mechanical advantage = load force/
effort force

mechanical energy A collective term for the forms of energy studied in MECHANICS: KINETIC

ENERGY, GRAVITATIONAL POTENTIAL ENERGY and ELASTIC potential energy.

mechanical equivalent of heat A conversion factor, now obsolete in the SI system of units, for converting units of HEAT, measured in CALORIES, into units of WORK, measured in ERGS. It has the value 4.185×10^7 ergs/calorie.

mechanics The branch of physics that deals with the effect of FORCES on objects or structures, particularly gravitational and CONTACT FORCES. *See also* DYNAMICS, STATICS.

median The value that is midway between the largest and the smallest values found in a set of data. *See also* AVERAGE.

medical imaging The name given to a range of techniques that are used to obtain images of the inside of a human body (usually to diagnose illness) without resorting to surgery. Techniques used are often based on the transmission of X-rays, the reflection of ULTRASOUND and on NUCLEAR MAGNETIC RESONANCE. *See also* ENDOSCOPE, MAGNETIC RESONANCE IMAGING, RADIOGRAPH, TOMOGRAPHY, ULTRASOUND IMAGING.

medium wave RADIO WAVES with a wavelength from 200 m to 600 m.

mega- (M) A prefix indicating that the size of a unit is to be multiplied by 10^6. For instance, one megawatt (MW) is equal to one million WATTS.

melting point The temperature at which a solid turns into a liquid, or vice versa. More technically, the melting point is the one temperature for a given pressure at which the solid and liquid can exist in equilibrium together. Most materials increase in volume, though only slightly, when they melt, and the melting point increases with increasing pressure. An important exception to this is water, which takes up less volume as a liquid than as a solid, thus water tends to melt under pressure.

memory Any electronic device that can store information, usually in the form of BINARY digits. Modern memory circuits are based on various forms of BISTABLE, which is put into one of its two possible states to represent a binary 1 or 0. The memory can then be read by examining the state of the bistable. Modern memory devices are able to store several million bits in a single integrated circuit. Memory is traditionally divided into ROM, or read only memory, where information can only be stored once, and RAM, or random access memory,

which can be overwritten with new data. Modern use of the term memory tends to refer to RAM, which is further divided into static and dynamic memory. Dynamic memory involves the storage of bits as charges on what are effectively small CAPACITORS in the integrated circuit. These charges tend to leak away with time, so the memory needs to be 'refreshed' periodically. Static RAM does not have this disadvantage but currently holds less data on a given size of integrated circuit than dynamic RAM.

meniscus The name given to the curved shape of a liquid surface in a tube, or any similarly shaped object, such as a lens with one CONVEX and one CONCAVE surface. A meniscus lens may be either converging or diverging, depending on which surface is the more strongly curved. *See also* CAPILLARY EFFECT, CONVERGING LENS, DIVERGING LENS.

Mercury The closest planet to the Sun, with an orbital radius of 0.39 AU (58 million km). Similar in many ways to our Moon, it is too small to retain an atmosphere and is heavily cratered. Mercury has a diameter of 4,900 km (0.38 times that of the Earth) and a mass of 3.3×10^{23} kg (0.055 times that of the Earth). Mercury orbits the Sun in 88 days and rotates about its own axis in 59 days.

mercury barometer A BAROMETER in which a tube filled with mercury and closed at the upper end is placed vertically, with its lower end in a trough of mercury open to the atmosphere. If the tube is long enough, the pressure of the air will not be sufficient to support the weight of the mercury and a vacuum (called a Toricellian vacuum) will appear at the top end of the tube. The height of the mercury column that can be supported is a measure of the pressure of the atmosphere, which is sometimes quoted as millimetres or inches of mercury (mmHg or inHg). The pressure that will support one millimetre of mercury is called one TORR. *See also* FORTIN BAROMETER.

meson Any one of a family of unstable HADRONS, made up of a QUARK and an antiquark. They exist as positive, negative and neutral particles and they include the kaon, the PI-MESON and the psi particle.

mesosphere A layer in the upper ATMOSPHERE in which temperature falls with increasing height.

metal Any of a class of elements that are typically lustrous, MALLEABLE, DUCTILE solids (mercury is a liquid) that are good conductors of heat and electricity. About 75 per cent of the known elements are classed as metals.

The structure of a metal is typically a rigid lattice of positive ions, through which the outermost electrons, the VALENCE ELECTRONS, are free to move. The forces between the ions in the metal lattice are repulsive at short distances, but at larger separations the electrons screen the repulsion between ions, resulting in an attractive force that is strong enough to give most metals melting points well above room temperature.

The presence of FREE ELECTRONS in a metallic lattice make metals good CONDUCTORS of heat and electricity. The electrical CONDUCTIVITY allows metals to reflect electromagnetic radiation, including light, resulting in their characteristic shiny appearance. The interatomic forces in metals are generally weaker then those in IONIC SOLIDS, so metals are relatively soft, and in their pure forms are usually malleable and ductile. *See also* ALLOY, METALLIC BONDING, THERMAL CONDUCTION.

metallic bonding The bonding that holds together atoms in a METAL. The overlapping of the ORBITALS in a metallic solid produces an energy band that is effectively a DELOCALIZED ORBITAL, so that VALENCE ELECTRONS can move freely from one atom to the next, allowing the metal to conduct heat and electricity. A simple model of a metallic structure is of a lattice of positive ions surrounded by a 'sea' of electrons. The relatively weak nature of metallic bonding compared to ionic bonding (*see* IONIC BOND) accounts for the softness and low melting points of metals when compared to IONIC SOLIDS.

metalloid, *semi-metal* Any element that has some of the properties of a METAL, but which is not completely metallic. Many of them are SEMICONDUCTORS, or form compounds that are semiconductors. Typical examples are arsenic, boron, germanium, silicon and tellurium.

metallurgy The study of the properties of METALS, and more usually their ALLOYS, particularly in respect of their engineering properties.

metastable equilibrium Any state that is a state of STABLE EQUILIBRIUM for small displacements but which is unstable for larger displacements, such as a ball lying in a small hollow in an upturned bowl.

metastable state A state of some physical system that is unstable, but in which the system will remain for an unusually long time, or a time sufficiently long for the state to be regarded as stable. SUPERHEATED water is an example of a metastable state, and some ENERGY LEVELS in atomic nuclei are also metastable.

meteorology The study of the Earth's ATMOSPHERE, especially WEATHER SYSTEMS.

method of mixtures A way of measuring HEAT CAPACITIES. A substance is heated to a measured temperature and then mixed with a second substance at a lower temperature. The mixture will be at a temperature between those of the two substances before they were mixed. If there are no heat losses, the heat energy given out by the hotter substance as it cools down will equal the heat energy taken in by the cooler substance as it heats up. If the heat capacity of one of the substances is known, that of the other can then be calculated.

metre The BASE UNIT of length in the SI system (*see* SI UNIT). The metre was originally defined as the length of a standard metal bar, then in terms of a certain number of wavelengths of light, but now defined as the distance travelled by light in a vacuum in $1/299{,}792{,}458$ seconds.

metric system Any system of measurements based on the METRE as the unit of length, the GRAM as the unit of MASS and the SECOND as the unit of time, or on some multiple of these units. In science, the system almost universally used is the SI system of units, which defines units for all physical quantities, derived from seven BASE UNITS. *See* SI UNITS.

metric ton, *tonne* A unit of MASS. One tonne is equal to 1,000 kg.

metrology The science of measurement, in particular measuring quantities very precisely.

Michelson–Morley experiment An experiment, first performed in 1881, set up to measure the velocity of the Earth through the ether (the hypothetical medium that was thought to fill all space). The ether was thought to support the propagation of ELECTROMAGNETIC RADIATION through space, and the results of the Michelson–Morley experiment were to have led to a calculation of the velocity of light.

In the experiment, light waves were made to interfere with each other (*see* INTERFERENCE) after they had travelled along two paths set at right angles to one another. The interference pattern was unchanged when the whole experiment was rotated, suggesting that the motion of the Earth through space did not alter the speed of the light waves. This lead to an abandonment of the ether theory, and ELECTROMAGNETIC WAVES are now known to be able to travel through free space. The abandonment of the ether hypothesis also led indirectly to the formulation of the SPECIAL THEORY OF RELATIVITY.

micro- (μ) A prefix indicating that a unit is to be multiplied by 10^{-6}. For instance, the microampere (μA) is equal to one millionth of an AMPERE.

micrometer An instrument for the accurate measurement of small objects (generally up to 10 cm). Many forms of micrometer exist, but typically the object to be measured is placed between a fixed surface and a moving surface, which can be advanced by a screw thread until it is just touching the object. The screw thread is machined accurately to a known pitch and provided with a scale which enables fractions of one rotation to be measured accurately. Accuracies of 10^{-6} m are possible using a simple micrometer equipped with a VERNIER scale.

microprocessor A complex INTEGRATED CIRCUIT that is capable of performing a large number of logical functions on BINARY numbers. The microprocessor contains an ARITHMETIC LOGIC UNIT (ALU), which performs various operations on numbers that are stored in SHIFT REGISTERS, with the output also being stored in a REGISTER, usually called the ACCUMULATOR. Instructions can be given to the microprocessor to perform various operations on the numbers in the registers or to move data between the registers and memory connected to the microprocessor.

microscope An optical device that uses a system of lenses to magnify objects too small to be seen in fine detail with the naked eye. In 1665 Robert Hooke (1635–1703) was the first to record microscopic examination of cells in cork, and Anton van Leeuwenhoek (1632–1723) recorded bacteria in 1683.

A SIMPLE MICROSCOPE has a single lens but limited powers of magnification, whereas a COMPOUND MICROSCOPE uses two lenses and light passes from an object through the first lens (object lens) to produce a magnified image that is then magnified further by the second lens (eyepiece). The total magnification is the product of the magnification of

each lens and is maximally 1,500–2,000 times in a light microscope, achieved by oil-immersion objective lenses. The RESOLVING POWER of a light microscope is limited to two points 0.2 μm apart. The thinner the material being observed, the greater the clarity of image.

See also ELECTRON MICROSCOPE, PHASE CONTRAST MICROSCOPY.

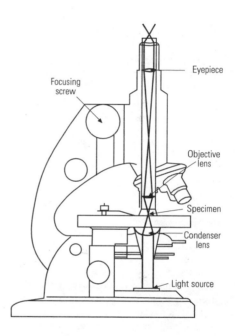

Light microscope.

microscopy The study or use of MICROSCOPES.

microwave ELECTROMAGNETIC RADIATION with a wavelength in the range 30 cm to 1 mm. The short wavelength means that microwaves can be formed into narrow beams using parabolic reflector dishes of a reasonable size without too much DIFFRACTION, thus they tend to be used for point-to-point communication rather than broadcasting. Microwaves are also often used in DBS (direct broadcasting satellite) systems. Their ability to form narrow beams with an aerial of reasonable size is also put to use in RADAR systems.

Microwave frequencies coincide with the RESONANT FREQUENCIES of some covalently bonded molecules (*see* COVALENT BOND). In particular, waves of 12.6 cm wavelength are resonantly absorbed by water molecules. This effect is used in microwave ovens, where microwave energy is resonantly absorbed and converted to heat by water in food.

See also KLYSTRON, MAGNETRON.

microwave spectroscopy The study of the absorption spectra (*see* ABSORPTION SPECTRUM) of MICROWAVES by gases, typically using microwaves with a wavelength of a few millimetres. Since these waves have frequencies that excite the natural frequencies of rotation of simple covalently bonded molecules (*see* COVALENT BOND), it is possible to study certain properties of these bonds, such as their natural length, and their angle to one another.

mild steel An alloy of iron with a few per cent of carbon. Mild steel is cheap and easy to machine and to weld, but rusts quickly.

Milky Way Our own GALAXY, which appears as a band of stars across the sky. Study of the distribution of stars in the Milky Way shows that the SOLAR SYSTEM is located close to the plane of the galaxy, in one of the spiral arms.

milli- (m) A prefix indicating a unit is to be multiplied by 10^{-3}. For instance, one milliampere (mA) is equal to one thousandth of an AMPERE.

millibar (mb) A unit of PRESSURE commonly used in METEOROLOGY. One millibar is equivalent to 100 PASCAL.

Millikan's oil drop experiment An experiment first performed by Robert Millikan (1868–1953) in 1909, in which the charge of an oil drop is measured by suspending it in an electric field. The charges of all the drops were found to be multiples of a single charge, the charge on the electron, thus showing the quantized (*see* QUANTUM) nature of electric charge.

minimum The smallest of a set of numbers, or the smallest value taken by a function. A local minimum is a value of a function that is smaller than the value of that function for values of independent variable adjacent to that giving the minimum value, but which may be exceeded when the independent variable takes on other, very different, values.

minor planet *See* ASTEROID.

mirage The illusion of water seen when looking at a hot surface, such as desert sand, at a shallow angle. The air just above the surface is heated by the surface, so is less dense and has a lower REFRACTIVE INDEX than the cooler air further from the ground. This leads to TOTAL INTERNAL REFLECTION of light from the sky as it approaches the surface. The unstable nature of the hot layer, resulting from CONVECTION, gives the reflected light an unsteady appearance, similar to light from the sky reflected in a pool of water. Unlike real water, the mirage remains at a fixed angle to the observer, so the mirage recedes as the observer approaches it.

mirror A device used to reflect some form of wave, usually visible light. Mirrors used for the REFLECTION of light are typically made from a thin layer of metal (usually aluminium) deposited on a layer of glass. For precision use, the metal is placed on the front surface of the glass (such mirrors are called front-silvered). Mirrors with the reflective coating behind the glass (back-silvered) are less easily damaged but suffer from multiple reflections off the front of the glass surface. *See also* CURVED MIRROR, PLANE MIRROR.

missing mass Mass that is not observed in the Universe, but which must exist if the density is equal to the CRITICAL DENSITY. *See* COSMOLOGY.

mist Droplets of water formed by condensation in the atmosphere close to ground level, but with the visibility remaining above 1,000 m. *See also* FOG, WEATHER SYSTEMS.

mixture A substance containing two or more elements or compounds, but where there is no chemical bonding between the constituents of the mixture. The constituents can be separated without a chemical reaction taking place. For example, in a mixture of iron and sulphur, the iron can be removed using a magnet, but once the iron and sulphur have combined to form iron sulphide (a compound) they cannot be separated by physical means.

m.k.s. system A system of physical units derived from the METRIC SYSTEM and based on the metre, kilogram and second. It has been replaced by SI UNITS.

mode The value in a set of data that occurs most often, or the range for which the frequency is greatest.

moderator A material that absorbs energy from fast-moving neutrons in a NUCLEAR REACTOR, without absorbing the neutrons themselves. Neutrons slowed in this way are called thermal neutrons. They are more likely to cause fission

[handwritten annotation: GAS CONSTANT for dry air 0.287053 kPa K⁻¹ or J mol⁻¹ K⁻¹]

in uranium–235 nuclei and less likely to be absorbed by uranium–238. DEUTERIUM and carbon–12 (as graphite) are often used as moderators. *See also* NUCLEAR FISSION.

modulation The process of changing some feature of a CARRIER WAVE, such as its amplitude or frequency, in order to convey some information. *See also* AMPLITUDE MODULATION, FREQUENCY MODULATION, PHASE MODULATION, PULSE-CODE MODULATION, SIDEBAND.

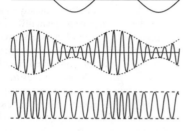

modulating signal

amplitude modulated carrier wave

frequency modulated carrier wave

Modulation.

modulus In mathematics the value of a number, taken as positive, regardless of whether the number is positive or negative. The modulus of a REAL NUMBER x is denoted by the symbol $|x|$. For a COMPLEX NUMBER, $a+ib$, the modulus is $\sqrt{(a^2 + b^2)}$.

modulus of rigidity, *shear modulus* A measure of the elasticity of a material subjected to a shearing force (*see* SHEAR). It is equal to the tangential force per unit area divided by the angular deformation in radians.

molar conductivity The CONDUCTIVITY of a solution divided by its concentration, in order to allow the contribution to the conductivity of each MOLE of solute ions to be determined. In a strong ELECTROLYTE (one that is fully IONIZED) the molar conductivity is essentially independent of concentration, except at very high concentrations. In a weak (only partially ionized) electrolyte, the molar conductivity rises only gradually with dilution, the material being fully ionized at very low concentrations.

molar gas constant, *universal gas constant* (R) The constant of proportionality linking the ABSOLUTE TEMPERATURE of one MOLE of gas to the product of its pressure and volume. R is equal to 8.3 JK⁻¹mol⁻¹. *See* IDEAL GAS EQUATION.

molar heat capacity The HEAT CAPACITY of a piece of material containing one MOLE of molecules. For an IDEAL GAS the molar heat capacity at constant pressure is greater than that at constant volume by R, the MOLAR GAS CONSTANT. For a material with a molar heat capacity at constant pressure of C_P and a molar heat capacity at constant volume of C_V:

$$C_P - C_V = R$$

See also DULONG AND PETIT'S LAW, RATIO OF SPECIFIC HEATS.

molar volume The volume occupied by one MOLE of a given material. All IDEAL GASES have the same molar volume at a given temperature and pressure. At STANDARD TEMPERATURE AND PRESSURE this is 22.4 dm³.

mole The SI UNIT of amount of substance. One mole is defined as being the amount of substance containing as many atoms (or molecules or ions or electrons) as there are carbon atoms in 12 g of carbon–12.

molecular mass *See* RELATIVE MOLECULAR MASS.

molecular orbital An ORBITAL formed by the overlapping of orbitals between two atoms that are linked together by a COVALENT BOND. It is the reduction in energy brought about by the formation of one of the possible molecular orbitals, as compared to the two separate atomic orbitals, that makes the bond stable. *See also* ANTIBONDING ORBITAL, BONDING ORBITAL, DELOCALIZED ORBITAL.

molecular weight An obsolete term for RELATIVE MOLECULAR MASS.

molecule The smallest part of a chemical COMPOUND that can exist without it losing its chemical identity. Molecules are made of one or more ATOMS held together by IONIC or COVALENT BONDS. *See also* MACROMOLECULE.

moment The turning effect of a FORCE. The magnitude of a moment is equal to the magnitude of the force multiplied by the shortest distance between the axis and the line along which the force acts. If an object is in EQUILIBRIUM, the sum of the moments of all the forces about any point will be zero.

moment of inertia A quantity that measures the distribution of MASS within an object. The moment of inertia of a body is the ratio of the MOMENT acting on the body to the ANGULAR

ACCELERATION caused by that moment. For a solid object, which can be thought of as made up of a series of points of mass m_i each located at a distance r_i from the axis, the moment of inertia I is given by

$$I = \Sigma m_i r_i^2$$

The ANGULAR MOMENTUM is $I\omega$, and the rotational KINETIC ENERGY $\frac{1}{2}I\omega^2$, where ω is the ANGULAR VELOCITY. See also ROTATIONAL DYNAMICS.

momentum The MASS of an object multiplied by its VELOCITY. Momentum is a measure of how easy or difficult it is to stop an object. To change the momentum of an object, an IMPULSE must be exerted equal to the change in momentum required. The impulse is the FORCE acting on the object multiplied by the time for which that force acts. For an object of mass m moving with velocity v, the momentum p will be

$$p = mv$$

See also LAW OF CONSERVATION OF MOMENTUM.

monochromatic (*adj.*) Describing ELECTROMAGNETIC RADIATION, in particular visible light, of a single wavelength. The term monochromatic is also used to describe a beam of particles where all particles have the same energy.

monoclinic (*adj.*) Describing a CRYSTAL structure where the UNIT CELL has two sets of faces at right angles to one another and the third is at some other angle. The size and shape of the unit cell are characterized by three lengths and one angle.

monomode (*adj.*) Describing an optical fibre (*see* FIBRE OPTIC) with a diameter of typically one micrometer (10^{-6} m), comparable to the WAVELENGTH of light for which the fibre is designed. Light travels down this fibre without bouncing off the sides, so avoiding spreading of light pulses arising from the different lengths of path along the fibre. See also GRADED INDEX, MULTIMODE, STEP INDEX.

moon Any natural body that ORBITS a PLANET. Many moons are known in the SOLAR SYSTEM, ranging in size from Ganymede, a moon of JUPITER, with a diameter of 5,250 km, which is larger than the planet MERCURY, to irregularly shaped objects only a few kilometres across, several of which have been discovered around the GAS GIANTS by the VOYAGER spacecraft and the HUBBLE SPACE TELESCOPE. The term Moon,

(capital M) is used to refer to the only such object in orbit around the Earth. It orbits the Earth once every 27.3 days at an average distance of 384,400 km. It has a mean diameter of 3,476 km and a mass of about 0.012 that of the Earth. The Moon has no atmosphere and very little surface water.

motor Any device that converts ELECTRICAL ENERGY into mechanical WORK, usually in the form of rotational motion. See also D.C. MOTOR, INDUCTION MOTOR.

motor effect The force produced by a current-carrying wire in a MAGNETIC FIELD.

moving-coil galvanometer An AMMETER in which the current flows through a coil suspended in a magnetic field. The result is a turning effect, which is opposed by a pair of flat coiled springs, called hairsprings. The hairsprings are also used to carry the current in and out of the coil. The coil rotates until the turning effect produced by the current is balanced by the turning effect of the springs. A pointer, or mirror with a beam of light shining onto it, is attached to the coil so the amount of motion, and thus the size of the current, can be read from a graduated scale.

MRI See MAGNETIC RESONANCE IMAGING.

multimode (*adj.*) Describing an optical fibre (*see* FIBRE OPTIC) with a diameter that is large compared to the WAVELENGTH of the light passing through it. In such a device, the light travels by a succession of REFLECTIONS off the walls of the fibre. Many different paths through the fibre are possible, with different lengths, leading to the disadvantage that pulses of light tend to become spread out as they pass along the system. This limits the rate at which the light can be modulated (*see* MODULATION) and hence the rate at which such a system can be used to convey information. See also GRADED INDEX, MONOMODE, STEP INDEX.

multiplexing Any technique for sending more than one set of data down a single TELECOMMUNICATIONS link. Multiplexing can be achieved either by using a range of different FREQUENCIES, or by sending data relating to different channels of communication one after another in sequence, a technique called TIME-DIVISION MULTIPLEXING. Different frequency bands are used to convey information about the left and right channels in stereo radio systems and for the brightness of the red, green and blue parts of the picture, plus sound, in colour television. Time-division multiplexing is mainly used with DIGITAL signals, such as digitized speech signals relating to many different telephone conversations sent down a single optical fibre.

multiplier A RESISTANCE connected in series with a VOLTMETER to enable it to read higher voltages. The use of a multiplier has the advantage of increasing the resistance of the voltmeter.

mu-meson *See* MUON.

muon, *mu-meson* A charged LEPTON with a mass about 210 times that of the electron, to which it decays with a HALF-LIFE of about 2×10^{-6} s.

mutual inductance A measure of the effect whereby a changing current in one coil will induce an ELECTROMOTIVE FORCE in another coil. It is equal to the induced e.m.f. divided by the rate of change of current. The SI UNIT of mutual inductance is the HENRY. If two coils have a mutual inductance M, a rate of change of current dI/dt in one will produce an e.m.f. in the other of

$$E = M dI/dt$$

See also LENZ'S LAW.

myopia *See* SHORT-SIGHTEDNESS.

N

NAND gate (NAND = not and) A LOGIC GATE in which the output is HIGH unless all the inputs are high. This gate is simple to manufacture and can be used as a building block for many more complex devices.

nano- (n) A prefix indicating that a unit is to be multiplied by 10^{-9}. For instance, one nanometre (nm) is one billionth of a metre.

nanotechnology The technology of building machines, such as electric motors, and eventually whole robots, on a very small scale. Nanotechnology devices are typically only a few nanometres in size, and approach the limits that can be achieved given the sizes of individual atoms. Although promising for the future, nanotechnology is still very much at the experimental stage.

natural frequency The FREQUENCY at which an oscillating system will oscillate if it is displaced from EQUILIBRIUM and then allowed to oscillate without any further external forces. *See also* FORCED OSCILLATION, RESONANCE.

natural logarithm A LOGARITHM to BASE e. The natural logarithm of x is denoted by the symbol $\ln x$.

natural number Any positive whole number, such as 1, 2, 3 etc.

neap tide The TIDE produced when the gravitational effects of the Moon and Sun oppose each other, producing the smallest rise and fall in tide level.

negative In physics, the name given to one of the two types of electric charges. Electrons are negatively charged particles. Objects that are negatively charged have more electrons than protons.

negative feedback FEEDBACK that is OUT OF PHASE with the original signal, thus tending to reduce the GAIN of an AMPLIFIER. Negative feedback usually has the advantages of increasing the stability of an amplifier against external changes, such as temperature or supply voltage fluctuations. It also reduces DISTORTION. *See also* POSITIVE FEEDBACK.

neon lamp A light source using a GAS DISCHARGE

through neon. It gives a bright orange-red light, commonly used in advertising signs.

nematic (*adj.*) Describing a material containing long molecules that can be aligned by the application of an electric field. *See* LIQUID CRYSTAL.

Neptune The eighth most distant planet from the Sun, with an orbital radius of 30.1 AU (4,500 million km). It is the smallest and most distant of the GAS GIANTS, with a diameter of 48,000 km (3.8 times that of the Earth) and a mass of 1.0×10^{26} kg (17 times that of the Earth). Neptune takes 165 years to orbit the Sun, and rotates on its own axis in 16 days.

Little was known about Neptune until it was visited by the VOYAGER spacecraft in 1989. Images taken then discovered six new satellites, bringing the total to eight, and a faint RING SYSTEM. The atmosphere is composed largely of methane and hydrogen. Strong winds were detected, as was a dark spot similar to Jupiter's GREAT RED SPOT, though this may be a temporary feature.

Historically, Neptune is important because its existence was predicted in 1846, ahead of its discovery, by calculations based on unexplained disturbances in the orbits of other planets.

Nernst heat theorem Any chemical change involving pure crystalline materials, and taking place at ABSOLUTE ZERO will involve no change in ENTROPY.

neutral 1. (*adj.*) Describing an object with no electric CHARGE or that contains equal amounts of POSITIVE and NEGATIVE charge.

2. In a mains electricity system, a wire that remains close to EARTH potential throughout the a.c. cycle (*see* ALTERNATING CURRENT).

neutral equilibrium A state where the displacement of a system produces no forces on it, for example a ball lying on a flat surface.

neutrino Any one of a family of three light (probably massless) neutral LEPTONS. They are the electron neutrino, the muon neutrino and the tau neutrino. Each carries the same

ELECTRON NUMBER, muon number, etc., as whichever charged lepton it is associated with. Neutrinos have no charge, probably zero REST MASS, and move at the speed of light. Each neutrino has an associated antineutrino, the most familiar of which, the antiparticle of the electron neutrino, is emitted from the nucleus during BETA DECAY.

Neutrinos interact only via the WEAK NUCLEAR FORCE, so even in solid rock, they have a MEAN FREE PATH of many LIGHT YEARS. The neutrino was originally proposed as a hypothetical particle to explain the apparent violation of the LAW OF CONSERVATION OF ENERGY in BETA DECAY. Subsequently, neutrino interactions have been observed, including those from beta decay in NUCLEAR REACTORS, from the Sun and from SUPERNOVA explosions.

neutron The neutral particle found in the nuclei of all elements except hydrogen. It is slightly more massive than the PROTON. Outside the nucleus, the neutron has a mean life of about 12 minutes, before decaying into a proton, an ELECTRON and an antineutrino (*see* NEUTRINO). The neutron was first discovered in 1930 in a series of experiments in which beryllium was bombarded with ALPHA PARTICLES. This process released neutrons that were not detected directly, but which would knock PROTONS out of paraffin wax in ELASTIC collisions. *See also* ATOM, HADRON, ISOTOPE, QUARK, MASS NUMBER.

neutron degeneracy pressure The mechanism that supports NEUTRON STARS against further GRAVITATIONAL COLLAPSE. It is caused by the PAULI EXCLUSION PRINCIPLE acting on NEUTRONS in much the same way as it produces ELECTRON DEGENERACY PRESSURE. The densities at which neutron degeneracy pressure acts are those associated with the atomic nucleus, about 10^{15} times the density of ordinary solid matter.

neutron diffraction A technique similar to X-RAY DIFFRACTION, but using low energy NEUTRONS, which have a DE BROGLIE WAVELENGTH similar to the X-rays they replace. As the neutrons are scattered by the nuclei rather than by the electrons of the crystal, this technique can be used to obtain further information about crystal structures. *See also* WAVE NATURE OF PARTICLES.

neutron star An extremely dense collapsed star. If the active core of a star (which is about 10 per cent of its total mass) exceeds 1.4 solar masses, the NUCLEAR FUSION produces heavy atomic nuclei and energy starts to be absorbed by further fusion rather than released. This accelerates rather than opposes further collapse and the core shrinks rapidly. Protons and electrons fuse to form neutrons and the collapse is only halted by NEUTRON DEGENERACY PRESSURE with a density roughly 10^{15} times that of ordinary matter. The outer layers rebound off this very rigid core and the resulting explosion, in which the star suddenly increases its brightness many millions of times, is called a SUPERNOVA. The remnant of the core is called a neutron star. Neutron stars are often observed as PULSARS.

newton The derived SI UNIT of FORCE. Force is equal to the mass times the acceleration, so one newton is defined as the force that will make a mass of one kilogram accelerate at one metre per second per second.

Newtonian mechanics MECHANICS based on NEWTON'S LAWS OF MOTION, and not taking account of the effects of QUANTUM MECHANICS or RELATIVITY. Newtonian mechanics generally works well provided the systems being considered are large compared to the size of an atom and move at speeds small compared to the speed of light.

Newton's law of gravitation The gravitational attraction between two objects is proportional to their masses and inversely proportional to the square of the distance between their respective centres of mass. The gravitational force F between two masses m and M separated by a distance r is

$$F = GMm/r^2$$

where G is the GRAVITATIONAL CONSTANT.

Newton's laws of motion Three laws that together sum up the behaviour of objects under the influence of FORCES acting upon them. These laws were first put forward by Isaac Newton (1642–1727) in his famous work the *Principia* (1687). They formed the basis of the study of motion until the early years of the 20th century, when the SPECIAL THEORY OF RELATIVITY suggested that Newton's laws are an approximation to the truth, and are most valid for particles moving at speeds much less than the speed of light.

Newton's first law states that an object will remain at rest, or continue to move in a straight line at a steady speed, unless it is acted upon by an unbalanced external force. In other words if the vector sum of the forces acting on an object is zero, it will not accelerate.

Newton's second law states that the rate of change of MOMENTUM of a body is directly proportional to the force on it. If the object has a constant MASS, this is equivalent to saying that the force is directly proportional to the mass times the ACCELERATION. This introduces the idea of INERTIAL MASS as a measure of the amount of matter in an object: the greater the mass the less the motion of the object will be changed by a given force. In other words, mass is a measure of the INERTIA of a body, or its resistance to having its motion changed. For an object of mass m experiencing a force F, the acceleration will be *a*, with

$$F = ma$$

Newton's third law is usually stated as 'for every action there is an equal and opposite reaction', i.e. if object A exerts a force on object B , then object B will exert an equal but opposite force on object A.

Newton's rings A THIN-FILM INTERFERENCE effect seen when a CONVEX lens is placed on a flat glass plate, producing circular interference fringes. The effect was first described in 1704 by Isaac Newton (1642–1727). It arises from INTERFERENCE between light which is partially reflected from the bottom of the lens and the top of the glass plate.

Nicad *See* NICKEL-CADMIUM CELL.

nickel-cadmium cell (Nicad) A high capacity rechargeable electrochemical CELL, often used to power portable electrical appliances.

Nicol prism A device for polarizing light (*see* POLARIZED), comprising two triangular PRISMS made of a birefringent material, usually calcite (*see* BIREFRINGENCE). The two prisms are cemented together with a material which has a REFRACTIVE INDEX between the refractive indices of the birefringent material for the ordinary and extraordinary rays. The ordinary ray is totally internally reflected (*see* TOTAL INTERNAL REFLECTION) and passes out of the prism at a large angle to the original (unpolarized) ray, whilst the extraordinary ray passes out of the second prism parallel to its original direction.

node 1. A point at which two or more lines meet, or where a curve intersects itself.

2. A point of minimum AMPLITUDE in a STANDING WAVE. The distance from one node to the next is half the WAVELENGTH of the wave. *See also* ANTINODE.

noise Any unwanted sound and, by extension, any unwanted signal, such as in an electronic system. Noise that covers a broad spread of frequencies is called white noise.

non-Euclidean space *See* EUCLIDEAN SPACE.

non-inertial frame of reference Any one of two or more CO-ORDINATE systems where an observer stationary with respect to one of the co-ordinate systems would see an observer stationary with respect to any of the other co-ordinate systems as accelerating.

non-inverting amplifier An AMPLIFIER with a positive GAIN.

non-inverting input One of the inputs of a DIFFERENTIAL AMPLIFIER. Signals applied to this input will be amplified with a positive GAIN.

non-renewable (*adj.*) A term used to describe FOSSIL FUELS and other energy sources that are being consumed at a rate which far exceeds the production of new reserves. *See also* RENEWABLE RESOURCE.

NOR gate (NOR = NOT OR) A LOGIC GATE that implements the NOR function. A NOR gate has two or more inputs and a single output that is HIGH only if all the inputs are LOW.

normal An imaginary line at right angles to a surface.

normal distribution, *Gaussian curve* In statistics, a bell-shaped curve obtained when the

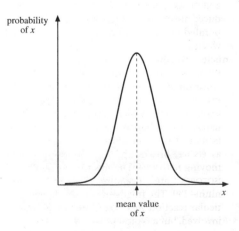

Normal distribution.

frequency distribution for a characteristic that shows continuous variation is plotted on a graph. It is the PROBABILITY distribution for random events centred on a MEAN $<x>$ and having a STANDARD DEVIATION σ, such that the probability of finding a value x is proportional to $\exp[(x-<x>)^2/\sigma]$. An example is the height of individuals in a population; most individuals are of intermediate height, with a few at each extreme. *See also* AVERAGE.

normal reaction The COMPONENT of the force of contact between the two surfaces that is at right angles to those surfaces.

northern lights *See* AURORA.

NOT gate *See* INVERTER.

npn transistor A JUNCTION TRANSISTOR in which the EMITTER and COLLECTOR are made of N-TYPE SEMICONDUCTOR, whilst the BASE is made of P-TYPE SEMICONDUCTOR.

n-type semiconductor A SEMICONDUCTOR in which the charge is carried predominantly by electrons rather than HOLES.

nuclear binding energy The energy released when an atomic NUCLEUS is formed from its constituent PROTONS and NEUTRONS. The binding energy is largest for the largest nuclei, but the binding energy per NUCLEON is greatest for iron–56. *See also* MASS DEFECT.

nuclear energy The energy released in processes involving the atomic NUCLEUS, particularly radioactive decay (*see* RADIOACTIVITY), NUCLEAR FUSION and NUCLEAR FISSION.

nuclear fallout The radioactive material produced after the NUCLEAR FISSION of material in a NUCLEAR WEAPON. Many of the elements produced are chemically active and may be incorporated into the food chain. *See also* NUCLEAR WINTER.

nuclear fission The splitting of a heavy atomic NUCLEUS into two smaller nuclei, with the emission of energy. Some nuclei show SPONTANEOUS FISSION, a form of radioactive decay (*see* RADIOACTIVITY), which takes place as an alternative to ALPHA DECAY. A more useful process is INDUCED FISSION, in which a nucleus, described as FISSILE, breaks up after absorbing a slow moving NEUTRON. The most important fissile ISOTOPES are uranium–238 and plutonium–239. The fragments produced in a particular reaction vary according to the isotopes involved, but a typical process would be

$$^{235}_{92}\text{U} + ^1_0\text{n} \rightarrow ^{148}_{57}\text{La} + ^{85}_{35}\text{Br} + 3^1_0\text{n}$$

Because lighter nuclei contain a smaller proportion of neutrons than heavy nuclei, a number of neutrons, usually two or three, are released in each fission. Lighter nuclei are also more tightly bound than heavy nuclei (*see* NUCLEAR BINDING ENERGY) and thus some energy is released in the process. Most of this energy is in the form of the KINETIC ENERGY of the neutrons.

In a NUCLEAR REACTOR, the energy of the neutrons is absorbed when they collide with a material called a MODERATOR, which contains light nuclei that do not absorb neutrons. The moderator becomes heated by the energy absorbed from the neutrons which, now moving more slowly, can go on to produce further fission and release more neutrons. The result is a CHAIN REACTION, where each step produces neutrons that can go on to produce further neutrons. In its uncontrolled form, this is the principle of the ATOMIC WEAPONS that were used against the Japanese by the Americans at the end of the Second World War.

See also THERMONUCLEAR REACTION.

nuclear fusion The bringing together of two light atomic nuclei to form a heavier one. Since the heavier nuclei, up to iron–56, are more tightly bound (*see* NUCLEAR BINDING ENERGY), energy will be released in this process. However, nuclei are positively charged so must be very fast moving to overcome the ELECTROSTATIC repulsion that tends to keep them apart. Thus nuclear fusion can only take place at extremely high temperatures. The right conditions exist for nuclear fusion in the cores of stars, and the fusion of hydrogen to helium, by a process called the PP CHAIN, is the mechanism by which the Sun produces the energy necessary for life on Earth. This fusion process involves the conversion of protons to neutrons, and therefore involves the WEAK NUCLEAR FORCE, so it happens much too slowly to be suitable as a source of energy for use on Earth.

In a HYDROGEN BOMB extreme temperatures are produced by triggering a NUCLEAR FISSION reaction. This fission explosion then triggers the fusion reaction, which releases energy so quickly that most of the nuclei have fused before the material of the bomb has time to expand. This method of triggering fusion is called INERTIAL CONFINEMENT.

A more likely method for the peaceful use of nuclear fusion is MAGNETIC CONFINEMENT. At the temperatures needed for fusion, matter is in the form of a PLASMA, a gas of electrons and nuclei. As this gas is made up of charged particles, they can be confined by magnetic fields in what is known as a magnetic bottle. Various arrangements of magnetic fields have been tried, but the most successful seems to be the TOKAMAK.

In the D-T REACTION, much of the energy released by the fusion is carried by the neutron, which, being uncharged, will escape the magnetic field and deposit its energy in the surroundings of the machine. One proposal is to surround the tokamak with a blanket of lithium – the neutron will react with the lithium nuclei to produce further TRITIUM, thus 'breeding' new fuel. The energy released by the neutron would be used to heat water to drive a steam turbine to generate electricity in the conventional manner.

Another form of fusion experiment which has had some success in recent years is pellet fusion, in which a small metal pellet containing deuterium and tritium is heated by bombardment on all sides with high energy laser beams. Whilst this appears promising on a small scale, it is less clear how the arrangement could be scaled up to give the continuous power output that is proposed for larger tokamaks. *See also* COLD FUSION.

nuclear magnetic resonance (NMR) The absorption of energy from an oscillating electromagnetic field by the nucleus of an atom, such as hydrogen. All atomic nuclei have a MAGNETIC MOMENT resulting from their CHARGE and SPIN. Aligning this moment in a strong, constant magnetic field produces a number of ENERGY LEVELS for the nucleus. If a weaker, oscillating field is then applied, the nuclei will absorb energy, provided the frequency of oscillation corresponds to the difference in energy between two energy levels. Once this energy has been absorbed, the nuclei 'relax', or fall back to lower energy levels, at a rate dependent on the chemical environment of the atom, again emitting electromagnetic radiation. Nuclear magnetic resonance is used with hydrogen nuclei both as a tool for discovering the structure of organic molecules, and as an imaging technique in medicine. *See also* MAGNETIC RESONANCE IMAGING.

nuclear magneton A unit of MAGNETIC MOMENT equal to 5.05×10^{-27} Am^2. *See* MAGNETON.

nuclear physics The branch of physics that concerns itself with the atomic NUCLEUS, in particular the ENERGY LEVELS of NUCLEONS within the nucleus, collisions between nuclei and the processes of RADIOACTIVITY.

nuclear reaction Any process in which an atomic NUCLEUS is changed in some way. In particular, a process triggered by a nucleus being struck by some incoming particle, often a PROTON, NEUTRON or PHOTON of GAMMA RADIATION. Nuclear reactions differ from chemical reactions in that the energies involved are often a significant fraction of the REST MASS energies of the nuclei, so the total rest mass present before and after the reaction may be significantly different. *See also* EQUIVALENCE OF MASS AND ENERGY, NUCLEAR FISSION, NUCLEAR FUSION, THERMONUCLEAR REACTION.

nuclear reactor A sealed vessel containing FISSILE material together with CONTROL RODS and a MODERATOR and some means of extracting the heat produced in the moderator as it absorbs the KINETIC ENERGY of the neutrons released in NUCLEAR FISSION.

A CHAIN REACTION is usually controlled by using control rods made of a neutron-absorbing material such as boron or cadmium. If the nuclear reactor is cooled by passing a suitable fluid through it, the heat extracted can be used to make steam for an electrical generator – this is the basis of the nuclear power station. The energy released by each fission is many million times greater than the energy released in any single chemical reaction, thus the energy yield for a given mass of fuel is far greater than could be obtained from conventional fuels such as coal or oil.

The nuclei produced by the fission process tend to have too many neutrons to be stable so are often radioactive. In the case of a NUCLEAR WEAPON this material is called NUCLEAR FALLOUT, in a power station it is the RADIOACTIVE WASTE.

See also FAST BREEDER REACTOR.

nuclear weapon Any device that uses the energy released from NUCLEAR FISSION or (more commonly) NUCLEAR FUSION, for destructive purposes. Nuclear weapons were originally designed as bombs to be dropped from aircraft, but now also exist in the form of missiles and artillery shells. *See also* NUCLEAR FALLOUT, NUCLEAR WINTER.

nuclear winter A hypothetical period of prolonged cold weather following a war in which widespread use is made of NUCLEAR WEAPONS. It has been suggested that such a war would put sufficient dust into the atmosphere to have significant effects on the climate.

nucleon A particle found in an atomic NUCLEUS: a NEUTRON or a PROTON.

nucleon number *See* MASS NUMBER.

nucleus (*pl. nuclei*) The positively charged massive centre of an ATOM. It is made up of particles called PROTONS, which are positively charged, having a charge equal in size to the negative charge on an ELECTRON, and NEUTRONS, which have no charge and have slightly more mass than the protons. Protons and neutrons are collectively called nucleons. The neutrons and protons are both far more massive than the electrons that surround the nucleus, but the nucleus is far smaller than the atom itself. A typical atom is 10^{-10} m in diameter, whilst a nucleus is 10^{-14} m across.

The number of protons in the nucleus determines the number of electrons needed to produce a neutral atom. It is the arrangement of these electrons that determines the chemical properties of an ELEMENT. Thus different numbers of protons in the nucleus produce atoms of different elements. The number of neutrons in a nucleus has no affect on the chemical properties, but does affect the mass of the atom. The number of protons in a nucleus is called the ATOMIC NUMBER (Z). The total number of nucleons is called the MASS NUMBER (A). A NUCLIDE may therefore be represented by the notation $_Z^A X$ where X is the element.

The existence of a nucleus was first proposed by Ernest Rutherford (1871–1937) as an alternative to the CURRANT-BUN MODEL of the atom, and was proved by the RUTHERFORD SCATTERING EXPERIMENT. Although most nuclei found in nature are stable, Marie Curie (1867–1934) discovered some elements, several of them previously unknown, which are radioactive. It is now known that RADIOACTIVITY arises from changes in the nucleus. More recently the development of the NUCLEAR REACTOR has led to the discovery of many new unstable nuclei.

See also LIQUID DROP MODEL, SHELL MODEL.

nuclide An atomic NUCLEUS identified as having a particular number of NEUTRONS and PROTONS. Thus nuclei of different ISOTOPES or different elements are different nuclides.

O

objective *See* OBJECT LENS.

object lens, *objective* In a REFRACTING TELESCOPE, MICROSCOPE or BINOCULARS, the lens that collects light from the object being viewed.

obtuse angle Any angle between 90 and 180 degrees.

oceanography The study of the physical, chemical, geological and biological features of the ocean.

octahedral (*adj.*) Having the shape of an octahedron; that is, a figure with eight triangular faces, each side having the same length.

octahedron A POLYHEDRON with eight plane faces.

octave An interval in PITCH corresponding to a doubling in frequency of the sound.

oersted (Oe) Unit of MAGNETIC FIELD strength in the C.G.S. SYSTEM. One oersted is equal to 79.58 AMPERES per metre (Am^{-1}).

ogive A graph showing how the CUMULATIVE FREQUENCY of a set of data increases over the range of the data. For example, in a statistical analysis of the heights of individuals in a given population, the ogive will be the graph of the total number of individuals having a height less than the specified value. For a

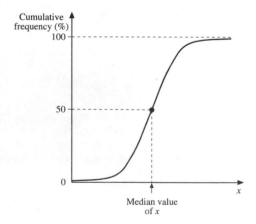

Ogive.

NORMAL DISTRIBUTION, the ogive climbs slowly at first, then more rapidly before levelling out as it approaches 100 per cent of the population.

ohm (Ω) The SI UNIT of RESISTANCE. One ohm is the resistance of a conductor that requires a POTENTIAL DIFFERENCE of one VOLT across it to make a CURRENT of one AMPERE flow.

ohmic (*adj.*) Obeying OHM'S LAW.

ohmmeter A device for measuring RESISTANCE. A simple ohmmeter contains a source of ELECTROMOTIVE FORCE in series with a MOVING-COIL GALVANOMETER and a VARIABLE RESISTOR, which is adjusted so the galvanometer reads full scale when there is no other resistance in the circuit. Any additional resistance reduces the current, leading to a non-linear scale reading in the opposite sense to a normal scale. More modern instruments provide a direct digital read-out.

Ohm's law For some materials, under constant physical conditions, the CURRENT flowing is proportional to the POTENTIAL DIFFERENCE, i.e. the RESISTANCE is constant. Many conductors, particularly metals, follow this law provided they are kept at a constant temperature. Such materials are said to be ohmic. Non-ohmic conductors have a resistance that varies with the current flowing through them.

oil-drop experiment *See* MILLIKAN'S OIL DROP EXPERIMENT.

omega-minus particle A BARYON containing three STRANGE QUARKS. It has a mass 1.8 times that of the proton and a HALF-LIFE of 8.2×10^{-9} s, decaying by the WEAK NUCLEAR FORCE. Discovered in 1964, the omega-minus particle was an important success for the model of particle physics which eventually led to the discovery of quarks.

Oort cloud A region around the edge of the SOLAR SYSTEM, from where COMETS are believed to originate.

op amp *See* OPERATIONAL AMPLIFIER.

opaque (*adj.*) Describing a material which does not permit light (or some other specified form of electromagnetic radiation) to pass through it. *Compare* TRANSPARENT, TRANSLUCENT.

open circuit A circuit in which there is some break, either introduced deliberately – such as by a switch – or as the result of a fault.

open loop gain The GAIN of an AMPLIFIER, particularly an OPERATIONAL AMPLIFIER, which has no FEEDBACK.

operational amplifier (*abbrev. op amp*) The key building-block of many ANALOGUE electronic circuits. An operational amplifier is a type of INTEGRATED CIRCUIT that forms a high gain DIFFERENTIAL AMPLIFIER, with an OPEN LOOP GAIN of typically several million. The operational amplifier also has a very high input resistance, so virtually no current flows into either input. The output voltage cannot exceed the supply voltage, so in many applications, the two inputs are at virtually the same ELECTRIC POTENTIAL. The NON-INVERTING INPUT is often connected to EARTH (zero potential), so the INVERTING INPUT is close to earth potential, and forms what is called a VIRTUAL EARTH, a point in a circuit that although not connected to earth, can always be taken as being at earth potential.

Most operational amplifier circuits also involve NEGATIVE FEEDBACK, in which some of the output is fed back to the inverting input. This reduces the gain of the circuit but produces a system with properties that do not depend on the detailed performance of the operational amplifier itself, and so are not affected by changes caused by changes in temperature, ageing, etc.

See also AMPLIFIER.

optical activity The ability of certain substances to rotate the plane of polarized light (*see* POLARIZATION). An optically active substance may be a crystal, liquid or solution. Optical activity arises from a lack of symmetry in the three-dimensional structure of the molecules (or crystal lattices) concerned. *See also* POLARIMETER.

optical axis *See* PRINCIPAL AXIS.

optical binary *See* BINARY STAR.

optical centre The point at which the PRINCIPAL AXIS meets a lens. It is the thickest point in a CONVERGING LENS; the thinnest point in a DIVERGING LENS.

optical fibre *See* FIBRE OPTIC.

optic nerve A large nerve between the eye and the brain. It carries information from the

sensory cells in the RETINA to the visual centres in the brain.

optics The study of visible light. Optics is usually divided into ray, or geometrical, optics, which deals with the phenomena of REFRACTION and REFLECTION, and wave optics, which deals with DIFFRACTION and INTERFERENCE.

orbit The path taken by an object, called a SATELLITE, that is moving under the influence of the GRAVITY of some larger object, such as a star or planet. If the total energy of the orbiting satellite is positive, it will have enough KINETIC ENERGY to escape completely from the gravitational influence and the orbit will be open, with the satellite curving around the source of gravity before travelling on into space. If the total energy is negative, so the GRAVITATIONAL POTENTIAL ENERGY more than outweighs the kinetic energy, the orbit will be closed and the satellite will trace out a closed path, either an ELLIPSE (a flattened circle) or a circle.

Orbital motion was first studied by Johannes Kepler (1571–1630), who put forward three laws to describe the orbits of planets around the Sun (*see* KEPLER'S LAWS). Kepler did not fully understand the force of gravity that was responsible for these orbits; this was explained by Isaac Newton (1642–1727), who studied the orbit of the Moon, realizing that, like an apple falling from a tree, it was in FREE FALL. By calculating the CENTRIPETAL ACCELERATION of the Moon, Newton was able to find how the force of gravity varied with distance. *See also* CIRCUMPOLAR ORBIT, GEOSTATIONARY ORBIT.

orbital A region of space in an ATOM or MOLECULE that can be occupied by a maximum of two electrons. An orbital is often thought of as the volume that encloses the space within which the probability of finding an electron is greater than a specified figure, since in principle, electron WAVEFUNCTIONS extend out infinitely far from the atom.

Two electrons in a single orbital often form a stable configuration called a LONE PAIR, whilst a single electron is often active in the formation of chemical bonds. Orbitals may overlap and share electrons between atoms to form a COVALENT BOND, or an orbital may gain or lose an electron so that the atom or molecule forms an ION.

Atomic orbitals are labelled by a number in the sequence 1, 2, 3 etc., called the PRINCIPAL QUANTUM NUMBER, with higher numbers

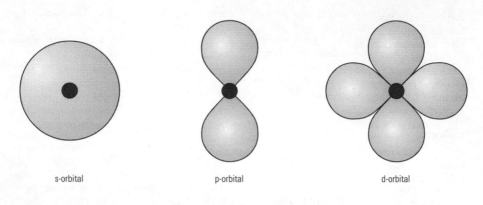

s-orbital   p-orbital   d-orbital

Electron orbitals.

indicating higher energies. The principal quantum number indicates the size of the orbital. They are also labelled by a letter, s, p, d, f, called the SUBSIDIARY QUANTUM NUMBER. This denotes the shape of the orbital and the ANGULAR MOMENTUM of an electron in that orbital. For a principal quantum number of 1, only an S-ORBITAL exists. For principal quantum number of 2 there is one s-orbital and three P-ORBITALS. With a principal quantum number of 3 there is a single s-orbital, three p-orbitals and five D-ORBITALS, and so forth. The energy of the orbitals, and hence the sequence in which they are filled by electrons is basically

1s, 2s, 2p, 3s, 3p, 4s, 3d, 4p, 5s, 4d, 5p,
6s, 4f, 6p, 7s, 5f

The order in which electrons go into these orbitals to form neutral atoms is responsible for the structure of the PERIODIC TABLE.

See also HUND'S RULE OF MAXIMUM MULTIPLICITY, HYBRID ORBITAL, MOLECULAR ORBITAL, PAULI EXCLUSION PRINCIPLE, SHELL.

order 1. (*physics*) The value of the number n in DIFFRACTION and DIFFRACTION GRATING equations, such as

$$d\sin\theta = n\lambda$$

where d is the width of the diffraction slits, θ is the angle of the diffraction and λ is the wavelength of light).

2. (*mathematics*) In an equation containing DERIVATIVES of a function, the largest number of times any function is differentiated (*see* CALCULUS). Thus an equation in which some quantity is differentiated twice, but no quantity is differentiated more than twice, is called a second order DIFFERENTIAL EQUATION.

ordinary ray *See* BIREFRINGENCE.

OR gate A LOGIC GATE with two or more inputs and an output that is high if any of the inputs is high.

origin In a GRAPH, or any system of CO-ORDINATES, the point at which the axes intersect, where the value of all the co-ordinates is zero.

orographic (*adj.*) Describing the formation of cloud and sometimes rain when moist air is forced upwards by a land mass. Orographic effects account for the higher than average rainfall on the side of any hills that face the PREVAILING WIND, and the dry area, called the RAIN SHADOW on the lee (downwind) side of the prevailing wind.

orthorhombic (*adj.*) Describing a crystal structure in which the UNIT CELL has all its faces at right angles to one another, with all the faces rectangular, but none square, so there are three different lengths characterizing the size of the unit cell.

oscillation A motion that repeats itself at regular intervals. The time taken for one complete oscillation is called the period, whilst the number of oscillations in one second is called the frequency. The unit of frequency is the HERTZ (Hz).

In order to oscillate, a system must have a position of STABLE EQUILIBRIUM, so that when displaced from the equilibrium position, the system will experience a force tending to return it to equilibrium. The mass of the system means that it will overshoot its equilibrium position and move away from equilibrium in the opposite direction and so on.

For many systems it is the case, at least to some degree of approximation, that the restoring force tending to return the system to equilibrium is directly proportional to the displacement from the equilibrium position. The motion is said to be ISOCHRONOUS, i.e. the period does not depend on the amplitude. In this case the system is said to exhibit SIMPLE HARMONIC MOTION, with the displacement from equilibrium varying with the SINE or COSINE function.

In any oscillation there is an interchange between KINETIC ENERGY and POTENTIAL ENERGY – GRAVITATIONAL POTENTIAL in the case of the SIMPLE PENDULUM and ELASTIC potential in the SPRING PENDULUM. In each case the kinetic energy is a maximum and the potential energy a minimum as the pendulum passes through its equilibrium position, and the kinetic energy is zero and the potential energy a maximum at the extremes of the oscillation. The total energy is constant throughout the oscillation.

For simple harmonic motion with a displacement x from equilibrium, moving with a velocity v and an acceleration a:

$$a = -\omega^2 x$$

where ω is the ANGULAR FREQUENCY. If the motion is started by releasing the system from rest, then after time t:

$$x = A\cos\omega t$$

where A is the amplitude. The velocity at this point is:

$$v = -\omega A\sin\omega t$$

In simple harmonic motion, the maximum velocity is ωA and is reached as the system passes through its equilibrium position. The acceleration after time t is given by:

$$a = -\omega^2 A\cos\omega t$$

The maximum acceleration is $\omega^2 A$. *See also* DAMPING, FORCED OSCILLATION, RESONANCE.

oscillator A circuit for generating ALTERNATING CURRENT, particularly at high frequency for the generation of a CARRIER WAVE in a radio system. The frequency of the oscillator is fixed by some resonant device (*see* RESONANCE), which

may be a PIEZOELECTRIC quartz crystal with a mechanical resonance at the desired frequency, or a TUNED CIRCUIT containing an INDUCTANCE and a CAPACITANCE. In the latter case, the current in the inductance and the capacitance will be exactly OUT OF PHASE, and at one particular frequency they will also be equal in magnitude. Thus the circuit will appear to have a very high resistance at a single frequency. An AMPLIFIER circuit with POSITIVE FEEDBACK is used, so some of the output is fed back and further amplified. Such a circuit will oscillate if it has sufficient GAIN, the tuned circuit can be used to fix the frequency at which this feedback will be greatest. *See also* CRYSTAL OSCILLATOR.

oscilloscope *See* CATHODE RAY OSCILLOSCOPE.

osmosis The movement of a liquid solvent, usually water, from a less concentrated solution to a more concentrated solution through a SEMIPERMEABLE MEMBRANE (one that is permeable in both directions to the solvent but varying in permeability to the solute) until the two concentrations are equal or isotonic.

If external pressure is applied to the more concentrated solution, osmosis is prevented and this provides a measure of the OSMOTIC PRESSURE or osmotic potential of the more concentrated solution, which is measured in pascals (Pa). The osmotic pressure is greater the more concentrated the solution.

The less dilute solution is called hypotonic, and the more concentrated solution is called hypertonic. The passage of water by osmosis will occur across a semipermeable membrane from any solution of weaker osmotic pressure to one of higher osmotic pressure, regardless of whether the dissolved substance on both sides of the membrane is the same or not. Osmosis is vital in controlling the distribution of water in living organisms, for example in the transport of water from the roots up to the stems of plants.

osmotic potential *See* OSMOTIC PRESSURE.

osmotic pressure, *osmotic potential* The pressure difference that can occur across a SEMIPERMEABLE MEMBRANE as a result of OSMOSIS. It is defined as the pressure that needs to be applied across a semipermeable membrane to prevent osmosis. KINETIC THEORY shows that the osmotic pressure is proportional to the concentration of the solute and to the ABSOLUTE TEMPERATURE of the liquid.

Otto cycle *See* FOUR-STROKE CYCLE.

out of phase Describing two OSCILLATIONS that are not exactly in step with one another or that are exactly out of step with one another.

overdamped (*adj.*) Describing a system that could oscillate if allowed to move freely, but where the level of DAMPING is too large to permit this. *See also* UNDERDAMPED.

overtone Frequencies of STANDING WAVES higher than the FUNDAMENTAL frequency for the system in question. The next highest frequency after the fundamental is called the first overtone, the next frequency above this is the second overtone, etc. With a standing wave on a stretched string, there is a NODE at each end, and the overtones will include all the HARMONICS.

In a pipe that is open at only one end, such as in a wind instrument, there is a node at one end and an ANTINODE at the other end. The node in air pressure is at the open end, but the node in air displacement is at the closed end. In each case there is an antinode at the opposite end. Such a pipe will produce overtones containing only odd harmonics and thus a sound which is musically different from that produced by a string. A pipe that is open at both ends, however, will produce all the harmonics.

ozone hole An area of lower than usual ozone concentration in the OZONE LAYER above the Earth's poles. *See* ATMOSPHERE.

ozone layer A protective layer consisting of the gas ozone, O_3, 15 to 40 km above the Earth's surface. It is formed by the effect of ultraviolet (UV) radiation on oxygen molecules. UV light splits oxygen (O_2) molecules into two atoms, one of which then combines with oxygen to create ozone. The ozone layer prevents harmful UV radiation reaching the Earth's surface, but in recent years it has become clear that the layer is being damaged by human activities.

P

Pangaea A single large land mass that is believed to have broken up into the TECTONIC PLATES that form the continents we observe today.

parabola A CONIC SECTION formed when the plane intersecting the cone is parallel to a straight line drawn on the surface of the cone. A parabola can also be represented by any equation of the form

$$y = ax^2 + bx + c$$

where a, b and c are constants.

parabolic dish An AERIAL that includes a metal or wire mesh surface with a cross-section which is part of a PARABOLA. This surface acts as a reflector, focusing incoming MICROWAVES onto a smaller detector or forming a narrow beam of outgoing radiation. Large dishes are used in RADIO TELESCOPES, whilst smaller ones are often found in RADAR and DIRECT BROAD-CAST SATELLITE systems. To be effective, the radius of the dish must be large compared to the WAVELENGTH of the waves used, otherwise the beam will spread as a result of DIFFRACTION. For this reason, such systems can only be used with wavelengths shorter than about 1 m.

parabolic mirror A MIRROR with a reflecting surface having a shape with a cross-section that is part of a PARABOLA (this shape is called a PARABOLOID of revolution). See CURVED MIRROR.

paraboloid A three-dimensional surface that has a parabolic (see PARABOLA) cross-section in one direction and a circular cross-section in the perpendicular direction.

parallel (*adj.*) **1.** (*electricity*) Describing electric devices that are connected in such a way that current can return to the power supply by flowing through one device or another. The voltage across each element in a parallel circuit will be the same, whilst the total current flowing will be equal to the sum of the currents flowing in the individual elements.

2. (*geometry*) Describing two lines or planes, such that the closest distance from a point on one line or plane to the other line or plane is always the same, regardless of which point is chosen.

parallel axis theorem An equation that enables the MOMENT OF INERTIA of an object about a given axis to be calculated from its moment of inertia about a second axis parallel to the first axis and passing through the CENTRE OF MASS of the object. If the moment of inertia about the axis passing through the centre of mass is I, and the moment of inertia about the parallel axis is I' with this axis passing a distance r from the centre of mass, then if the mass of the object is M,

$$I' = I + Mr^2$$

parallelogram A geometrical figure with four straight sides, with pairs of sides parallel to one another and with parallel sides having equal lengths. Unlike a rectangle, the two pairs of sides are not at right angles to one another. The area of a parallelogram is equal to one half the length of one of the sides multiplied by the perpendicular distance between the two sides having this length.

paramagnetism A weak form of magnetism found in some elements and molecules (such as O_2), causing these materials to have a RELA-TIVE PERMEABILITY slightly greater than 1. A paramagnetic material will align parallel to any applied magnetic field.

Paramagnetic materials contain electrons with a total ANGULAR MOMENTUM that is not zero. This causes each atom or molecule to behave as a MAGNETIC DIPOLE and these dipoles will line up in an external magnetic field. The alignment is destroyed by thermal vibrations, so paramagnetism generally decreases with increasing temperature.

parameter A quantity in a mathematical equation on which the other variables depend, but which is kept constant while the other variables are being investigated. The calculation may later be repeated with a different value for the parameter. See also PARAMETRIC EQUATION.

parametric equation A mathematical equation

in which a number of variables are expressed in terms of one or more PARAMETERS. Thus for a particle moving in a circle, the position of the particle may be expressed in terms of the parameter t, with

$$x = r cos\omega t, y = r sin\omega t$$

Calculating x and y for just one value of t gives a single point on the circle. By using all values of t, the whole of the path is plotted.

parity 1. A description of how a physical system or quantity compares with its own mirror image. If a system is unchanged on reflection it is said to have even parity. If it is reversed, the parity is odd. For example, the RIGHT HAND GRIP RULE would become a left hand grip rule if reflected, and therefore has odd parity. It was once believed that all the laws of physics would be basically unchanged under such a rule, though some quantities, such as magnetic fields, called pseudovectors, might reverse their direction. However, in 1956 Lee and Yang discovered that this is not true in the case of BETA DECAY.

2. In computing, a scheme for detecting and correcting errors. The sum of all the BITS in each piece of data must be an even or an odd number. An extra bit, called the parity bit, is added, which is zero if the sum is even and one if the sum is odd. The scheme enables single-bit errors to be detected.

parsec A unit of distance used in astronomy. One parsec is the distance of a star that makes an angle of one arc second ($\frac{1}{3600}$ degrees) between the Earth and the Sun. One parsec is equivalent 3.09×10^{13} km.

partial pressure In a mixture of gases, that part of overall pressure that can be attributed to the presence of one specified gas in the mixture; that is, the pressure that it would exert if it were alone. See also DALTON'S LAW OF PARTIAL PRESSURES.

particle accelerator A device designed to accelerate charged particles to high energies so that they can be fired at targets or made to collide with particle beams moving in the opposite direction. An electric field is used to accelerate the particles. In modern machines the particles will pass through this electric field many times over. Magnetic fields are used to focus the particle beam, and in some machines to steer the beam around a circular path. See also COLLIDING BEAM EXPERIMENTS, CYCLOTRON, LINEAR ACCELERATOR, SYNCHROCYCLOTRON, SYNCHROTRON.

particle detector A device used in particle physics to detect and track the paths of SUBATOMIC PARTICLES. They are used to identify new particles produced when fast-moving electrons or protons hit a target or an oncoming beam of POSITRONS or antiprotons. Many such experiments use a combination of detectors combined with shielding through which only certain types of particle (such as MUONS) can pass. A common feature of most detectors is that they operate in a magnetic field so the CHARGE and MOMENTUM of the particles can be deduced from the curvature of the tracks they leave in the detector. See also BUBBLE CHAMBER, CLOUD CHAMBER, DRIFT CHAMBER, PARTICLE ACCELERATOR, SPARK CHAMBER.

particle physics The study of ELEMENTARY PARTICLES. See also GRAND UNIFIED THEORY, QUANTUM THEORY, STANDARD MODEL.

pascal (Pa) The SI UNIT of PRESSURE. One pascal is equivalent to a pressure of one NEWTON per metre squared ($N m^{-2}$).

Pascal's law In any confined fluid, pressure is transmitted uniformly throughout the fluid and acts at right angles on any surface in contact with that fluid. See also HYDROSTATIC PRESSURE.

passive device An electronic device, such as a RESISTOR, CAPACITOR or INDUCTOR, that cannot amplify a current. The behaviour of passive devices can be described fully by simple mathematical equations. See also ACTIVE DEVICE.

path difference The difference in the length of two paths by which light beams travel from a source to a point where they overlap and interfere (see INTERFERENCE). Path difference is often expressed as a number of wavelengths of the light in the material concerned (this is sometimes called an optical path difference). One important example of interference is the case where two waves start in PHASE and interfere after travelling along different paths. For CONSTRUCTIVE INTERFERENCE the path difference must be either zero or a whole number of wavelengths. For DESTRUCTIVE INTERFERENCE, the path difference must be a whole number of wavelengths plus an odd half-wavelength.

Pauli exclusion principle A consequence of QUANTUM THEORY, formulated by Wolfgang Pauli (1900–58), which states that no two fermions may occupy the same quantum state. This has applications in the arrangements of electrons in atoms, where each ORBITAL contains only two electrons, each in one of the two possible SPIN states, described as spin up and spin down, depending on the direction of the spin ANGULAR MOMENTUM relative to some reference direction.

p.d. *See* POTENTIAL DIFFERENCE.

pendulum A system in which a point-like mass exhibits oscillatory motion. *See* SIMPLE PENDULUM.

perigee For an object in orbit around the Earth, the position in the orbit that is closest to the Earth.

perihelion For an object in orbit around the Sun, the position in the orbit that is closest to the Sun.

period 1. The time taken to complete an ORBIT or OSCILLATION.

2. A horizontal row of elements in the PERIODIC TABLE, containing a sequence of elements of consecutive atomic number and steadily changing chemical properties.

periodic function Any mathematical function that takes on the same value repeatedly for regular intervals of the independent variable. In particular the trigonometric functions, SINE, COSINE, TANGENT etc., have a period of 360° (or 2π radians). So

$$\sin(\theta + 360°) = \sin\theta$$

for any θ.

periodic table An arrangement of the ELEMENTS in order of increasing ATOMIC NUMBER that brings out the similarities between them. Atomic number increases along the rows of the table (called periods) and the table is arranged so that elements with similar properties are placed in the same column or group. Thus the left-hand column of the table, group 1 (formerly group I), contains the alkali metals, lithium, sodium, potassium etc., each of which is a metal that reacts violently with water to form a compound containing a singly charged positive metal ion. The next column, group 2 (formerly group II), contains the alkaline earths – metals that are reactive enough to form oxides on exposure to air, but which react only slowly with water. To the right of the table are group 17 and group 18 (formerly groups VII and VIII). Group 17 contains the halogens – reactive elements that tend to form singly charged negative ions, whilst group 18 contains the NOBLE GASES, with a full outer

SHELL of electrons, which are the least reactive of the elements.

In the middle of the table are the transition metals, which show only slight changes in properties with increasing atomic number, as electrons fill the D-ORBITALS.

The periodic table also contains two long series of elements, the lanthanides and the actinides. Within each of these series the chemical properties are generally very similar, with each element differing only in the number of electrons in f-orbitals.

periscope A device that enables objects to be viewed over or around some obstacle preventing direct viewing – such as the ocean surface in the case of a submarine periscope. A pair of totally internally reflecting prisms (*see* TOTAL INTERNAL REFLECTION) are used to bend the light through 90° at the top and bottom of the periscope tube.

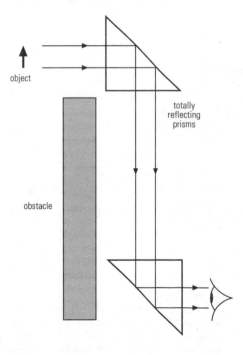

Periscope.

permanent gas A GAS that cannot be turned into a liquid by the action of pressure alone. Oxygen and nitrogen are examples.

permanent magnet A MAGNET that retains its magnetism indefinitely. The atoms in a permanent magnet each behave like a small BAR MAGNET, due to the imbalance in the rotational motion of their electrons – effectively an electric current within the atom. In FERROMAGNETIC materials these atoms are all aligned in a single direction over a region of space called a DOMAIN. In an unmagnetized piece of a ferromagnetic material, neighbouring domains are magnetized in different directions, so the total magnetic effect is zero. Under the influence of an external magnetic field, the domains which are aligned in the same direction as the field tend to grow, producing INDUCED MAGNETISM. Once all the domains are aligned the material cannot be magnetized any further – it is then described as saturated. In some materials, such as pure iron, that are described as magnetically soft, the domains do not retain their alignment when the applied field is removed.

The amount of magnetism remaining when the applied field is removed is called the REMANENCE, whilst the amount of reverse magnetic field needed to demagnetize a magnetized material is called the COERCIVE FORCE. Magnetically soft materials are used in ELECTROMAGNETS and TRANSFORMER cores, where the direction of magnetization must constantly be reversed without any unnecessary waste of energy. The alignment of domains can be destroyed by heating – the temperature at which a material loses its ferromagnetic properties is called the CURIE POINT.

The removal of any permanent magnetism, called DEGAUSSING, can also be achieved by moving the material through a SOLENOID in which an ALTERNATING CURRENT is flowing. As the sample moves past the solenoid, it is magnetized first in one direction then the other, with successive magnetizations becoming weaker and weaker.

permeability (μ) A measure of the degree to which a material may be magnetized. The absolute permeability μ is the ratio of MAGNETIC FLUX density B induced in a medium to the MAGNETIC FIELD strength H of the external field inducing it:

$$\mu = B/H$$

The RELATIVE PERMEABILITY μ_r is the ratio of the magnetic flux density of a material to that induced in free space μ_0 by the same magnetic field strength:

$$\mu_r = \mu/\mu_0$$

The quantity μ_0 is known as the permeability of free space and has the value $4\pi \times 10^{-7}$ HENRY per metre. *See also* AMPERE.

permittivity (ε) The ratio of the ELECTRIC FLUX density induced a material to the external ELECTRIC FIELD strength inducing it. The permittivity of free space ε_0 is a fundamental constant that measures the strength of the ELECTROSTATIC force in a vacuum. It is given in the equation:

$$F = Q_1 Q_2 / 4\pi\varepsilon_0 r^2$$

where F is the force between two charges Q_1 and Q_2 separated by a distance r in a vacuum. This equation is a statement of COULOMB'S LAW. ε_0 has the value 8.85×10^{-12} Fm^{-1}. *See also* RELATIVE PERMITTIVITY.

perpendicular (*n., adj.*) A line at an angle of 90° (at right angles) to a specified plane or line. The point at which the perpendicular meets the line or plane is sometimes called the foot of the perpendicular.

perpendicular bisector A line that is at right angles to a line joining two points and which divides that line into two sections of equal length.

perpetual motion The state of some hypothetical machine which, once set in motion, would continue to move for ever without any further energy input.

The existence of friction and other dissipative forces make perpetual motion impossible on a macroscopic scale, though electrons in atoms, for example, do represent a form of perpetual motion. Systems such as these are sometimes described as showing perpetual movement rather than perpetual motion, since the original idea was that perpetual motion machines might be used as sources of energy. Once the LAW OF CONSERVATION OF ENERGY was understood, interest in such devices declined.

petrol engine An INTERNAL COMBUSTION ENGINE burning petrol. Petrol engines are mostly used for smaller vehicles, such as cars, small vans and light aircraft.

A mixture of petrol and air is drawn into the engine by a PISTON moving down a CYLINDER during the INDUCTION STROKE. It is then compressed, and thus heated, as the piston

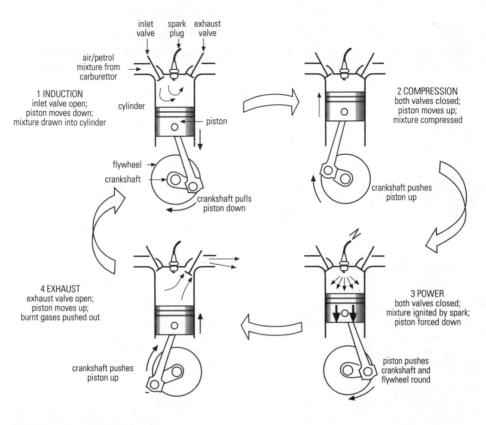

Cycle of operation of the petrol engine.

moves back up the cylinder during the COM-PRESSION STROKE. During this time the fuel vaporizes, and as the piston reaches the top of the stroke it is ignited by a spark plug, to which a high voltage from an INDUCTION COIL is connected. The expanding, burning gases force the piston back down the cylinder for the POWER STROKE, and as the piston returns up the cylinder, the burnt gases are pushed out of the cylinder in the EXHAUST STROKE. The cycle then repeats. This sequence is called the four-stroke cycle, since the piston moves up and down the cylinder four times for each load of fuel burnt. Many engines contain several cylinders, typically four for a car engine, to produce a more even supply of power.

See also COMPRESSION RATIO, DIESEL ENGINE, KNOCKING, PRE-IGNITION, TURBOCHARGING, TWO-STROKE CYCLE.

petrology The branch of GEOLOGY that deals with rocks, in particular their formation and chemical and physical structure.

phase 1. A measure of the stage that an OSCILLA-TION has reached at a given instant, particularly when comparing two oscillations. Phase is usually expressed as an angle, with the complete oscillation being represented by 360°, or 2π radians. Two oscillations moving together have a phase difference of 0 and are said to be in phase, whilst two oscillations exactly out of step with one another, so they are always moving in opposite directions, are said to be exactly out of phase, or to have a phase difference of 180° or π radians.

2. Any one of the different arrangements in which the molecules of a certain substance may exist, as a GAS, LIQUID or as one or more SOLID forms.

3. The proportion of the Moon or a planet, such as Venus or Mercury, that is illuminated by the Sun as seen from the Earth. The phase is usually expressed as a percentage, or using the terms New Moon (0 per cent), Full Moon (100 per cent), First Quarter (50 per cent, between New and Full) and Last Quarter (50 per cent, between Full and New).

phase contrast microscopy A modification of a light MICROSCOPE that utilizes the different REFRACTIVE INDEX of features within an object so that a transparent object can be seen in detail without the need for a coloured stain.

phase diagram A graph of temperature against pressure that shows changes in melting and boiling points with pressure. Three lines on this diagram indicate the combinations of pressure and temperature at which two states can exist together in equilibrium. Where the three lines meet is the TRIPLE POINT. At pressures below that of the triple point, the liquid state does not exist and a solid when heated will sublime (turn from solid to gas). Carbon dioxide at normal (atmospheric) pressures is an example of this: solid carbon dioxide (called dry ice) will turn directly into a gas.

phase modulation A method of transmitting information in which the relative phase of a CARRIER WAVE is varied with the amplitude of the signal. *See also* MODULATION.

phase rule A rule that determines the number of DEGREES OF FREEDOM for any system containing one or more PHASES in equilibrium. If the number of phases present is P, the number of degrees of freedom is F and the number of chemically distinct components present in the system is C, then

$$F + P = C + 2$$

phase space A multi-dimensional space that has one dimension for every independent variable of position and MOMENTUM possessed by the particles in a system. Thus any particular state of a system is represented by a single point in phase space and the development of a system with time can be expressed in terms of the path it traces out in phase space.

phase velocity In a travelling wave, the speed at which a point of a given PHASE, for example a crest in a TRANSVERSE WAVE, travels through a material. It is related to the frequency and

wavelength of the wave by the equation

$$\text{speed} = \text{frequency} \times \text{wavelength}$$

In many materials the speed of the wave does not depend on the frequency or the wavelength but only on the material through which the wave is travelling. The stiffer the material the faster the wave will travel as the motion of one part of the material is passed on more effectively to neighbouring regions. Waves also travel more rapidly in less dense materials, as the smaller mass accelerates more rapidly. Materials in which the speed of the wave does depend on the wavelength of the wave are called DISPERSIVE.

For a solid of YOUNG'S MODULUS E and density ρ, the speed of the wave is given by

$$v = \sqrt{(E/\rho)}$$

For sound travelling in a gas at a pressure p and density ρ, with a RATIO OF SPECIFIC HEATS γ,

$$v = \sqrt{(\gamma p/\rho)}$$

For a transverse wave on a string stretched by a tension T, and with a mass per unit length of μ,

$$v = \sqrt{(T/\mu)}$$

phon A unit of loudness which, unlike the DECIBEL, takes into account the differing response of the ear to sounds of differing frequencies. The loudness of a sound measured in phons is the same as the loudness in decibels of a 1 kHz tone perceived as being of equal loudness to the sound being measured.

phonon A QUANTIZED vibration in a crystal lattice, analogous in some ways to a photon of electromagnetic radiation. The SPECIFIC HEAT CAPACITY of a non-metal and its THERMAL CONDUCTIVITY, for example, can be expressed in terms of phonons and the way they interact with each other. Because the lattice has a finite size, there is a minimum wavelength which a phonon can possess, equal to $2d$, where d is the lattice spacing.

phosphor Any material that, when struck by electrons, converts some of the kinetic energy of the electrons into visible light.

phosphorescence The production of visible light as a result of a chemical reaction, electron bombardment or other process in which electrons move from one ENERGY LEVEL to a lower energy level. Also, the emission of visible light

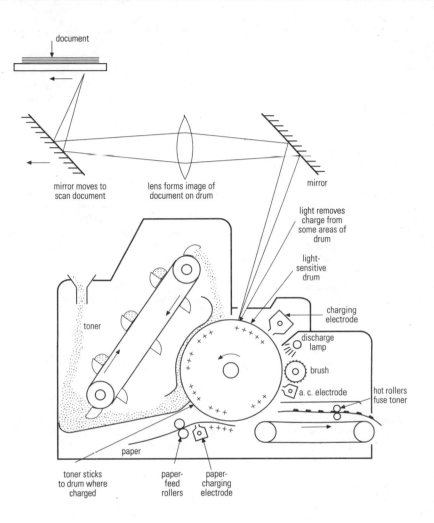

Photocopier.

some time after light of a shorter wavelength (e.g. ultraviolet) has been absorbed. *Compare* FLUORESCENCE.

photocathode A negative ELECTRODE designed to release electrons when struck by light. *See* PHOTOELECTRIC CELL, PHOTOELECTRIC EFFECT.

photocell *See* PHOTOELECTRIC CELL.

photoconductive (*adj.*) Describing a material, such as cadmium sulphide, that conducts electricity when light falls upon it. The CONDUCTIVITY of the material increases with the intensity of the light. The action of light cre-

ates ELECTRON-HOLE PAIRS in an otherwise insulating material.

photoconductive cell *See* PHOTOELECTRIC CELL.

photocopier A machine for copying printed or written documents. A light illuminates the document to be copied and a lens forms an image of this document on a charged drum with a PHOTOCONDUCTIVE surface. The action of the photoconductive layer causes electric charge to escape from the drum in those regions which are to be white on the final document. Ink, in the form of a dry powder known as toner, sticks to the charged portions of the drum. A piece of paper is given an electrostatic charge (*see* STATIC ELECTRICITY) and then rolled onto the drum, transferring ink from the drum to the paper. The paper then passes though a heated roller, which melts the ink, allowing it to soak into the paper to form a permanent image.

photodiode A REVERSE BIASED PN JUNCTION DIODE in which incoming PHOTONS create ELECTRON-HOLE PAIRS, allowing a current to flow. The size of the current is dependent on the intensity of the light. Photodiodes are used to measure the intensity of light or as switches to detect light.

photoelasticity The property of many transparent plastics, which exhibit BIREFRINGENCE to an extent dependent on the degree of STRESS to which the material is subjected. By building models of a complex structure in a photoelastic material and viewing this in polarized light (*see* POLARIZATION), the effects of stresses on the structure can be made visible.

photoelectric cell, *photocell* One of several devices for detecting light and other forms of ELECTROMAGNETIC RADIATION. A photoemissive cell consists of a negative ELECTRODE (the photocathode) and a positive-collecting electrode (anode) in a vacuum. PHOTONS striking the photocathode liberate electrons, in a process known as the PHOTOELECTRIC EFFECT. The electrons are attracted to the anode, and the resulting electric current is a measure of the light intensity.

In a photovoltaic cell, a POTENTIAL DIFFERENCE is set up between two layers as a result of irradiation by light (*see* SOLAR CELL). In a photoconductive cell, the conductivity of a SEMICONDUCTOR increases on exposure to light.

photoelectric effect The emission of electrons from the surface of a metal on exposure to ELECTROMAGNETIC RADIATION. Electrons are emitted only if the wavelength is below some minimum which depends on the metal used. If light of too long a wavelength is used, the effect is not observed no matter how bright the light.

The particle-like nature of light first became apparent from studies of the photoelectric effect. It contradicts the wave model of light, which holds that the energy in a light beam can be increased either by increasing the intensity or by having a shorter wavelength.

Albert Einstein's (1879–1955) explanation

of the photoelectric effect (1901) suggested that light comes in quanta (*see* QUANTUM), or individual particles, called PHOTONS. Each electron gains the energy needed to escape by absorbing the energy of one photon. The energy of each photon is related to the frequency, f, of the light by

$$E = hf$$

where E is the photon energy and h is a FUNDAMENTAL CONSTANT called PLANCK'S CONSTANT. The minimum energy needed for an electron to escape from the surface of a metal is called the WORK FUNCTION, and the photoelectric effect cannot take place for a given metal if the photon frequency is so low (i.e. the wavelength so long) that the photon energy is smaller than the work function.

The theory is further justified by experiments in which the energy of the ejected electrons is measured by collecting them on a plate held at a potential negative to the metal plate from which they are produced. The maximum POTENTIAL DIFFERENCE that can exist with photoelectrons still arriving is called the stopping potential and is proportional to the maximum energy with which photoelectrons are released. The photoelectric effect is the basis of many light-detecting devices.

See also CCD, PHOTOELECTRIC CELL, PHOTOMULTIPLIER, PHOTOVOLTAIC EFFECT.

photoemission The release of electrons from a metal ELECTRODE by the PHOTOELECTRIC EFFECT.

photoemissive cell *See* PHOTOELECTRIC CELL.

photographic film A light-sensitive surface that relies on a chemical reaction caused by light striking grains of a silver halide, which is made visible by chemical reactions when the film is developed.

photometer Any device for measuring light intensity. Modern photometers generally comprise a SOLAR CELL in conjunction with a digital VOLTMETER calibrated to give a direct reading of light intensity.

photometry The science of measuring light intensity.

photomultiplier A sensitive light-detecting device based on the PHOTOELECTRIC EFFECT. Light strikes a PHOTOCATHODE and releases electrons. A series of collecting ELECTRODES then multiplies the number of electrons. The first electrode is sufficiently positive that electrons reach it with sufficient energy to knock out further electrons, in a process known as SECONDARY EMISSION. This is repeated several times, with each collecting electrode being more positive than the previous one, so that a single PHOTON releasing a single electron will result in many thousands of electrons being produced. In this way, individual photons can be detected with high efficiencies.

photon The QUANTUM of ELECTROMAGNETIC RADIATION, having an energy related to the frequency, f, of the light by

$$E = hf$$

where E is the photon energy and h is PLANCK'S CONSTANT. The concept of the photon was introduced by Albert Einstein (1879–1955) to explain the PHOTOELECTRIC EFFECT and also removed the ULTRAVIOLET CATASTROPHE from calculations of BLACK-BODY RADIATION. The theory was not immediately accepted, as there was strong evidence from YOUNG'S DOUBLE SLIT EXPERIMENT that light behaved as a wave rather than a particle (*see* WAVE-PARTICLE DUALITY). The modern theory of photons and their interaction with matter is called QUANTUM ELECTRODYNAMICS.

photovoltaic cell *See* SOLAR CELL.

photovoltaic effect A PHOTOELECTRIC EFFECT in which a POTENTIAL DIFFERENCE is created between two layers as a result of light falling on the boundary between the two.

physical binary *See* BINARY.

physics The branch of science that deals with MATTER, ENERGY and their interactions. Physics attempts to find laws, usually mathematical in form, that accurately describe a wide range of phenomena throughout the Universe.

pi (π) An irrational number equal to the circumference of any circle divided by its diameter. Pi is equal to 3.1416 (to 4 decimal places).

pico- (p) A prefix placed in front of a unit to denote that the size of that unit is to be multiplied by 10^{-12}. For instance a picofarad (pF) is one million millionth of a FARAD.

pie chart A method for displaying data using a circle as the whole sample, divided into segments. The angles of the segments show the percentage of the whole represented by each portion. It provides a clear, simple representation of proportions but does not give precise information.

piezoelectric The effect by which a POTENTIAL DIFFERENCE (p.d.) appears between the faces of

certain materials when they are subject to STRESS. The effect arises from distortions in the arrangement of charges in some ANISOTROPIC crystals, notably quartz, and some plastics. The reverse effect also occurs, with the material deforming in response to an applied p.d. This effect is used in many electronic TRANSDUCERS. *See also* CRYSTAL OSCILLATOR.

pi-meson, *pion* (π^+, π^-, π^o) The lightest MESON, with a mass 0.15 times the proton mass, and the one with the longest HALF-LIFE. The positive and negative pi-mesons have half-lives of 2.6×10^{-8} s. The half-life of the neutral pion (or pi-zero) is far shorter, 8×10^{-17} s.

pinion A small GEAR, the smaller of a pair of gears. *See also* RACK AND PINION.

pion *See* PI-MESON.

piston A round metal plate, usually with a flat surface, which moves up and down in a CYLINDER in response to pressure changes. The piston is a key part in the INTERNAL COMBUSTION ENGINE and in many HYDRAULIC and PNEUMATIC systems.

pitch A musical description related to the FREQUENCY of a sound; a doubling in frequency represents an increase in pitch of one OCTAVE.

pitchblende The principle ore of uranium, consisting mainly of uranium(IV) oxide, UO_2. Pitchblende is also a source of radium, thorium and polonium.

pitot head A forward-facing open tube fitted to an aircraft as part of a PITOT-STATIC SYSTEM.

pitot-static system In aircraft, a system used to measure the difference between the pressure of the air striking the front of the aircraft and the static pressure of the surrounding air, in order to measure the speed of the aircraft through the air.

pivot A fixed point about which an object, such as a LEVER, can rotate. To reduce the effects of friction, some form of lubrication or BEARING is often used at a pivot.

pixel Any one of the large number of small single elements from which an image is formed on a computer screen or other computer image-forming system.

Planck's constant (h) A constant that determines the scale of the effects of WAVE-PARTICLE DUALITY. Since h is small, the effects of QUANTUM MECHANICS only become apparent at short distances or over short times. Planck's constant has a value of 6.6×10^{-34} Js. *See also* PHOTOELECTRIC EFFECT.

plane A flat surface in space. Any vector normal to a particular plane always has the same direction, and any straight line joining any two points on the plane lies entirely in the surface.

plane mirror A flat MIRROR. When objects are viewed by looking at light reflected from a plane mirror, a VIRTUAL IMAGE is seen, which is as far behind the mirror as the object was in front of the mirror, with the line joining object and image being at 90° to the plane of the mirror.

plane-polarized (*adj.*) Describing an ELECTRO-MAGNETIC WAVE in which the electric and magnetic fields each oscillate in a single plane, perpendicular to one another and to the direction of wave propagation. *See* POLARIZATION. *Compare* CIRCULARLY POLARIZED.

planet A large, roughly spherical, celestial object composed mainly of rock or gas and in orbit around the Sun. The smaller planets, including the Earth, are composed mostly of rock and are called TERRESTRIAL PLANETS. The larger planets are made mostly of gas and are called GAS GIANTS. The periods of the orbits of the planets increase with distance from the sun (*see* KEPLER'S LAWS), ranging from 88 days for Mercury to 240 years for Pluto.

planetary nebula A stage in the life cycle of some stars in which the outer layers of a RED GIANT are blown off into space over a period of typically one million years, revealing a WHITE DWARF at the core of the star.

planetary satellite *See* SATELLITE.

plasma A highly IONIZED gas in which the number of FREE ELECTRONS approximately equals the number of positive IONS. As a whole, a plasma is electrically neutral. They are produced at such high temperatures that collisions between particles are violent enough to produce almost complete IONIZATION.

Most of the matter in the Universe exists as a plasma, which is sometimes described as the fourth state of matter. Plasmas occur in stars and in interstellar space, and are induced in thermonuclear reactors and in GAS DISCHARGE tubes.

The formation of a plasma requires a large amount of energy, either from an electric field or by the absorption of short wavelength electromagnetic radiation, usually ULTRAVIOLET. Unless this energy supply is maintained, a plasma will cool rapidly, releasing energy when ions and electrons recombine. A plasma behaves in much the same way as a gas, with the additional property that, since it contains charged particles, it is influenced by magnetic fields. *See also* MAGNETIC BOTTLE, MAGNETIC CONFINEMENT, MAGNETOHYDRODYNAMICS.

plastic 1. (*n.*) Any of the stable synthetic materials that are liquid at some stage in their manufacture. Plastics may be shaped when in their liquid stage and then set to a rigid or semi-rigid solid.

2. (*adj.*) Describing a material or process in which the material is deformed by a force and does not return to its original shape once the force is removed.

plate tectonics The study of the motion of TECTONIC PLATES.

Platonic solid Any of the five regular POLYHEDRONS – the cube, dodecahedron, icosahedron, octahedron, tetrahedron.

plotting compass A small MAGNETIC COMPASS used for determining the direction of a MAGNETIC FIELD in the laboratory.

Pluto The ninth, and last, planet from the Sun, with an orbital radius of 39.4 AU (5.9 billion km). Relatively little is known about Pluto, which was not discovered until 1930, except that it is small, and probably composed of rock, ice and frozen methane, with a mass of 6.6×10^{23} kg (0.11 times that of the Earth), and a diameter of 2,700 km (0.21 times that of the Earth). It has a satellite called Charon, which is almost as large as the planet itself. Pluto has an orbital period of 248 years and rotates about its own axis every 6.9 days.

pneumatic (*adj.*) Describing a system in which air is used to transmit forces from one place to another or to drive a machine. Because no sparks are produced, pneumatic tools are often used in places where there is a danger of explosion, such as coal mines.

pn junction diode A simple electronic device, produced by forming a junction between P-TYPE and N-TYPE SEMICONDUCTORS. Electrons from the n-type material, and HOLES from the p-type, diffuse across the boundary and cancel each other out, forming a DEPLETION LAYER in which there are very few CHARGE CARRIERS. When the p-type material is made negative and the n-type positive, the diode is said to be reverse biased – the holes in the p-type material and the electrons in the n-type layer are pulled away from the junction and no current will flow. If the polarity is reversed, with the

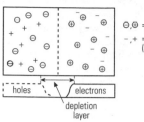

Θ,⊕ = fixed ions
−,+ = charge carriers
(electrons and holes)

pn junction diode.

p-type material positive and the n-type negative, charge carriers are attracted into and across the depletion layer and a current flows. The diode is then said to be forward biased.

Diodes can be used for RECTIFICATION, the conversion of alternating to direct current, and for this purpose a BRIDGE RECTIFIER circuit is often used. They are also used for DEMODULATION in radio circuits. If the junction is made sufficiently large, a reverse biased diode will have a significant CAPACITANCE, which can be used in a TUNED CIRCUIT. Such diodes are called varicap diodes.

pnp transistor A JUNCTION TRANSISTOR in which the EMITTER and COLLECTOR are made of P-TYPE SEMICONDUCTOR, whilst the BASE is N-TYPE.

point In mathematics, a position in space, on a graph for example, usually defined by its CO-ORDINATES, and having no physical extent, so it can be fully represented by a single set of co-ordinates rather than by a range of co-ordinates.

Poiseulle's equation An equation for finding the rate of volume flow of a viscous fluid in LAMINAR FLOW down a tube. If the VISCOSITY of the fluid is η and the tube has radius r and length l, the volume flow per second, V, caused by a pressure difference p between the ends of the tube is

$$V = \pi p r^4 / 8 l \eta$$

polar (*adj.*) **1.** Relating to the POLE. A polar AIR MASS is one that has come from regions close to the one of the Earth's poles.

2. Describing a molecule or a substance, particularly a solvent, that contains POLAR BONDS.

polar bond A COVALENT BOND in which the electrons spend a higher proportion of their

time closer to one atom in the bond than the other, thus one atom in effect carries a partial negative charge whilst the other has an equal positive charge. These are sometimes indicated on diagrams by $\delta+$ and $\delta-$. Typical examples are hydrogen chloride, HCl, and water H_2O, each of which have polar bonds with the hydrogen atom forming the positive end of the bond.

H – Cl
δ+   δ–

The extreme version of a polar bond, with the electrons spending all their time attached to one atom, is an IONIC BOND. Polar bonds tend to lead to a greater attraction between molecules and materials with higher boiling points than would be expected from VAN DER WAALS' BONDING.

See also HYDROGEN BOND.

polar co-ordinates CO-ORDINATES that in two dimensions define the location of a point in terms of its distance from the ORIGIN and the angle between the direction to the point and a fixed direction. These co-ordinates are generally denoted as r and θ and if the x-axis is taken as the $\theta = 0$ direction, they are related to CARTESIAN CO-ORDINATES by the equations:

$$x = r\cos\theta$$
$$y = r\sin\theta$$

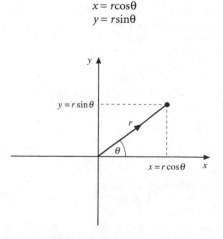

Polar co-ordinates.

In three dimensions, cylindrical polar co-ordinates or spherical polar co-ordinates may be used. Cylindrical polar co-ordinates use a second distance, z, to measure the position of the point above or below the plane of the origin. Spherical polar co-ordinates measure the distance r of the point from the origin together with a polar angle θ, above or below the plane perpendicular to an axis called the polar axis, and an azimuthal angle ϕ found by projecting the point into the plane perpendicular to the polar axis.

polarimeter A device for analysing polarized light (*see* POLARIZATION), particularly for measuring the OPTICAL ACTIVITY of a solution. The sample to be analysed is placed between two pieces of POLAROID. The first (the polarizer) polarizes the light and the second (the analyser) can be rotated until no light passes through the polarimeter. The degree of rotation required to prevent any light passing through the polarimeter is a measure of the optical activity of the sample.

polarization 1. The direction of motion of the material through which a TRANSVERSE WAVE is travelling. In the case of an ELECTROMAGNETIC WAVE, the direction of polarization is the direction of the electric field, since it is this, rather than the magnetic field, that is responsible for most of the physical effects of the wave. Electromagnetic waves may be PLANE-POLARIZED, with the electric field in a single direction, or CIRCULARLY POLARIZED, in which case the electric field direction follows a circular path, which may be either left- or right-handed, as seen by an observer facing in the direction in which the wave is travelling. A circularly polarized wave will have its direction of polarization reversed on reflection. Radio aerials transmit and receive waves that are plane-polarized in the direction of the wire elements of the aerial. *See also* BIREFRINGENCE, BREWSTER'S LAW, OPTICAL ACTIVITY, POLARIZATION BY REFLECTION, POLAROID.

2. A measure of the extent to which molecules have been polarized (*see* POLARIZE) by an electric field.

polarization by reflection Unpolarized light reflected from a non-metallic surface will become partially polarized, with the electric field predominantly in the direction parallel to the reflecting surface. This reflected light is completely polarized when the ANGLE OF INCI-DENCE is equal to the BREWSTER ANGLE for the material, at which point the reflected and refracted light (*see* REFRACTION) are at right angles. *See also* BIREFRINGENCE, BREWSTER'S LAW.

polarize (*vb.*) **1.** Of light, to transmit, reflect or scatter one direction of POLARIZATION in unpolarized light more strongly than another.

2. Of an ELECTRIC FIELD, to pull a molecule or atom away from symmetry so the centres of positive and negative charge no longer coincide; or to align POLAR MOLECULES.

polar molecule A molecule in which the electrons in a COVALENT BOND are not evenly distributed, so they have a higher probability of being found at one end of the molecule than the other. *See also* HYDROGEN BOND, POLAR BOND.

Polaroid Trade name for a transparent plastic material that plane-polarizes light passing through it (*see* POLARIZATION). The plastic contains long-chain molecules along which electrons can travel, absorbing the energy from electromagnetic radiation. During manufacture, the plastic is stretched, aligning the molecules. Light that is polarized perpendicular to the direction of alignment will be transmitted through the Polaroid, whilst light polarized parallel to the molecules is absorbed. Polaroid filters are used in some sunglasses and are sometimes fitted to cameras to reduce the amount of reflected (rather than scattered) light and so reduce glare.

pole 1. The point on the surface of an object, a planet for example, at which the axis of rotation passes through that surface.

2. The region on the surface of a magnet at which the MAGNETIC FIELD LINES enter and leave the magnet.

pole and barn paradox A RELATIVISTIC PARADOX in which a man runs into a barn carrying a pole which is the same length as the barn. According to an observer at rest in the barn, the pole is now shorter than the barn (LORENTZ CONTRACTION) so it would be possible to shut the doors at each end of the barn and trap the pole in the barn. According to the runner, the barn is now shorter than the pole, so it would not be possible to trap the pole. The resolution of this paradox comes from the idea that the shock wave of the leading edge of the pole running into the barn door cannot travel along the pole faster than the speed of light, thus the pole cannot be perfectly rigid, and so the runner can, just, be trapped in the barn.

polycrystalline (*adj.*) Describing any solid material that occurs in pieces with no regular shape, but on examination can be seen to be made of many small CRYSTALS. Within each crystal the molecules are arranged regularly, but the direction of alignment of this structure varies from one small crystal to the next, and the boundaries between one crystal and the next (GRAIN BOUNDARIES) are often irregular.

polygon A closed two-dimensional figure having straight sides. The type of polygon is defined by the number of sides: 3 sides is a triangle; 4 a quadrilateral; 5 a pentagon; 6 a hexagon; etc. If a polygon has sides all of the same length it is described as regular.

polyhedron A closed three-dimensional figure having at least four faces, each of which is flat. In a regular polyhedron, all sides have the same length. There are only five such regular polyhedrons, called the Platonic solids. These are the tetrahedron (four triangular faces), the cube (six square faces), the octahedron (eight triangular faces), the dodecahedron (12 faces each in the shape of a pentagon) and the icosahedron (20 triangular faces).

population inversion A state in which more atoms are in an EXCITED STATE than in some lower energy state, such as the GROUND STATE. *See also* LASER.

p-orbital The second lowest energy ORBITAL for a given PRINCIPAL QUANTUM NUMBER. p-Orbitals exist only for principal quantum numbers of two or greater. There are three p-orbitals for a given principal quantum number, each formed of two lobes with each of the three orbitals at right angles to the other two.

porous (*adj.*) Describing any material that contains many small cracks or holes, able to absorb water, air or some other fluid.

position vector A VECTOR whose length and direction are equal to the distance and direction of a specified point from the ORIGIN.

positive In physics, the name given to one of the two types of electric CHARGE. PROTONS are positively charged. Objects that are positively charged normally reach that state by losing electrons rather than by gaining protons.

positive feedback FEEDBACK that is IN PHASE with the original signal. This generally has the effect of destabilizing a circuit and may lead to OSCILLATION. *See also* NEGATIVE FEEDBACK.

positron The ANTIPARTICLE of the ELECTRON. The positron is a stable ELEMENTARY PARTICLE with the same mass as an electron, but the opposite charge and LEPTON number. Although positrons are stable in isolation, when a positron meets an electron, they will annihilate one another, producing two or more gamma rays.

potential difference (p.d.) A more technical term for VOLTAGE, the p.d. between two points being the amount of energy converted from electrical energy to other forms when one COULOMB of charge flows between the two points.

$$\text{Potential difference} = \frac{\text{energy transformed}}{\text{charge flow.}}$$

potential divider A pair of RESISTORS with a supply voltage connected across both of the resistors so that a fraction of this voltage appears across each of the resistors. This arrangement is used in volume controls in radios and AMPLIFIERS. For a potential divider with resistors R_1 and R_2 and an input voltage V_{in} and an output voltage V_{out}

$$V_{out} = V_{in}R_2/(R_1 + R_2)$$

potential energy The ENERGY possessed by an object or system as a result of its position or state, such as a stretched spring or a mass in a GRAVITATIONAL FIELD. It is the amount of work done by the object or system moving from a state at which it is said to have no potential energy to a higher state. *See also* GRAVITATIONAL POTENTIAL ENERGY.

potentiometer An alternative name for a VARIABLE RESISTOR, particularly when used as an adjustable POTENTIAL DIVIDER.

pound (lb) A unit of mass, now obsolete in science but still in everyday use in the US. One pound is approximately 0.454 KILOGRAMS.

poundal A unit of force, now obsolete in science, equal to the pull of gravity on Earth on a mass of one POUND. One poundal is approximately 4.5 NEWTONS.

powder coating A way of giving a plastic surface to a metal object, preventing corrosion, in a way similar to the application of paint. The object to be coated is given an ELECTROSTATIC charge and then sprayed with an oppositely charged powder. The opposite charges ensure that the powder sticks to the object. The object is then heated to melt the powder into a solid layer.

power 1. (*mechanics*) The rate at which WORK is done. The SI UNIT of power is the WATT.

$$\text{power} = \text{energy/time} = \text{force} \times \text{speed}$$

2. (*arithmetic*) The number of times a quantity must be multiplied by itself in an algebraic expression, thus a^5, which means $a \times a \times a \times a \times a$, is referred to as 'the fifth power of a' or 'a to the power (of) 5' or just 'a to the fifth'. If the power is 2 or 3, the number is said to be 'squared' or 'cubed' respectively.

power factor For a device through which an ALTERNATING CURRENT is flowing, the COSINE of the PHASE between the voltage across the device and the current through it. The electrical power delivered to the device is found by multiplying together the ROOT MEAN SQUARE current and voltage and the power factor. For a pure RESISTANCE the power factor is 1, for a pure REACTANCE, it is 0.

power series See SERIES.

power stroke The stage in the operation of a PETROL ENGINE where fuel burns and expands to drive the engine.

pp chain The NUCLEAR FUSION reactions by which a MAIN SEQUENCE STAR fuses hydrogen into helium. The details are complicated, but the first step in the chain involves two protons fusing to form a hydrogen–2 nucleus, with the release of a POSITRON and a NEUTRINO. Various steps are then possible, but most commonly the hydrogen–2 nucleus fuses with another proton to form a helium–3 nucleus with the emission of a gamma ray. Two helium–3 nuclei then fuse and release two protons to give a helium–4 nucleus.

precession The slow change in alignment of the axis of a more rapidly spinning object, such as a GYROSCOPE, under the influence of an external force.

pre-ignition The condition in a PETROL ENGINE where the fuel starts to burn before the spark has been generated in the SPARK PLUG. Until the 1980s, lead-based compounds were routinely added to petrol to enable the petrol-air mixture to be compressed to a greater extent without igniting. Concerns over the high levels of toxic lead compounds in the air of some large cities have lead to the introduction of unleaded petrol, which has no such additives, but can only be burnt in engines designed to compress the fuel a little less before igniting it. See also COMPRESSION RATIO.

preon A hypothetical ELEMENTARY PARTICLE from which QUARKS and LEPTONS are made, postulated to explain the large number of different types of such particles. No workable model of particle physics based on preons has yet been constructed and there is no experimental evidence for their existence.

pressure The FORCE acting on each square metre of area. The unit of pressure is the PASCAL (Pa), this being a pressure of one NEWTON per square metre.

The amount of pressure acting on an object is related to the amount it is deformed or damaged, thus shoes with small heels do more damage to a soft floor than larger heels worn by a person of the same weight. Knives are made with sharp edges, and nails with sharp points, to minimize the contact area and make the pressure produced as large as possible. Padded seats are more comfortable than hard ones, because they deform to produce a large contact area and less pressure.

On a molecular level, the pressure of a fluid can be thought of in terms of the pressure exerted on a surface by the molecules of the fluid. The pressure is equal to the average change in MOMENTUM per molecular collision multiplied by the average number of collisions per second per square metre. Because there are a very large number of collisions, they are felt as a constant pressure, rather than a series of separate impacts.

See also BAROMETER, BOURDON GAUGE, DALTON'S LAW OF PARTIAL PRESSURES, DYNAMIC PRESSURE, HYDROSTATIC PRESSURE, HYDROSTATICS, KINETIC THEORY.

pressure law A GAS LAW used to define the IDEAL GAS TEMPERATURE SCALE. It states that, for a fixed mass of gas held in a constant volume, the pressure is proportional to the ABSOLUTE TEMPERATURE; that is, the pressure divided by the absolute temperature is a constant. For a fixed mass of an ideal gas with a pressure p and absolute temperature T held at constant volume,

$$p/T = \text{constant}$$

See also BOYLE'S LAW, CHARLES' LAW, CONSTANT VOLUME GAS THERMOMETER, IDEAL GAS EQUATION.

pressure relief valve See VALVE.

prevailing wind The most common wind direction at a given location. See also AIR MASS.

Prévost's theory of exchanges The absorption and emission of THERMAL RADIATION from a body is equal when it is in EQUILIBRIUM with its surroundings. In this state, the temperature of the body remains constant. If the body and its surroundings are at different temperatures there is a net flow of energy. *See also* HEAT, THERMAL EQUILIBRIUM.

primary cell Former name for a non-rechargeable electrochemical CELL.

primary coil In a TRANSFORMER or INDUCTION COIL, the coil that is connected to the power supply.

primary colour Red, green or blue, the three colours of light that when mixed together in equal proportions produce white light. When mixed in other proportions they can produce any other colour (except black).

primitive cell *See* UNIT CELL.

primordial (*adj.*) Referring to the state of some system at its earliest stage, in particular the conditions in the Universe before the formation of stars and galaxies, and of conditions on the Earth before the origin of life.

principal axis, *optical axis* The line joining the centre of a LENS or MIRROR to the PRINCIPAL FOCUS, at right angles to the FOCAL PLANE. It is the line along which a ray of light can travel through an optical system without any change of direction, except for a 180° reflection at a mirror.

principal focus A point associated with a LENS or CURVED MIRROR, through which all rays originally parallel to the PRINCIPAL AXIS will pass after REFRACTION or REFLECTION at the lens or mirror. For a DIVERGING LENS or a CONVEX mirror, this is a VIRTUAL FOCUS – the rays diverge as if they had originated from an imaginary point on the other side of the lens or mirror.

principal quantum number The QUANTUM NUMBER used to label an ORBITAL to give a broad indication of its energy. Thus in describing a 2s orbital, 2 is the principal quantum number.

principle of equivalence One of the basic premises of the GENERAL THEORY OF RELATIVITY, which states that there is no difference between a system that is undergoing a steady acceleration and one that is in a constant GRAVITATIONAL FIELD. Thus an observer in a lift who suddenly finds himself weightless will not know whether the lift cable has broken and the lift has entered FREE FALL or if the Earth's gravity has mysteriously been turned off.

This equivalence is a consequence of two ways of thinking about MASS. Mass can be thought of as the resistance to acceleration,

INERTIAL MASS in Newton's second law (*see* NEWTON'S LAWS OF MOTION) and also as the quantity that controls the size of gravitational interactions in NEWTON'S LAW OF GRAVITATION – this is GRAVITATIONAL MASS. The equivalence of these two masses leads to the idea that all objects fall with the same acceleration in a given gravitational field, and thus to the principal of equivalence.

principle of least time *See* FERMAT'S PRINCIPLE.

principle of superposition The total displacement of two or more WAVES arriving at a point is equal to the sum of the displacements of the individual waves. *See* INTERFERENCE.

prism 1. A three-dimensional shape formed by extending a two-dimensional shape, especially a triangle, into the third dimension, so it has two faces that are POLYGONS with all the others being rectangular. The volume of a prism is equal to the length of its parallel sides multiplied by the area of the face perpendicular to these sides.

2. A triangular block of glass or some other transparent material used to refract light (*see* REFRACTION) in optical systems.

probability A mathematical measure of how likely an event is. If an event cannot happen, it is given a probability of 0, if it is certain, the probability is 1. If an event is expected to happen n times out of N, the probability is n/N. If an event has a probability p, the probability of it not happening is $1 - p$. For two events that are independent, with probabilities p and q, the probability of both happening is pq. For two alternative events, one or the other of which can happen but not both, the probability that one or the other will occur is $p + q$.

probability amplitude A quantity used in QUANTUM MECHANICS to find the probability of a system being in a particular state or a particle in a particular position. Probability amplitudes obey the PRINCIPLE OF SUPERPOSITION and so can interfere constructively or destructively (*see* INTERFERENCE), giving wave-like properties to the particles they describe. The probability of finding a particular state is the square of the probability amplitude. *See also* WAVE-PARTICLE DUALITY.

progression A SERIES of algebraic expressions added together, each one differing from the previous one in some way, typically either by the addition of a certain number (an ARITH-

METIC PROGRESSION) or by multiplication by a certain number (a GEOMETRIC PROGRESSION).

progressive wave, *travelling wave* Any WAVE that transmits ENERGY in one direction through a material, as opposed to a STANDING WAVE.

projectile Any object that is launched and then allowed to move freely under the influence of gravity. *See* FREE FALL. *See also* BALLISTICS.

projector A device designed to produce an enlarged image on a screen. In a motion picture projector the object is a transparent piece of photographic film. Because the image is enlarged, the light falling on any one part of the screen is likely to be rather dim, so the light falling on the object being projected is usually made as bright as possible using CONDENSERS.

PROM (programmable read-only memory) A MEMORY device consisting of an INTEGRATED CIRCUIT that is programmed after manufacture and can hold the data permanently.

proportional (*adj.*) Describing two quantities that vary together in such a way that the ratio of one to the other is a constant, shown by the symbol $\propto$. If $y \propto x$ then y/x will be a constant, called the constant of proportionality. If x and y are plotted on a graph, the result will be a straight line passing through the ORIGIN, with a gradient equal to the constant of proportionality.

proton The positively charged particle found in the NUCLEUS of an ATOM. The number of protons in a nucleus is called the ATOMIC NUMBER and fixes the number of ELECTRONS needed to produce a neutral atom, which in turn determines the chemical properties of the element.

Protons were originally believed to be one of the fundamental constituents of matter, together with electrons and NEUTRONS. However, they are now thought of as the lightest member of the family of BARYONS, objects containing three QUARKS. The proton is the only baryon that is stable in isolation (neutrons are only stable when combined with protons in nuclei), though some modern theories suggest that protons may decay very slowly into LEPTONS (*see* PROTON DECAY).

The mass of a proton is 1.6×10^{-27} kg. It has a charge of 1.6×10^{-19} C, equal but opposite to the charge on an electron.

See also HADRON.

proton decay The decay of nuclear matter, particularly PROTONS into LEPTONS, predicted by GRAND UNIFIED THEORIES, but not confirmed experimentally. If the proton is unstable, its lifetime must be many orders of magnitude greater than the age of the Universe.

protostar A cloud of gas, mostly hydrogen and helium, that is collapsing under its own GRAVITY and may eventually become hot enough at its centre for NUCLEAR FUSION to take place, leading to the formation of a STAR.

psi particle *See* J/ψ.

Ptolemaic model A model of the SOLAR SYSTEM which places the Earth at the centre with the Sun, Moon and PLANETS all moving around the Earth. In order to explain the detailed motions of the planets, they were pictured as moving on epicycles, circles each with a planet rotating around its circumference, whilst the centre of the epicycle travelled around the Earth in a circular path. This complex model was later superseded by the far simpler COPERNICAN MODEL.

p-type semiconductor A SEMICONDUCTOR in which the charge is predominantly carried by HOLES rather than electrons.

pulley A wheel with a groove in it around which there is a rope supporting a LOAD, or a belt transmitting a load to another pulley. It is possible for an EFFORT to be applied to a rope which loops around several pulleys, so that the TENSION is used several times over in supporting the load. If two wheels of different sizes are used, the different MOMENTS of the tension in a belt around the two pulleys produces a machine with a VELOCITY RATIO that is equal to the inverse ratio of the pulley sizes.

pulsar A STAR that flashes with a well defined period, typically ranging from a few hundred times per second to once every few seconds. Pulsars are now recognized as NEUTRON STARS, and the flashing is caused by the interaction between the radiation from the neutron star and its intense magnetic field. This causes radiation to be emitted only along the magnetic poles of the neutron star. If the magnetic poles are not aligned with the axis of rotation of the star, a beam of radiation will sweep through space and an observer will see a pulse whenever this points towards their telescope.

pulse A wave or electrical signal of short duration.

pulse-code modulation A method of transmitting information in which the AMPLITUDE of a signal is sampled and converted into a BINARY

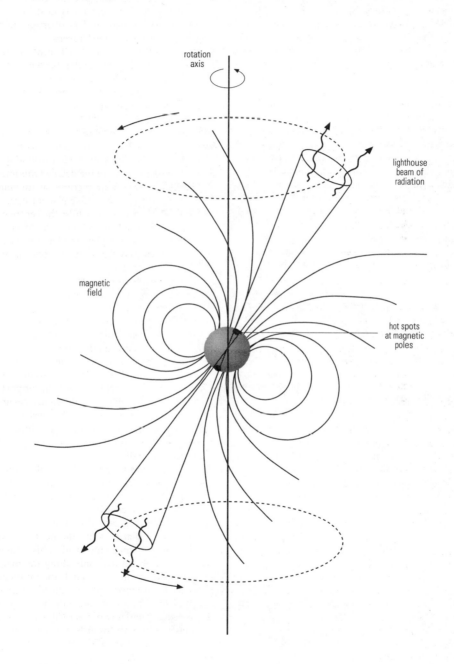

rotation
axis

lighthouse
beam of
radiation

magnetic
field

hot spots
at magnetic
poles

Pulsar.

code, which is transmitted as a series of on and off pulses. *See also* MODULATION.

pulse-jet An early form of JET ENGINE that operated by admitting air through a series of slats which were then closed by the pressure of the burning gas. As this left the rear of the engine, the slats opened again to admit a fresh supply of air.

pump A device for moving or changing the pressure of a fluid. A pump normally consists of a chamber together with two valves through which fluid can flow in one direction only, or some other mechanism for ensuring that the fluid can flow in one direction only. Fluid enters the chamber through one of the valves. The volume of the chamber is then reduced and fluid expelled through the other valve.

pumped storage An energy storage scheme that uses electricity at times of low demand to pump water from a low-level reservoir to one at a higher level. At times of high electricity demand, the process is reversed to generate electricity. Although some energy is inevitably lost in the process, it has economic benefits as there is no way of storing surplus electrical energy in sufficient quantities and it enables the most cost-efficient generating plant to be used more continuously.

pupil An aperture at the centre of the IRIS of the EYE. The size of the pupil can be adjusted by the muscles of the iris to control the amount of light entering the eye.

p-wave In SEISMOLOGY, the LONGITUDINAL wave that arrives first after an EARTHQUAKE. *See* SEISMIC WAVE.

pyramid A three-dimensional geometrical shape having a rectangular (usually square) base, and three triangular sides (usually EQUILATERAL) meeting at a point called the apex. The volume of a pyramid is $Ah/3$, where A is the area of the base, and h is the height of the apex above the base.

pyroelectricity The production of a charge on the surface of certain materials as a result of electrical POLARIZATION produced by a change in temperature.

pyrometer Any device for measuring high temperatures, especially one that operates by analysing visible light or infrared radiation given off by a hot object. One simple form of pyrometer works by viewing the hot object behind the filament of a light bulb. The current through the filament is adjusted until the filament disappears from view, being neither brighter nor less bright than the object being viewed. At this point the filament and the hot object are at the same temperature. A VARIABLE RESISTOR controlling the filament temperature is calibrated to read temperature directly.

pyrometry The measurement of high temperatures. *See* PYROMETER.

Q

Q factor A number that describes the amount of DAMPING in an oscillating system. The Q factor is 2π times the total energy in the OSCILLATION divided by the energy lost to heat due to the damping forces in one oscillation. The larger the Q factor, the longer a system can continue to oscillate on its own. *See also* RESONANCE.

QCD *See* QUANTUM CHROMODYNAMICS.

QED *See* QUANTUM ELECTRODYNAMICS.

quadrant One quarter of a circle, or some other angular range of 90°.

quantized Describing a quantity that occurs only in fixed amounts, often multiples of a basic unit (or QUANTUM). For example, electric CHARGE always occurs in multiples of the ELECTRON charge.

quantum (*pl. quanta*) A particle, or the group of waves associated with such a particle, which has a fixed (quantized) value of some quantity. Thus, for example, the PHOTON is a quantum of ELECTROMAGNETIC RADIATION with a fixed amount of energy. The term quantum also refers to the minimum amount by which certain properties, such as the energy or ANGULAR MOMENTUM, of a system can change.

quantum chromodynamics (QCD) The QUANTUM THEORY of the STRONG NUCLEAR FORCE. QCD explains that the force between two QUARKS is carried by particles called GLUONS and increases with their separation until there is sufficient energy to create a quark-antiquark pair. Thus quarks are only seen in COLOUR neutral states made up of three quarks (BARYONS), three antiquarks (an antibaryon) or a quark and an antiquark (a MESON).

quantum electrodynamics (QED) A description of the interactions between charged particles, combining the principles of QUANTUM THEORY and the SPECIAL THEORY OF RELATIVITY. QED explains the forces between charged particles in terms of the exchange of PHOTONS. The ideas of QED were subsequently applied to the STRONG NUCLEAR FORCE and the WEAK NUCLEAR FORCE to produce the STANDARD MODEL of particle physics.

quantum mechanics A system of mechanics developed from QUANTUM THEORY. Quantum mechanics explains the properties of atoms, molecules and ELECTROMAGNETIC RADIATION.

quantum number A number that represents the value of some quantity such as CHARGE, which is conserved in certain types of interactions and which is only found in whole number multiples of some basic quantity. *See also* PROBABILITY AMPLITUDE.

quantum physics *See* QUANTUM THEORY.

quantum theory, *quantum physics* The theory that energy is absorbed or released in discrete, indivisible units called quanta. According to quantum theory, there is no real distinction between effects traditionally described in terms of WAVES, such as light, and objects that are more usually thought of as particles, such as ELECTRONS. This double nature of waves and particles is referred to as WAVE-PARTICLE DUALITY. The strange nature of quantum theory has given rise to a number of GEDANKENEXPERIMENTS (thought experiments) carried out with idealized apparatus, to illustrate some of the ideas. *See also* BAND THEORY, ENERGY LEVEL, GAUGE BOSON, HEISENBERG'S UNCERTAINTY PRINCIPLE, HYDROGEN SPECTRUM, PAULI EXCLUSION PRINCIPLE, PHOTOELECTRIC EFFECT, QUANTUM CHROMODYNAMICS, QUANTUM ELECTRODYNAMICS, SPIN, VIRTUAL PARTICLES, WAVE NATURE OF PARTICLES.

quark Any member of the fundamental family of particles from which all HADRONS, including the PROTON and NEUTRON are made. Quarks are held together in hadrons by the STRONG NUCLEAR FORCE and come in six varieties called FLAVOURS. Protons and neutrons are composed of up and down quarks. Heavier quarks called strange, charm, top and bottom quarks are also known. Each flavour of quark has a corresponding antiparticle.

Each quark has charge of +⅔ or −⅓ in units of the electron charge. Each also carries BARYON NUMBER (a quantity conserved in all interactions) ⅓. The heavier quarks carry quantum numbers called strangeness, charm, etc., which are conserved in interactions involving the strong nuclear force but not in those involving the WEAK NUCLEAR FORCE. Hadrons containing these quarks have relatively long HALF-LIVES, the times involved (typically 10^{-10} s) are short by conventional standards, but long compared with the 10^{-23} s lifetime typical of hadrons that decay by way of the strong force.

See also ASSOCIATED PRODUCTION, QUANTUM CHROMODYNAMICS, STANDARD MODEL.

quarter-wave plate A sheet of birefringent material (see BIREFRINGENCE) with a thickness such that the difference in delay for ordinary and extraordinary rays is equal to one quarter of the PERIOD of the wave. A quarter-wave plate can be used to convert plane polarized light into circularly polarized light and vice versa (see POLARIZATION). Two such plates, or a single plate of twice the thickness, called a half-wave plate, can be used to reverse the direction of circular polarization or to rotate the direction of polarization of plane polarized light through 90° without any loss of INTENSITY.

quasar Contraction of quasi-stellar object. An object that appears as a star even when viewed through a powerful telescope, but which has an unusually high RED-SHIFT. It is now believed that many quasars are in fact extremely distant galaxies. See also DOPPLER EFFECT, HUBBLE'S LAW.

quasi-stellar object See QUASAR.

quenching A process in which a material is heated and then cooled suddenly, usually using oil or water. The result is that the large numbers of DISLOCATIONS produced by thermal vibrations at high temperatures become 'frozen' in place and tangled with one another, so are unable to move through the metal. This results in a material that is very hard, but brittle. See also ANNEALING.

R

rack and pinion A device for converting rotational motion into linear, side-to-side motion, such as in the steering mechanism of a car. The rotation of the steering wheel rotates a small gear (the pinion) which meshes with a toothed bar (the rack), which moves the car's wheels from side to side.

rad 1. Abbreviation for RADIAN.

2. A former unit of absorbed DOSE of IONIZING RADIATION. One rad is equivalent to 10^{-2} GRAY.

radar A system based on the reflection of MICROWAVES for the detection of objects that reflect the microwaves, particularly ships and aircraft. Radar systems use a rotating AERIAL to transmit a series of microwave pulses. A receiver then picks up any reflected signal. The time taken for the pulse to return enables the distance to the reflecting object to be calculated and the direction of the aerial at the time the signal is received provides information on the direction of the reflecting object. Radio waves reflected by a moving object will have a frequency which differs slightly from the transmitted waves, as a result of the DOPPLER EFFECT. This shift in frequency can be used to measure the speed of an object, or to distinguish moving objects such as aircraft from a stationary background of radar reflections from hills, buildings, etc. *See also* SONAR.

radian (*abbrev. rad*) The SI UNIT of plane ANGLE. The length of an arc which subtends an angle θ is equal to the radius of the arc multiplied by the angle in radians. Whenever an angle (as opposed to a trigonometrical function of the angle) appears in an equation, it is essential that the angle be expressed in radians. One radian is equal to 57.3°, a complete circle is 2π radians.

radiance The total amount of ELECTROMAGNETIC RADIATION emitted by an object, per square metre of receiving area as measured at a distance of one metre from the object.

radiant energy RADIATION, particularly ELECTROMAGNETIC RADIATION, emitted by an object. *See* BLACK-BODY RADIATION.

radiation In general, the emission of rays, waves or particles from a source, with intensity falling off according to an INVERSE SQUARE LAW. In particular, the term is used for IONIZING RADIATION and ELECTROMAGNETIC RADIATION. *See also* BACKGROUND RADIATION, BLACK-BODY RADIATION, IRRADIATION, RADIOACTIVITY, THERMAL RADIATION.

radiation detectors Any device for detecting IONIZING RADIATION by the IONIZATION it produces. Types of radiation detectors include the spark counter, Geiger–Müller tube and solid-state detector. In each case the ionization produced by the particle occurs in a region where there is an ELECTRIC FIELD.

In the case of a spark counter or Geiger–Müller tube, two ELECTRODES in a gas are used to produce an electric field that is strong enough for the ions to be accelerated and gain sufficient energy to create further ions when they collide with other gas molecules. The spark counter operates in air at ATMOSPHERIC PRESSURE and produces a spark that can be seen and heard.

In the Geiger–Müller tube, lower pressures are used so the ions accelerate for longer before colliding with other molecules. In this way, a lower voltage can be used and the detector is more sensitive. Only ALPHA PARTICLES are sufficiently ionizing to be detected by spark counters; a Geiger–Müller tube will detect BETA PARTICLES and GAMMA RADIATION but can only detect alpha particles if the window through which the radiation must pass to reach the low pressure gas of the tube is sufficiently thin. The pulse of current that is produced when an ionizing particle enters a Geiger–Müller tube is counted electronically and can be fed to a loudspeaker to produce a clicking sound.

In a solid-state detector, which is essentially a REVERSE BIASED PN JUNCTION DIODE, electrons and HOLES are produced allowing a pulse of current to flow in proportion to the amount of energy deposited in the detector by the ionizing radiation.

radiation fog FOG formed by the rapid cooling of air close to ground that radiates its heat away into space on a clear night. *See also* ADVECTION FOG.

radio The transmission of information, particularly speech or music by use of RADIO WAVES. To do this, a CARRIER WAVE is used, a radio wave of some specified frequency, generated by an electronic circuit called an OSCILLATOR.

To send any information, the carrier wave must be modulated, or changed in some way. The simplest way to do this is simply to switch it on and off in accordance with some agreed code, such as Morse code. More sophisticated communication systems involve changing the amplitude (AMPLITUDE MODULATION, AM) or frequency (FREQUENCY MODULATION, FM) of the carrier wave in accordance with the ANALOGUE or DIGITAL information that is to be transmitted.

At the receiving end, a radio receiver needs firstly to select the required signal from the many radio waves arriving, and a TUNED CIRCUIT is used for this. The system then must extract the information conveyed by the carrier wave. This process is called demodulation. The received information can then be converted back to its original form.

See also LONG WAVE, SHORT WAVE, UHF, VHF.

radioactive series, *decay series* A series of NUCLIDES, all but the last radioactive, where the decay of the first nuclide gives rise to the second, which decays into the third and so on. In general, the first member of such a series has a much longer HALF-LIFE than the others, so that a sample of this nuclide will create the other nuclides in the series with an equilibrium between the rates of production and decay. Thus the amounts of each of the members of the decay series, apart from the final product, will be proportional to their half-lives. A typical radioactive series is that produced by the decay of uranium–238.

$$^{238}_{92}U \rightarrow {}^{234}_{90}Th \rightarrow {}^{234}_{91}Pa \rightarrow {}^{234}_{92}U \rightarrow$$

$$^{230}_{90}Th \rightarrow {}^{226}_{88}Ra \rightarrow {}^{222}_{86}Rn \rightarrow {}^{218}_{84}Po \rightarrow$$

$$^{214}_{82}Pb \rightarrow {}^{214}_{83}Bi \rightarrow {}^{214}_{84}Po \rightarrow {}^{210}_{81}Tl \rightarrow$$

$$^{210}_{82}Pb \rightarrow {}^{210}_{83}Bi \rightarrow {}^{210}_{84}Po \rightarrow {}^{208}_{82}Pb$$

See also RADIOACTIVITY.

radioactive waste The radioactive material produced in the NUCLEAR FISSION of material in a NUCLEAR REACTOR. This material is extracted from the spent fuel and has to be stored carefully until its RADIOACTIVITY has fallen to a safe level – for some ISOTOPES in the waste, this takes several thousand years.

radioactivity The spontaneous decay of unstable atomic nuclei with the emission of IONIZING RADIATION, either ALPHA PARTICLES, BETA PARTICLES or GAMMA RADIATION. Radioactivity occurs spontaneously in many naturally-occurring radioisotopes without any external influence, and may also be induced in certain unstable nuclei by bombarding with NEUTRONS or other particles.

Radioactivity can be harmful to living tissue because of the damage done to living cells by the ionizing radiation. However, radioactivity can also be used to kill cancerous cells. To ensure that healthy tissue does not receive a dose which may lead to further cancers, the radiation source is either implanted in the patient or is in the form of a beam aimed at the patient from several directions, overlapping to form a large dose at the location of the tumour.

See also ACTIVITY, ALPHA DECAY, BACKGROUND RADIATION, BETA DECAY, DECAY CONSTANT, HALF-LIFE, RADIOACTIVE SERIES, RADIOACTIVE WASTE, TRACER TECHNIQUE.

radiocarbon dating An important example of the use of the decrease in ACTIVITY of a radioactive ISOTOPE to find the age of some objects. All living organisms extract carbon from their surroundings. Some of this carbon will be the unstable isotope carbon–14, which is produced in the atmosphere by the interaction of COSMIC RADIATION with nitrogen nuclei. When the organism dies, it stops taking in carbon from its surroundings, and since carbon–14 has a HALF-LIFE of about 5,700 years, measurement of the proportion of carbon–14 compared to stable carbon–12 enables the age to be determined. One technique is to measure the level of RADIOACTIVITY, but since this is very small, greater precision can be achieved and smaller samples are needed if a MASS SPECTROMETER is used to detect the nuclei.

Radiocarbon dating assumes that the proportion of carbon–14 in the atmosphere has remained constant. This is not quite true, as has been shown by the comparison of radiocarbon dates with those obtained by other methods, such as examining the growth rings of trees – a technique called dendrochronology.

radiograph An X-RAY shadow picture, often used in medical diagnosis, taken by placing the patient between an X-RAY TUBE and a suitable detector, such as a fluorescent screen or photographic film. The amount of X-radiation absorbed depends on the density of the tissue through which the X-rays pass. Since bone absorbs far more radiation than soft tissue, radiographs can be used to examine bone structures. In other cases, a CONTRAST ENHANCING MEDIUM, such as barium sulphate, may be introduced into the patient as a BARIUM MEAL to examine the stomach for example, or injected into an artery to produce an image of the arteries (arteriogram). A related form of imaging involves the GAMMA CAMERA. *See also* MEDICAL IMAGING.

radioisotope Any radioactive ISOTOPE.

radio telescope A radio receiver connected to a large dish AERIAL and used to detect RADIO WAVES from space. Since radio waves have a far longer wavelength than visible light, they are diffracted more (*see* DIFFRACTION) so a radio telescope will produce an image with much less RESOLUTION than a visual telescope, despite its larger diameter. *See also* INTERFEROMETRY.

radio waves ELECTROMAGNETIC WAVES with a wavelength greater than about 1 mm. They are produced by oscillating electric charge. Radio waves are used in communications – in radar, television and RADIO broadcasting. Radio waves are also emitted by stars and can be detected by RADIO TELESCOPES.

Radio waves can be generated by producing oscillating currents in electric circuits and feeding this current into a wire or pattern of wires called an AERIAL or antenna. The oscillating charges in the aerial set up an oscillating electromagnetic field which then spreads out into space as an electromagnetic wave. A similar aerial can be used to detect the wave, with either the electric or magnetic field producing a current in the aerial. A RESONANT CIRCUIT can be used to separate currents of different frequencies so that the receiving system can be tuned to respond to signals at one frequency only. Although there is no theoretical upper limit on the wavelength of a radio waves, wavelengths longer than 10 km are of little practical use. *See also* LONG WAVE, MEDIUM WAVE, SHORT WAVE, UHF, VHF.

rain Droplets of water that fall to the Earth's surface from clouds. Droplets are formed when water vapour in the cloud condenses on CONDENSATION NUCLEI in the atmosphere. When they grow too large to be supported by any upward movement of air into the cloud, they fall to the ground as rain.

rainbow An arc of colours, seen when water droplets, usually from falling rain, are illuminated with the Sun roughly behind the observer. The effect is caused by sunlight being refracted (*see* REFRACTION) on entering each raindrop, undergoing TOTAL INTERNAL REFLECTION at the rear of the drop, and then being refracted again on leaving the drop. DISPERSION causes the different colours in the sunlight to be refracted at slightly different angles.

rain shadow A dry area on the side of a hill or mountain that faces away from the PREVAILING WIND. OROGRAPHIC uplifting of incoming moist air causes rain on the upwind side of the hills, leaving dry air on the downwind side.

RAM (**random access memory**) Electronic memory for storing and retrieving data, essentially an array of many thousands of BISTABLES that can each be set to a HIGH or LOW state in which they then remain until reset. The disadvantage of RAM is that it loses the information stored once the power supply is removed.

ramjet A type of JET ENGINE in which the air is forced into the engines by the motion of the aircraft through the air. Ramjets have been used in some high-speed experimental aircraft, but as they can only be started when already moving at high speed, they are unsuitable for routine use.

ramp generator A circuit for producing a VOLTAGE that increases steadily with time. A typical circuit is based on the charging of a CAPACITOR from a constant current supply. A counter circuit driven by a series of regular pulses with its output connected to a DIGITAL-TO-ANALOGUE CONVERTER may also be used, provided the steps are small enough for the output to vary in a reasonably smooth way.

random (*adj.*) Describing a quantity or outcome that cannot be predicted beforehand, but only expressed as a PROBABILITY. For example, when two dice are thrown, it is impossible to predict the total score, but the probabilities of various scores can be calculated.

random access memory *See* RAM.

range The set of values that a FUNCTION can take

on. For example, the range of the function $\sin x$ is $-1 \leq \sin x \leq 1$.

rarefaction A reduction in density, as between compressions when a sound wave travels through a gas.

raster A scanning pattern in which an electron beam covers the whole face of a CATHODE RAY TUBE in a series of lines. *See* TELEVISION.

ratio A description of the relative size of two quantities, often expressed as a pair of integers (a simple ratio). For example if two quantities are in the ratio 3:1 it means that the first quantity is three times the second one. If the ratio is 3:2, it means that the first quantity divided by 3 is equal to the second quantity divided by two, i.e the first quantity is ½ times the second one.

ratio of specific heats The HEAT CAPACITY of a substance at constant pressure divided by the heat capacity at constant volume. For air, this ratio, usually given the symbol γ, is about 1.4. For a material with a MOLAR HEAT CAPACITY at constant pressure of C_P and a molar heat capacity at constant volume of C_V

$$\gamma = C_P / C_V$$

ray An infinitely narrow, parallel-sided beam of RADIATION, such as light, or a line drawn to represent such a beam. The term is also loosely used to denote radiation of any type.

Rayleigh's criterion An EMPIRICAL rule for discovering the RESOLUTION of a telescope, microscope or DIFFRACTION GRATING. Rayleigh's criterion states that two objects of equal brightness can just be resolved as separate if the central maximum of the DIFFRACTION pattern from one object falls in the same place as the first minimum of the diffraction pattern from the other object.

Rayleigh wave In SEISMOLOGY, the surface wave that produces a rolling wave-like motion of the surface in an EARTHQUAKE. *See* SEISMIC WAVE.

RCD *See* EARTH-LEAKAGE CIRCUIT BREAKER.

reactance That part of the IMPEDANCE of a circuit which is due to its CAPACITANCE or INDUCTANCE. The total impedance Z of a circuit containing a CAPACITOR or INDUCTOR is given by

$$Z^2 = R^2 + X^2$$

where R is the RESISTANCE and X is the reactance of the capacitor or inductor. The reactance is the peak value of the POTENTIAL DIFFERENCE across the inductor or capacitor, divided by the peak value of the current through it (equivalently, the ROOT MEAN SQUARE (r.m.s.) voltage divided by the r.m.s. current). There is a PHASE difference between the voltage and the current of 90°, with the current leading the voltage for a capacitance and the voltage leading the current for an inductance.

The SI UNIT of reactance is the OHM. For an ANGULAR FREQUENCY ω, the reactance of a capacitance C is

$$X = \omega C$$

and for an inductance L it is

$$X = 1/\omega L$$

reactor Any vessel in which some kind of reaction takes place, in particular a vessel for containing NUCLEAR FISSION. *See also* NUCLEAR REACTOR.

read only memory See ROM.

real gas Any gas that shows significant departures from the behaviour predicted by the IDEAL GAS EQUATION. The behaviour of such gases can be modelled fairly well by VAN DER WAALS' EQUATION, which takes account of the volume taken up by the gas molecules themselves and of the attractive forces between the molecules (VAN DER WAALS' FORCES).

If the density of the gas is high enough and the temperature low enough, van der Waal's equation predicts a region where volume would increase with increasing pressure. This is unrealistic, and this region is in fact the area where the gas would exist as a SATURATED VAPOUR along with some liquid. The temperature at which this is first seen is called the critical temperature – above this temperature there is no distinction between the liquid and gas states.

See also IDEAL GAS.

real image An IMAGE in which the light rays from any given point on the object actually do meet at the corresponding point on the image. If a screen is placed at the place where a real image is formed, the rays will be scattered into the eyes of the observer and a representation of the object will be seen. *Compare* VIRTUAL IMAGE.

real number A number that contains no imaginary part. In other words, any number that is not a COMPLEX NUMBER.

rechargeable (*adj.*) Describing an electrochemical CELL in which the chemical reactions can be reversed by forcing a current through the cell against its ELECTROMOTIVE FORCE, enabling the cell to be re-used.

reciprocal 1 divided by the quantity concerned. For instance the reciprocal of x is $1/x$, which may be written x^{-1}.

rectangle A geometrical figure having four straight sides at right angles to one another. If the lengths of the sides of a rectangle are a and b, the area is ab.

rectangular hyperbola See HYPERBOLA.

rectification The process of converting ALTERNATING CURRENT to DIRECT CURRENT.

rectifier Any device designed to convert ALTERNATING CURRENT to DIRECT CURRENT. Nowadays a rectifier usually comprises a PN JUNCTION DIODE or an array of such diodes forming a BRIDGE RECTIFIER.

rectify (*vb.*) To convert ALTERNATING CURRENT to DIRECT CURRENT.

red giant A large cool STAR, a late stage of a star's life cycle. The LUMINOSITY of a red giant is high despite the relatively low temperature, because the surface area is large. Red giants lie to the top right of the HERTZSPRUNG–RUSSELL DIAGRAM.

Stars with masses greater than about 0.5 SOLAR MASSES are able to fuse helium and some are able to fuse further elements. In the case of a star of about 8 solar masses or more, this fusion will continue to produce heavier elements such as silicon and iron. During this stage of its lifetime the star's outer layers swell and cool. Eventually the outer layers are thrown off in a cloud of gas called a PLANETARY NEBULA revealing a WHITE DWARF at the core.

See also CEPHEID VARIABLE.

red-shift The apparent increase in wavelength of an ELECTROMAGNETIC WAVE by the DOPPLER EFFECT. This effect may be seen in the light reaching us from distant galaxies, leading to the theory that the Universe is expanding and that it began in a BIG BANG. *See also* HUBBLE'S LAW.

reed relay A pair of contacts made of a FERROMAGNETIC material and enclosed in a glass tube. In a magnetic field, INDUCED MAGNETISM holds the contacts together; when the field is removed they spring apart. A reed relay can be operated by switching on and off an ELECTROMAGNET or by the movement of a PERMANENT MAGNET. *See also* RELAY.

reflectance, reflectivity The proportion of the ELECTROMAGNETIC RADIATION falling on a surface that is reflected by that surface. A perfectly reflecting surface will have a reflectance of 1, whilst a perfectly black surface will have a reflectance of 0.

reflecting telescope A TELESCOPE in which light is gathered and focused to form an image by a CURVED MIRROR. Modern telescopes for astronomical use are almost always of the reflecting type as it is easier and cheaper to build a large accurately curved mirror than a LENS of the same size. This is because a lens has two surfaces which must be accurately machined whilst a mirror has only one. A lens must also be supported only around its edge, so large lenses tend to sag under their own weight, whilst mirrors can be supported over their entire surface.

Telescopes designed for visual observation need a second mirror to reflect the beam of light that has reflected off the main mirror to a position at which the observers' head will not be in the way of incoming light. Two designs are in common use, the Newtonian and the Cassegrain telescopes. A Newtonian telescope has a small secondary mirror placed at 45° to the main mirror and reflects the light out the side of the telescope, where it is viewed with an eyepiece. In a Cassegrain telescope, the secondary mirror is parallel to the main mirror and light is reflected back down the length of the telescope and through a hole in the main mirror, where an eyepiece is located.

Both Newtonian and Cassegrain telescopes tend to be small; most professional astronomy is carried out with telescopes that are not used for direct viewing, but have a photographic plate, CCD, or other light detector located at the focus of a single curved mirror. Earth-based telescopes of more than about 1 m in diameter are limited in the amount of detail they can see by the effects of REFRACTION from the Earth's turbulent atmosphere. This has led to the launch of telescopes into space (*see* HUBBLE SPACE TELESCOPE), and also to experiments with adaptive optics, where the mirror is distorted under electronic control in an attempt to compensate for the distortions introduced in the atmosphere.

See also REFRACTING TELESCOPE.

reflection The process by which a wave strikes the boundary between one medium and another, and leaves the boundary travelling in a new direction but through the same medium as it started in.

A beam of light reflected from a smooth surface such as a mirror leaves in a single direction. This is called regular or specular reflection. If the surface is less smooth the light will be scattered, being reflected in many different directions.

The direction at which light strikes a reflecting surface is called the angle of incidence, this angle being measured from an imaginary line called the normal, which is at right angles to the surface. The angle at which the light leaves the reflecting surface, again measured from the normal is called the angle of reflection.

The behaviour of light striking a mirror is described by the laws of reflection. These state (i) the angle of reflection is equal to the angle of incidence and (ii) the incident ray (the ray of light arriving at the mirror), the reflected ray (the ray leaving the mirror), and the normal all lie in the same plane.

See also FERMAT'S PRINCIPLE, TOTAL INTERNAL REFLECTION.

reflectivity See REFLECTANCE.

reflex angle An angle greater than 180° but less than 360°.

reflex magnetron See KLYSTRON.

refracted ray The ray of light leaving a boundary between two transparent materials, the light having passed from one material into the other. See REFRACTION.

refracting telescope A TELESCOPE in which light is gathered and focused to form an image by a lens. See also REFLECTING TELESCOPE.

refraction The process by which a wave, such as a light wave, changes speed as it passes from one medium to another. Unless the wave meets the boundary between the two materials at right angles, one side of the WAVEFRONTS will meet the boundary before the other side and will thus experience the change of speed sooner, the result being a change in the direction of the wave. If a wave passes into a medium where it is slowed down (generally a denser medium) the wave will be refracted in such a way that it travels more closely aligned with the direction of the normal (the imaginary line at right angles to the boundary between the two surfaces). If the wave passes into a medium in which it travels more quickly

(generally a less dense material), it will be refracted so as to diverge from the direction of normal.

The angle at which an incoming ray of light (called the incident ray) hits the boundary between the two materials is measured from the normal and is called the angle of incidence. The angle at which the outgoing ray (the refracted ray) leaves the boundary is called the angle of refraction and is also measured from the normal. The amount of refraction is given by SNELL'S LAW.

See also APPARENT DEPTH, BIREFRINGENCE, FERMAT'S PRINCIPLE, REFRACTIVE INDEX, REFRACTIVITY.

refractive index In REFRACTION, the ratio of the sine of the ANGLE OF INCIDENCE to the sine of the ANGLE OF REFRACTION. It is also equal to the ratio of the speeds of the wave in the two materials. When a refractive index is quoted for a single medium, the light is taken as entering that medium from a vacuum. Since the refractive index varies with wavelength, the wavelength is taken as that for yellow light (589 nm) unless otherwise stated. See also REFRACTIVITY, SNELL'S LAW.

refractivity A measure of the extent to which a medium refracts light.

$$refractivity = refractive\ index - 1$$

See also REFRACTIVE INDEX.

refractometer Any device for measuring REFRACTIVITY.

regelation The process of the refreezing of ice after it has been melted by the application of pressure, rather than by any increase in temperature. This effect is commonly demonstrated in an experiment in which a weighted copper wire passes through a block of ice, the pressure of the wire melting the ice beneath it, with the ice re-freezing after the wire has passed. The LATENT HEAT of fusion released by the freezing ice is conducted through the thickness of the copper wire and melts the next layer of ice.

register The part of a MICROPROCESSOR that holds the data currently being worked on. The size of the register determines how much data can be worked on at once and therefore helps to determine the size and capacity of the microprocessor. See also SHIFT REGISTER.

regular (adj.) Of reflection, see SPECULAR.

reheat system, afterburner A system used in GAS

TURBINE engines to increase power by burning extra fuel behind the turbines.

relative atomic mass, *atomic weight* The mass of an ATOM measured in ATOMIC MASS UNITS; that is, on a scale where a single atom of carbon–12 has a mass of 12 exactly. One MOLE of atoms will have a mass in grams that is equal to the relative atomic mass of the atom in atomic mass units.

In the case of an element that occurs with several ISOTOPES, the relative atomic mass is normally given as an average of the isotopes, weighted by their natural abundances. For example, chlorine occurs with two isotopes with relative atomic masses of approximately 35 and 37. However, the lighter isotope is roughly three times more common than the heavier one, so any compound made from a naturally occurring chlorine sample will have this ratio of isotopes. Thus the relative atomic mass of chlorine is 35.5.

relative density (r.d.), *specific gravity* The DENSITY of a substance at a specified temperature, divided by the maximum density of water (its density at 4°C). Since the density of water is close to 1 g cm^{-3}, the relative density is close to the density in grams per centimetre cubed. If its relative density is less than 1, a substance will float on water; if its density is greater than 1, it will sink.

The relative density of a gas is usually quoted relative to dry air – both at the same temperature and pressure.

relative humidity HUMIDITY expressed as a fraction of the maximum amount of water vapour that air can hold at that temperature. *See also* SATURATED.

relative molecular mass (rmm), *molecular weight* The mass of a MOLECULE measured in ATOMIC MASS UNITS. The relative molecular mass of a molecule is equal to the sum of the RELATIVE ATOMIC MASSES of the atoms from which the molecule is composed.

relative permeability The factor by which a material changes the strength of any MAGNETIC FIELD in which it is placed. It is the ratio of the magnetic field in a toroidal (doughnut shaped) SOLENOID with a core made of a specified material to the field in a similar solenoid with no core, ignoring the effects of SATURATION and HYSTERESIS. *See also* PERMEABILITY, SUSCEPTIBILITY.

relative permittivity The amount by which the CAPACITANCE of a CAPACITOR is increased by the introduction of a DIELECTRIC into the space between the plates. *See also* PERMITTIVITY.

relative velocity The VELOCITY of an object as seen from another moving object. *See also* GALILEAN RELATIVITY.

relativistic (*adj.*) Travelling at speeds close to the SPEED OF LIGHT. *See* SPECIAL THEORY OF RELATIVITY.

relativistic paradox A paradox, or apparently nonsensical result, arising from the SPECIAL THEORY OF RELATIVITY. All such paradoxes can be resolved by careful consideration of the physics involved. *See* POLE AND BARN PARADOX, TWIN PARADOX.

relativity A blanket term covering the RELATIVITY PRINCIPLE, SPECIAL THEORY OF RELATIVITY and GENERAL THEORY OF RELATIVITY. *See also* GALILEAN RELATIVITY.

relativity principle The idea that there is no observer who has a privileged viewpoint and can claim to be truly at rest. Thus a view of events seen by any observer is equally valid, and all velocities are relative to a given observer. This idea led to the SPECIAL THEORY OF RELATIVITY.

relay An electrical switch operated by the current flowing in an ELECTROMAGNET. Relays are used to control circuits using the flow of current in a separate circuit. The currents and voltages controlled may be far larger than those in the electromagnet. Car starter motors, for example, require currents of several hundred AMPERES. A relay is used to operate these motors, with the current in the electromagnet, which is only a few amperes, controlled by a key-operated switch. *See also* REED RELAY.

relay station An installation, normally on a hilltop or high building, which receives RADIO signals and amplifies and re-transmits them, increasing the coverage of a VHF transmitter.

reluctance In a MAGNETIC CIRCUIT, the MAGNETO-MOTIVE FORCE divided by the MAGNETIC FLUX. It is analogous to RESISTANCE in electrical circuits. As with electrical resistance, the reluctance of any part of a magnetic circuit is proportional to its length, and inversely proportional to its cross-sectional area. *See also* RELATIVE PERMEABILITY.

rem (Abbreviation for **Röntgen equivalent man**) A former unit for effective dose of IONIZING RADIATION. One rem is equivalent to 10^{-2} SIEVERT.

remanence The amount of magnetism retained by a FERROMAGNETIC material when the magnetizing field is removed. *See also* PERMANENT MAGNET.

renewable resource Any natural resource that can be replaced in a reasonable amount of time, for example wood, soil, water and fish. Although they are renewable, the continued supply of such resources relies on their proper use and conservation. Renewable energy is energy obtained from a renewable source, for example SOLAR ENERGY, WAVE POWER, HYDROELECTRIC POWER, GEOTHERMAL ENERGY and wind. NON-RENEWABLE resources cannot be replaced, for example coal, oil and metal ores. Some resources, for example metal used in motor cars and tin cans, can be recycled but it is often uneconomical to do so.

residual current device (RCD) *See* EARTH-LEAKAGE CIRCUIT BREAKER.

resistance A measure of the difficulty with which a CURRENT flows through a CONDUCTOR. It is equal to the VOLTAGE across the object whose resistance is being measured divided by the current through that object. The unit of resistance is the OHM.

$$\text{Resistance} = \text{potential difference}/\text{current}$$

When resistors are connected in SERIES, the total resistance is larger than any of the resistors in the circuit, it will be equal to the sum of the resistances in the circuit. When resistors are connected in PARALLEL, the total resistance is smaller than the resistance of the smallest resistor – connecting another resistor in parallel, no matter how large, can only make it easier for the current to flow. The total resistance is the reciprocal of the sum of the reciprocals of the individual resistors.

For resistances R_1, R_2, R_3 connected in series, the total resistance is R where

$$R = R_1 + R_2 + R_3$$

If these resistances are connected in parallel,

$$1/R = 1/R_1 + 1/R_2 + 1/R_3$$

See also CONDUCTANCE, INTERNAL RESISTANCE, OHMMETER, RESISTIVITY, THERMAL RESISTANCE, WHEATSTONE BRIDGE.

resistance thermometer A thermometer that uses a WHEATSTONE BRIDGE circuit to measure temperature by the change in resistance of a length of platinum wire. It is important to compensate for any changes in the resistance of the wires connecting the sensor to the rest of the bridge circuit. This is done by incorporating a pair of DUMMY LEADS, which run from the opposite arm of the bridge to the point where the sensor is located, but are connected together without being connected to the sensor. In this way, any changes in the resistance of the connecting wires, due to temperature changes for example, affect both arms of the bridge equally, so do not affect the balance of the bridge.

resistance wire Wire made of a metal alloy, deliberately designed to have a high RESISTANCE.

resistivity The RESISTANCE between opposite faces per one metre cube of material. Resistivity is measured in OHM metres (Ωm). The resistance of a conductor is proportional to its length and inversely proportional to the cross-sectional area through which the current flows. It also depends on the material used and the temperature of the material, and these are the two factors accounted for by the resistivity of the material.

For a conductor of length l with cross-sectional area A and resistivity ρ, the resistance R is given by

$$R = \rho l/A$$

See also CONDUCTIVITY.

resistor Any device designed to have a constant RESISTANCE, in order to control the current in a circuit. Small resistors are usually made from a film of metal or metal oxide on a ceramic former, coated with a layer of hard varnish to prevent moisture or dirt from altering the resistance. The value of the resistance is usually marked on the resistor by a series of coloured stripes.

Larger resistors, capable of dissipating more power are made from a coil of metal alloy wire called resistance wire, designed to have a larger resistance than normal wire. Resistors made in this way are called wire wound resistors.

See also VARIABLE RESISTOR.

resolution, *resolving power* The ability of an optical system, such as a camera, microscope telescope or SPECTROMETER to show fine detail in an image. It is usually measured in terms of the closest separation of two objects that can just be shown to be separate by the system under consideration. The resolution is inversely proportional to the WAVELENGTH of light being used. Thus the resolution of a light microscope is limited to two points about 0.2 μm apart, compared to typically 1 nm for the ELECTRON MICROSCOPE.

The resolution of an instrument is ultimately limited by DIFFRACTION. The light rays must pass through an APERTURE or lens at some stage, and will be diffracted as they do so. Resolution is often calculated in terms of RAYLEIGH'S CRITERION.

Experiments on the resolution of the human eye suggest that it comes close to the figure suggested by diffraction, at least near the centre of the field of vision. The density of cells on the retina is not quite high enough for the theoretical resolution to be achieved, and there would be no evolutionary advantage in it being higher. In smaller mammals, such as mice, the brain lacks the processing power to cope with this level of information, and the resolution is much lower than might be expected from diffraction considerations alone.

resolve (*vb.*) The process of finding the COMPONENT of a vector in a specified direction, or a pair of directions, usually at right angles to one another. *See also* RESULTANT.

resolving power *See* RESOLUTION.

resonance The state of an oscillating system when its NATURAL FREQUENCY of OSCILLATION is close to the frequency of some periodically varying driving force (*see* FORCED OSCILLATION). Resonance can occur in electrical circuits, mechanical systems, atoms and molecules.

At resonance the amplitude of the oscillating motion is a maximum and the PHASE difference between the oscillation and the driving force is 90°, with the motion lagging behind the force by one quarter of an oscillation. The size and sharpness of the resonance depends on the Q FACTOR of the oscillating system: the larger this is, the greater the response at resonance and the more rapid the change in phase as the resonance is approached.

Resonance can be useful in that it can be used to select one frequency and reject others, in tuning a radio for example. Resonance can also be a problem in systems which are subject to vibration – such as a car driving over an uneven road. In such systems, DAMPING is usually applied to ensure that the oscillation of the system does not reach dangerous levels. *See also* TUNED CIRCUIT.

resonant cavity A hollow space, often cylindrical in shape, with a particular RESONANT FREQUENCY; particularly such a cavity with metal walls and resonant at a frequency in the MICROWAVE region of the ELECTROMAGNETIC SPECTRUM.

resonant circuit *See* TUNED CIRCUIT.

resonant frequency The frequency at which the response to a periodic DRIVING FORCE reaches a maximum. *See* RESONANCE.

rest energy *See* REST MASS ENERGY.

rest mass (m_0) The MASS of an object, especially an ELEMENTARY PARTICLE, when at rest relative to the observer measuring the mass. Since mass increases with velocity, this is an unambiguous way of specifying the mass of a particular type of elementary particle, regardless of its motion. *See also* REST MASS ENERGY, SPECIAL THEORY OF RELATIVITY.

rest mass energy The ENERGY an object has by virtue of its REST MASS. It is equal to m_0c^2, where m_0 is the rest mass, and c is the speed of light in a vacuum. Rest mass energies, expressed in ELECTRON-VOLTS (eV), are often used in preference to masses in particle physics.

resultant In mathematics, the VECTOR obtained by adding a number of vectors together, particularly the vector sum of the forces acting on an object.

retina The light-sensitive layer at the back of the EYE. In humans, the retina contains sensory cells called rods and cones, which convert the light energy they receive into nerve impulses that travel along the optic nerve to the brain. At the point where the optic nerve leaves the retina, there are no rods or cones and this is called the blind spot. The cones are sensitive to colour and are used mostly for day vision. Rods are mostly used for night vision and cannot distinguish colour.

Colour vision is explained by the trichromatic theory, in which it is thought that three different types of cones exist. These respond to three different colours of light, green, blue and red, and other colours are perceived by a combined stimulation of these.

retrograde (*adj.*) Describing the motion of a planet against the background of stars which is in the opposite direction to the normal west-to-east motion. For the outer planets, this apparent backward drift occurs when the planet is overtaken in its orbit by the Earth, which moves in a smaller, faster orbit. This simple explanation of retrograde motion was one of the chief successes of the COPERNICAN MODEL of the SOLAR SYSTEM.

reverberation The repeated REFLECTION of sounds from the walls of an enclosed space.

reverberation time The time taken for the volume of echoing sound in a room, such as a concert hall, to fall by 60 DECIBELS.

reverse biased Describing the state of a PN JUNCTION DIODE when a voltage is applied that tends to increase the size of the DEPLETION LAYER, i.e. with the P-TYPE SEMICONDUCTOR negative and the N-TYPE SEMICONDUCTOR positive. In this state, no current can flow. *See also* FORWARD BIASED.

Reynolds number A dimensionless number used in HYDRODYNAMICS and AERODYNAMICS particularly to examine the validity of tests performed on scale models.

$$R = \rho v l / \eta$$

where R is the Reynolds number, v the speed of flow and ρ the density and η the VISCOSITY of the fluid. It can be shown that results from a scale model can only be reliably reproduced in the full size object if the two operate at the same Reynolds number. In practice, this is often difficult to achieve and limits the usefulness of tests with scale models.

rheology The study of the deformation and flow of matter, especially plastic solids and viscous liquids.

rheostat Former term for a VARIABLE RESISTOR, particularly when used to control the brightness of a lamp.

Richter scale A logarithmic scale used to measure the magnitude of EARTHQUAKES. The magnitude is calculated as a function of the total energy released by the earthquake and expressed on a logarithmic scale of 0 to 10.

right hand grip rule A rule for finding the direction of a MAGNETIC FIELD around a current-carrying wire or SOLENOID. For a wire, imagine gripping the wire with the right hand, with the thumb pointing in the direction of CONVENTIONAL CURRENT flow; the fingers then wrap around the wire in the same direction as the field lines. For a solenoid, the fingers are made to curl around like the

conventional current and the thumb gives the field direction.

ring system In astronomy, a set of flat discs found around all the GAS GIANTS made of small particles in ORBIT in bands around the equator of the planet. JUPITER'S rings are too faint to be seen from Earth and URANUS' and NEPTUNE'S rings are too dark to be observed against the dark background of the sky, but SATURN'S rings have been known almost since the invention of the telescope.

rod A type of light-sensitive cell found in the RETINA of the eye. Rods are mostly used for night vision and cannot distinguish colour.

roentgen *See* RÖNTGEN.

roller bearing *See* BEARING.

ROM (read only memory) A data storage system into which data can only be placed once. Once the data is stored the memory is 'blown' and the state of each unit of memory is fixed. *See also* EPROM, PROM.

röntgen, *roentgen* Former unit for DOSE of IONIZING RADIATION. One röntgen is approximately 8.7×10^{-7} GRAY.

root A solution to a mathematical equation, a value of an independent variable for which a function takes on a particular value, often zero. In particular, a root is the value of x for which $x^2 = y$ (called the square root of y), or $x^3 = y$ (the cube root of y).

root mean square (r.m.s.) A form of average in which a quantity is squared, the mean value found and then the square root taken. If an ALTERNATING CURRENT flows through a resistor, the r.m.s. values of the voltage and current are the ones that would produce the same heating effect as that voltage or current flowing continuously. For a SINUSOIDAL current or voltage, the root mean square value is 0.707 times the peak value.

rotational dynamics The study of the way in which an object rotates, described in terms of an angle between a point on the object and a fixed direction. The line around which the object rotates is called the axis.

The rotational motion of a body is described in terms of its ANGULAR VELOCITY, the rate of change of its angular position with time. If there is a force acting on the system along a line that does not cross the axis, the angular velocity of the object will change. The rate of change of angular velocity with time, called the ANGULAR ACCELERATION, is proportional to the MOMENT of the force. The amount of angular acceleration also depends on the mass of the object and the way in which that mass is distributed; the acceleration will be less for massive objects or for objects where the mass is far from the axis. These two factors are taken account of in the MOMENT OF INERTIA.

For a point of mass m moving at a speed v at a distance r from the axis, with the motion at right angles to the line joining the object and the axis

$$\omega = v/r$$

where ω is the angular velocity.

Rutherford–Bohr atom A model of the ATOM, incorporating the idea of an atomic NUCLEUS and the idea that ELECTRONS only occupy certain ENERGY LEVELS with QUANTIZED values of ANGULAR MOMENTUM. *See also* BOHR THEORY, HYDROGEN SPECTRUM, RUTHERFORD SCATTERING EXPERIMENT.

Rutherford scattering experiment The experiment performed in 1909 that showed the existence of the atomic nucleus. ALPHA PARTICLES from a radium source were fired in a vacuum at a thin sheet of gold foil and the emerging particles were detected. Most particles passed straight through the foil, but a few were deflected through very large angles, suggesting that they had experienced a large force from a massive object.

Rydberg constant The constant R in the RYDBERG EQUATION, equal to $1.10 \times 10^7 \, \text{m}^{-1}$. The Rutherford–Bohr model of the atom showed that this constant is related to other FUNDAMENTAL CONSTANTS by the relationship

$$R = me^4/8\varepsilon_0^2 h^3 c$$

where m is the mass of an electron, e the charge on an electron, ε_0 the PERMITTIVITY of free space, h PLANCK'S CONSTANT and c the SPEED OF LIGHT.

Rydberg equation An equation which gives the wavelengths λ, of the lines in the EMISSION SPECTRUM of the hydrogen atom,

$$1/\lambda = R \left(1/n^2 - 1/m^2 \right)$$

where R is the RYDBERG CONSTANT and n and m are positive integers, with m greater than n.

S

safety valve *See* VALVE.

satellite Any body in ORBIT around another body. The term usually refers to a natural or artificial satellite that moves in a closed orbit around a planet under the influence of the GRAVITATIONAL FIELD of that planet.

Large numbers of natural satellites have been discovered around the larger planets, whilst the Earth and Pluto are exceptional in having satellites with sizes which are a substantial fraction of the parent planet. The Earth's Moon, the four large satellites of JUPITER (called the GALILEAN SATELLITES) and Titan, the largest satellite of SATURN, are all comparable in size to small planets, but most of the others are far smaller.

Since the 1960s there has been an increasing use of artificial satellites. These are mainly communications satellites, for relaying telephone, radio and television signals around the globe; navigation satellites; military satellites; astronomy satellites, including telescopes that monitor the radio, infrared, ultraviolet and gamma ray emissions from the Solar System and distant galaxies; weather-monitoring satellites; and land resources satellites, used for routine mapmaking and for geological, agricultural and oceanographic purposes. Two special types of orbit are often used, the GEOSTATIONARY ORBIT and the CIRCUMPOLAR ORBIT.

saturated (*adj.*) **1.** Describing a SOLUTION or VAPOUR that can hold no more dissolved or evaporated substance of a specified type.

2. Describing a MAGNETIC material in which all the DOMAINS are aligned, so the magnetism is as strong as possible.

saturated vapour The state of a VAPOUR whose PARTIAL PRESSURE is equal to its SATURATED VAPOUR PRESSURE, the maximum density that the vapour can have at that temperature. If a saturated vapour is cooled, the liquid will condense.

saturated vapour pressure The PRESSURE in the VAPOUR above a liquid at which the molecules leave and re-enter the liquid at the same rate. The saturated vapour pressure increases with temperature. Once the saturated vapour pressure reaches the pressure of the atmosphere above the liquid, bubbles can form and the liquid will be at its BOILING POINT. *See also* SUPERSATURATED VAPOUR.

Saturn The sixth planet in order from the Sun, with an orbital radius of 9.53 AU (1.4 billion km). Saturn is a GAS GIANT like JUPITER, and is the second largest planet in the SOLAR SYSTEM, with a diameter of 110,000 km (8.5 times that of the Earth), and a mass of 5.7×10^{26} kg (95 times that of the Earth). Saturn is most famous for its complex RING SYSTEM. It has an atmosphere of hydrogen and methane, similar to Jupiter's. Eighteen satellites are known, of which the largest, TITAN, is large enough to retain its own atmosphere. Saturn orbits the sun every 29 years and rotates on its own axis every 10 hours.

s-block element Any element in GROUPS 1 or 2 of the PERIODIC TABLE, with the outer electrons in an S-ORBITAL.

scalar A quantity that has no associated direction such as mass and temperature. *Compare* VECTOR, which is specified by its magnitude and direction.

scatter diagram A graph on which individual data points, each consisting of a measured value for two different quantities, are plotted to see how closely a variation in one quantity is related to a variation in the other. If the two quantities are completely dependent on one another, the points will lie on a single line or curve. If there is no relationship, they will be scattered randomly over the graph.

Schrödinger's cat A thought experiment designed to illustrate the strange behaviour of WAVEFUNCTIONS and the philosophical problems with the interpretation of QUANTUM THEORY in PROBABILITY terms. A cat is sealed in a box and is either killed or allowed to survive with equal probability depending on the outcome of some quantum mechanical process, such as the decay of a single radioactive particle within one HALF-LIFE. The outcome is not known until the box is opened, so for an external observer the cat has a wavefunction that is a mixture of alive and dead until the observation is made, but alive or dead thereafter. This raises the question of when and how the wavefunction changes and the artificial boundary between an experiment and its observer.

Schrödinger's equation An equation that provides information about the WAVE NATURE OF PARTICLES and describes the behaviour of a quantity called the WAVEFUNCTION. The wavefunction represents a PROBABILITY AMPLITUDE, the square of which is a measure of the probability of finding the particle at a given point.

Schwarzschild radius The distance from the centre of a BLACK HOLE to its EVENT HORIZON. The Schwarzschild radius effectively determines the size of a black hole, since once an object has passed the event horizon, it cannot escape from the black hole.

scintillation counter A PARTICLE DETECTOR based on materials, called scintillators, that are particularly efficient at converting the energy deposited by charged particles into visible light. This light can than be channelled along an OPTICAL FIBRE and detected by a PHOTOMULTIPLIER. Scintillation counters are often used, along with other particle detectors, to measure the tracks of particles produced in high energy elementary particle physics experiments such as those performed at CERN.

scintillator A material that is particularly efficient at converting the energy deposited by IONIZING RADIATION into visible light.

screen 1. A surface onto which an optical image is projected, designed to scatter the received light into the eyes of those viewing the image.

    2. A conducting layer designed to protect a device from the influence of nearby electromagnetic fields.

sea breeze A wind blowing from the sea to the land driven by the CONVECTION CURRENTS produced as the land heats up more quickly than the sea during the day. *See also* WEATHER SYSTEMS.

Searle's bar A method of measuring the THERMAL CONDUCTIVITY of a good conductor by heating one end of a bar of the material whilst keeping the other end at a fixed temperature. The TEMPERATURE GRADIENT can be measured by inserting thermometers into holes drilled in the sample a known distance apart. The rate of heat flow can be found either by heating the bar electrically at a known rate, or by measuring the rate at which energy has to be removed from the cool end of the bar to maintain the temperature at a steady level. If the bar is cooled with water, for example, the heat flow can be found by measuring the flow rate of the cooling water and its temperature rise. To ensure that heat only flows through the sample under test and not to or from the surroundings, the whole experiment should be thermally insulated, or LAGGED.

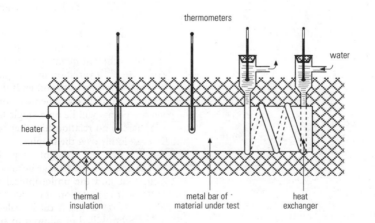

thermometers

water

heater

thermal insulation

metal bar of material under test

heat exchanger

Searle's bar.

secant A function of angle, the reciprocal of COSINE. In a right-angled triangle, the secant of an angle is equal to the hypotenuse of the triangle divided by the side of the triangle adjacent to the angle concerned.

second The SI UNIT of TIME. One second is defined as being equal to the time taken for 9,192,631,770 oscillations of the electromagnetic radiation produced in a transition between two specified ENERGY LEVELS of an atom of caesium–133.

secondary cell Former term for a rechargeable CELL.

secondary coil In a TRANSFORMER or INDUCTION COIL, the coil that supplies energy to a LOAD.

secondary colour Any colour that can be made by mixing two PRIMARY COLOURS. The secondary colours of light are cyan (blue and green), magenta (blue and red) and yellow (green and red).

secondary emission The release of electrons from an ELECTRODE that is itself struck by high energy electrons. Generally this effect is undesirable, but it is used to advantage in the PHOTOMULTIPLIER.

secondary wavelet A small section of WAVE-FRONT used to predict the position of the next wavefront in HUYGENS' CONSTRUCTION.

second law of thermodynamics Any change will bring about an increase in the total ENTROPY of the system if the change is irreversible, or will produce no entropy change if it is reversible. Systems that appear to produce order out of chaos, such as living organisms, do not actually result in a net decrease in entropy, as they give out sufficient heat to increase the entropy of their surroundings by an amount that more than compensates for the decrease in entropy of the living organism itself. *See also* ARROW OF TIME, CLAUSIUS STATEMENT OF THE SECOND LAW OF THERMODYNAMICS, KELVIN STATEMENT OF THE SECOND LAW OF THERMODYNAMICS.

Seebeck effect *See* THERMOELECTRIC EFFECT.

seismic wave A wave travelling through the Earth or along its surface, such as those produced during an EARTHQUAKE. Seismic body waves, which travel through the Earth, can be both longitudinal and transverse. The LONGITUDINAL WAVES travel faster than the TRANSVERSE WAVES and are called p-waves (for primary). P-waves are the first to arrive after an earthquake. The transverse body waves

travel more slowly and cannot pass through the molten core, which will not support this type of wave motion. They are called s-waves (for secondary).

There are two types of seismic surface waves, called Love waves (L-waves) and Rayleigh waves, and it is these vibrations that are responsible for most of the damage caused by an earthquake. L-waves are low frequency transverse waves that move along the upper part of the crust. Rayleigh waves travel a little deeper into the crust, causing a rolling wave-like motion of the surface. Seismic waves are detected with an instrument called a SEISMO-GRAPH.

seismograph An instrument for detecting EARTHQUAKES. Seismographs can also be used to detect the waves returning from man-made explosions, an important technique in geology. In particular they are used to detect reserves of oil or natural gas, which often become trapped in periclines (dome-shaped structures) of porous rock capped by impermeable (non-porous) rock.

seismology The study of EARTHQUAKES. Seismology gives a good deal of information about the structure of the Earth's crust as SEISMIC WAVES are reflected and refracted from various strata (layers) of rock, and sudden discontinuities (FAULTS) in these strata.

self-inductance A measure of the effect in which a changing current in a coil causes an ELECTROMOTIVE FORCE (e.m.f.) in the same coil, tending (by LENZ'S LAW) to oppose the change in current. Such an e.m.f. is sometimes called a BACK E.M.F. The self-inductance of a coil is equal to the induced e.m.f. divided by the rate of change of current which produces it. The SI UNIT of self-inductance is the HENRY. A rate of change of current dI/dt in a single coil with self-inductance L will produce a back e.m.f. of

$$E = L dI/dt$$

semi-circle One half of a circle, a SECTOR where the angle between the radii is 180°. The area of a semi-circle is $\pi r^2/2$.

semiconductor A material intermediate between a CONDUCTOR and an INSULATOR, which conducts electricity but not very well. By far the most important semiconductor is silicon, the material from which many electronic devices, particularly INTEGRATED CIRCUITS, are made.

Pure semiconductors do not conduct at all at very low temperatures, but thermal vibrations of the lattice, or the addition of certain impurities, can make the material conduct. A semiconductor in which the conduction is by charge carriers released by thermal vibrations is called an intrinsic semiconductor. If the charge carriers are released as a result of impurities the material is an extrinsic semiconductor. The addition of impurities, a process called DOPING, must be carefully controlled if the semiconductor is to have predictable properties. In the case of silicon, the lattice is held together by each atom forming four COVALENT BONDS with its neighbours. In such a structure there are no free CHARGE CARRIERS. The addition of an impurity with five VALENCE ELECTRONS, called a donor impurity, can release a FREE ELECTRON that can carry charge through the material. Such a material is called an n-type semiconductor. An impurity with only three valence electrons, called an acceptor impurity, will create a gap in the electron structure of the lattice, which may be filled by an electron from a neighbouring bond. In effect, the shortage of an electron, called a hole, moves through the lattice like a positive charge carrier. A semiconductor in which current is carried mostly by holes is called a p-type semiconductor.

The action of thermal vibrations or light will release electrons from the lattice structure, producing ELECTRON-HOLE PAIRS. This means that semiconductors conduct better at high temperatures, or on exposure to light.

The development of new semiconductors and the ability to produce different electrical characteristics in different parts of a single piece of material by careful doping has led to the evolution of integrated circuits, and is widely regarded as the most important technological development of the second half of the 20th century. Semiconductors are also used in a wide range of ACTIVE DEVICES and TRANSDUCERS, such as TRANSISTORS, DIODES and LIGHT-EMITTING DIODES.

See also BAND THEORY.

semi-metal See METALLOID.

semipermeable membrane A material through which one type of molecule can pass but not another. Typically, such membranes are considered in processes such as OSMOSIS, where the solvent molecules can pass through the membrane, but the solute cannot. One explanation of this effect is that the solute molecules are too large to pass through the membrane, which acts as a 'molecular sieve'. However, some membranes still work with IONIC solutions where the solute ions are smaller than the solvent molecules, suggesting that this cannot be the explanation for every case.

series 1. (*electricity*) In an electric circuit, devices are said to be in series if they are connected one after another, so that all the CURRENT flowing through one device has to flow through the next one. The current is the same in each element and the VOLTAGE across the combination is equal to the sum of the voltages across the individual circuit elements.

2. (*mathematics*) A sequence of mathematical quantities added together, each calculated by a set of rules determined by its position in the series. An example is a power series, where the nth term in the series contains the nth power of some variable. An example of a power series is

$$y = a_0 + a_1x + a_2x^2 + a_3x^3 + \ldots + a_nx^n$$

where all the a's are constants. See also PROGRESSION.

series-wound Describing a D.C. MOTOR in which the ARMATURE and the FIELD COILS are in SERIES.

shadowmask In a CATHODE RAY TUBE for a COLOUR TELEVISION system, a metal plate with many small holes through which electrons can pass to reach only certain areas on the screen. Electrons starting at one of the three ELECTRON GUNS can only hit one type of PHOSPHOR on the screen.

shear The deformation of a material in which two parallel surfaces move by differing amounts in a direction parallel to the surfaces; that is, the two surfaces slide over one another. See also SHEAR STRAIN, SHEAR STRESS.

shear modulus See MODULUS OF RIGIDITY.

shear strain A measure of the deformation produced by a SHEAR. The relative distance moved by two parallel planes in the material undergoing a shear divided by the separation between those planes. See also MODULUS OF RIGIDITY.

shear stress A measure of the strength of a force producing a SHEAR. The size of the force divided by the surface area parallel to the force over which the force acts. See also MODULUS OF RIGIDITY.

shell A series of atomic ORBITALS of roughly similar energies. Completing the filling of one shell and starting to fill the next results in a sudden change in chemical properties, atomic size, IONIZATION ENERGY, etc. This represents the start of a new PERIOD in the PERIODIC TABLE. The shells are labelled K, L, M, N, O, P. The K-shell is closest to the nucleus and can hold two electrons, the next shell is called the L-shell and holds a further eight electrons. The nth shell out from the nucleus can hold $2n^2$ electrons.

Elements with a full outermost shell of electrons are particularly stable; these are the noble gases. Elements that have a few electrons beyond a full shell will tend to lose them to form positive ions (for example sodium, which has 11 electrons, one beyond a full shell). Those with a few electrons too few, will tend to gain electrons to complete a shell, forming negative ions (for example fluorine with nine electrons, one short of a full shell). Those that are not close to a full shell tend to form COVALENT BONDS (carbon for example, with six electrons, four short of a full shell, will form four covalent bonds) or else exhibit more complex behaviour.

The shell model does not take full account of the energy differences within a single shell and so does not fully account for the behaviour of the elements of higher ATOMIC NUMBERS, where a full consideration of the orbitals (sometimes called sub-shells) must be used, though even for heavy elements it provides a useful description of their X-ray spectra (*see* SPECTRUM), which depend on the motion of electrons in the innermost shells.

shell model A model of the structure of the atomic NUCLEUS that treats individual PROTONS and NEUTRONS as existing in ENERGY LEVELS, similar in pattern to those occupied by ELECTRONS in an atom (*see* SHELL). The shell model works well for small nuclei, but the greater complexity of the interactions in the nucleus mean that the model breaks down for larger nuclei, for which the LIQUID DROP MODEL is more often used.

shift register A series of BISTABLES used with their clock inputs activated together and the set and reset inputs of one bistable connected to the outputs of the next bistable in the line. Digits can be fed in one after another, and the clock pulses cause them to move along the register from one bistable to the next. This is the way in which numbers are stored in computers for short-term memory, and how data can be entered step by step, as in a pocket calculator. *See also* CLOCKED BISTABLE.

SHM *See* SIMPLE HARMONIC MOTION.

shock wave A sudden change in pressure in a fluid, propagating as a LONGITUDINAL WAVE, such as that produced by an aircraft travelling though air at a SUPERSONIC speed. Such a shock wave in air is heard as a SONIC BOOM.

short-circuit A circuit with a low resistance, such that a dangerously large current may flow, possibly limited only by the INTERNAL RESISTANCE of the power supply.

short-sightedness, *myopia* A common defect of vision in which the eye lens is too strong for the size of the eyeball. Nearby objects can be seen clearly, but the sufferer cannot see distant objects clearly. It can be corrected using a DIVERGING LENS in front of the eye, effectively weakening the eye lens.

short wave RADIO WAVES with a wavelength from 200 m to 10 m. Short wave radio signals (and to some extent MEDIUM WAVE) can be reflected from an upper layer of the Earth's atmosphere called the IONOSPHERE, which is IONIZED by radiation from the Sun. Reflected radio waves can be used for long-distance communication, but the degree of reflection varies with the state of the ionosphere, which is in turn affected by the time of day, time of year and the activity of the Sun. Short wave communications systems are increasingly being replaced by systems based on communications satellites.

shower A sudden period of rain, often heavy and with large droplets, lasting only a few minutes and produced by convective activity (*see* CONVECTION) within CUMULUS or CUMULONIMBUS clouds. *See also* WEATHER SYSTEMS.

shunt Former term for PARALLEL as applied to electric circuits. Still used to denote a RESISTOR connected in parallel with a meter to produce an AMMETER of lower sensitivity and lower RESISTANCE.

shunt-wound (*adj.*) Describing a D.C. MOTOR in which the ARMATURE and the FIELD COILS are in PARALLEL.

shutter A device designed to allow light to enter for only a short period of time. Shutters are used in cameras to control the time for which light is able to reach the film.

sideband A range of frequencies on each side of the CARRIER WAVE frequency in any modulated radio wave (*see* MODULATION). Sidebands are produced by beats formed between the carrier frequency and the modulating frequency and account for the BANDWIDTH needed to transmit a radio signal. In an amplitude modulated signal (*see* AMPLITUDE MODULATION) the sidebands extend on each side of the carrier frequency by an amount equal to the highest frequency present in the modulating signal. *See also* SINGLE SIDEBAND.

sidereal day *See* DAY.

siemens (S) The SI UNIT of CONDUCTANCE. An object with a conductance of one siemens has a RESISTANCE of one OHM.

sievert (Sv) The SI UNIT of radiation DOSE EQUIVALENT. A dose equivalent of one sievert will have the same effect on living tissue as a dose of BETA PARTICLES depositing an energy of one JOULE per kilogram.

sigma particle (Σ^+, Σ^0, Σ^-) Any one of a family of three BARYONS, one positive, one neutral and one negative, all containing a strange QUARK (*see* STRANGENESS) and having masses around 1.3 times the mass of a PROTON. The charged particles decay via the WEAK NUCLEAR FORCE with a HALF-LIFE of 0.8×10^{-10} s for the positive particle and 1.5×10^{-10} s for the negative particle. The neutral SIGMA PARTICLE decays to the LAMBDA PARTICLE by the ELECTROMAGNETIC FORCE, with a half-life of just 5.8×10^{-20} s.

sign The expression of whether a number is greater or less than zero. Numbers greater than zero are termed positive (+), numbers less than zero are negative (−).

signal Any form of energy, particularly an electrical voltage or a modulated electromagnetic wave, used to convey information.

silicon chip A colloquial term for INTEGRATED CIRCUIT.

silicon controlled rectifier *See* THYRISTOR.

simple harmonic motion (SHM) An oscillating motion of an object or system about a fixed point such that the acceleration of the object is always directed to the fixed point and is proportional to the displacement from the fixed point.

If acceleration is a, the displacement is x and the ANGULAR FREQUENCY is ω, then

$$a = -\omega^2 x$$

The − sign indicates that the acceleration

is directed towards the fixed point. Simple harmonic motion is isochronous (period does not depend on the amplitude).

See also OSCILLATION.

simple microscope A CONVERGING LENS used as a magnifying glass to produce an enlarged VIRTUAL IMAGE. *See also* COMPOUND MICROSCOPE, MICROSCOPE.

simple pendulum A mass (called a bob) swinging on the end of a string or a pivoted rigid support. A simple pendulum is isochronous (period does not depend on amplitude) for small OSCILLATIONS – the period increases for larger swings. The period of oscillation of a simple pendulum is also independent of the mass of the bob. Historically, the simple pendulum was used as the timekeeping element in many clocks, though PIEZOELECTRIC quartz crystals are now used. For a simple pendulum of length l, the period is

$$T = 2\pi(l/g)^{1/2}$$

where g is the ACCELERATION DUE TO GRAVITY. *See also* CONICAL PENDULUM.

simplex (*adj.*) Describing a system, such as a two-way radio link using a single frequency, in which communications cannot be sent and received at the same time. *Compare* DUPLEX.

sine A function of angle. In a right-angled triangle, the sine of an angle is equal to the length of the side of the triangle opposite to the angle divided by the HYPOTENUSE.

sine wave A wave or oscillation the value of which varies smoothly with time in a way that can be described by the SINE function, such as SIMPLE HARMONIC MOTION.

single sideband (ssb) A form of amplitude modulated RADIO transmission in which the BANDWIDTH required is reduced by transmitting only one of the two SIDEBANDS produced when a CARRIER WAVE is amplitude modulated. *See also* AMPLITUDE MODULATION.

sinusoidal (*adj.*) Describing a SINE WAVE.

singularity A point at which a mathematical FUNCTION takes on an INFINITE value, for example the function $1/x$ has a singularity at $x = 0$. In physics, singularities are generally believed to be meaningless and represent a failure of a physical theory to fully describe a situation. A particular example of this is the infinite density of matter at the centre of a black hole.

SI unit Any one of the units of measurement in the internationally agreed METRIC SYSTEM

(*Système International*). In any physical equation, if all quantities are substituted into the equation in SI units, the result will also be in SI units. All units in the SI system are expressed in terms of seven base units and two supplementary units. The base units are METRE (length), SECOND (time), KILOGRAM (mass), AMPERE (current), KELVIN (temperature), MOLE (amount of substance) and CANDELA (LUMINOUS INTENSITY). The supplementary units are the RADIAN (angle) and STERADIAN (SOLID ANGLE).

Any quantity that cannot be expressed directly in terms of one of these units can be expressed in terms of a derived unit, such as the metre per second for velocity. Eighteen of the derived units are given special names. The derived units with special names are NEWTON (force), PASCAL (pressure), JOULE (energy), WATT (power), COULOMB (charge), FARAD (CAPACITANCE), OHM (RESISTANCE), SIEMENS (CONDUCTANCE), VOLT (POTENTIAL DIFFERENCE), HERTZ (frequency), TESLA (MAGNETIC FIELD STRENGTH), WEBER (MAGNETIC FLUX), HENRY (INDUCTANCE), LUMEN (LUMINOUS FLUX), LUX (ILLUMINANCE), BECQUEREL (ACTIVITY), GRAY (radiation DOSE) and SIEVERT (radiation DOSE EQUIVALENT).

slab magnet A short fat BAR MAGNET, with the POLES on the larger faces.

sleet SNOW that melts before it reaches the ground. *See also* WEATHER SYSTEMS.

slip plane In a metallic CRYSTAL, a surface along which the layers of atoms can move relatively easily. Thus a metal crystal is more DUCTILE when being pulled along a slip plane than at an angle to one. The POLYCRYSTALLINE nature of many metal samples makes the presence of slip planes less obvious than the CLEAVAGE PLANES in non-metallic crystals.

slip ring A metal ring attached to a rotating machine, such as the ARMATURE of an ALTERNATOR, to allow current to be carried in or out using brushes.

smoke detector A device for detecting small particles, particularly smoke, in the air. They are used in fire alarm systems. Many smoke detectors operate by measuring the IONIZATION current produced by a source of ALPHA PARTICLES. Smoke absorbs the alpha particles, reducing the ionization current.

smoothing The removal of changes in a DIRECT CURRENT supply. A RECTIFIER, for example, will provide a very unsteady voltage from an ALTERNATING CURRENT supply. A CAPACITOR can be used to even out these variations, charging during the peaks of the supply voltage and discharging to fill in the gaps.

Snell's law In REFRACTION, the SINE of the ANGLE OF INCIDENCE divided by the sine of the ANGLE OF REFRACTION is equal to a constant for a given pair of materials. This constant is called the REFRACTIVE INDEX:

$$\sin i / \sin r = n$$

where i is the angle of incidence, r is the angle of refraction and n is the refractive index of the material.

snow Frozen water, in a hexagonal crystal form, that falls to the Earth's surface from clouds when water condenses at temperatures below its freezing point. *See also* WEATHER SYSTEMS.

soft (*adj.*) In physics, describing a FERROMAGNETIC material in which the alignment of the DOMAINS is easily altered, so the material is easily magnetized and demagnetized.

solar (*adj.*) Associated with the Sun.

solar cell, *photovoltaic cell* A device used to convert light energy, usually from the Sun, into electricity. A solar cell typically consists of a glass or plastic plate with a very thin gold layer, through which light can pass, in contact with a thin layer of copper oxide on a copper base. A voltage is produced between the gold electrode and the base. Modern solar cells, based on semiconductors, are slightly more efficient than their predecessors, but solar cells remain an expensive and inefficient way of generating energy. They are used only where there are no alternatives, such as in spacecraft, and to power electrical devices in remote areas.

solar day *See* DAY.

solar energy Energy from the Sun. With the exception of NUCLEAR ENERGY and TIDAL POWER, all the energy currently available on Earth originated in the Sun. Many schemes have been proposed to convert energy from the Sun directly into heat or electricity. Direct conversion to electricity has found limited use in remote locations, such as providing the power for spacecraft. However, the technology is too expensive, and the amount of energy falling on each square metre too small, to make this energy form economically viable in competition with continued use of FOSSIL FUEL reserves. The use of solar energy for heating water is more widespread, but the temperatures

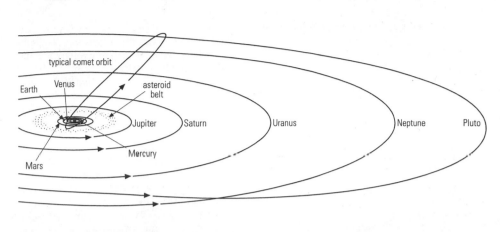

Above: Relative orbits of the planets in the Solar System.
Below: Relative sizes of the planets in the Solar System.

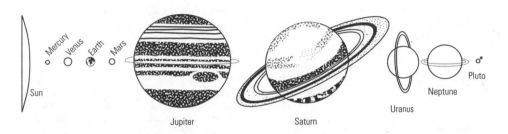

reached are not high enough for this to be used as the sole energy source when heating water.

solar mass The mass of the Sun, which is 2×10^{30} kg. The masses of other stars are often expressed in terms of the solar mass.

solar panel A BATTERY of SOLAR CELLS.

Solar System The name given to the Sun and the bodies held in ORBIT around it by its GRAVITY. The Sun appears to be a typical star, and there is some evidence that other stars may have similar systems. The bodies in orbit around the Sun include PLANETS, planetary SATELLITES, ASTEROIDS and COMETS.

solar wind The stream of high-energy particles, mostly PROTONS and ELECTRONS, given off by the Sun. The interaction between the solar wind and the Earth's atmosphere and magnetic field is responsible for AURORAE and can influence the IONOSPHERE and the state of the Earth's magnetic field.

solenoid A coil of wire used to produce a MAGNETIC FIELD. The magnetic field around a single wire is very weak unless a large current is flowing. To produce a larger field, wire is often wound into a coil. Solenoids often have iron cores to further increase the field strength – the field from the current aligns the DOMAINS in the iron, producing a far stronger field. In a solenoid without an iron core, the field is proportional to the current flowing in the coil, and to the number of turns per metre length of solenoid. The field at the ends of a long solenoid is half that at the middle, and the pattern of field lines is rather like that produced by a BAR MAGNET. The field direction can be found using the RIGHT HAND GRIP RULE. The magnetic field B near the centre of a long solenoid of n turns per metre, carrying a current I and having no core is

$$B = \mu_0 nI$$

where μ_0 is the PERMEABILITY of free space.

solid The state of matter in which a substance retains its shape. The molecules are closely packed – so a solid is not easily compressed –

and are rigidly held together. Solids may be either CRYSTALLINE, in which case the molecules are arranged in a regular lattice, or AMORPHOUS, in which there is no regular arrangement of the atoms. A crystalline solid has a definite melting point at which it becomes liquid, whereas an amorphous solid becomes increasingly soft over a range of temperatures until it assumes liquid properties.

Crystalline solids can occur as CRYSTALS, with the lattice ordering being maintained over long distances to produce pieces of material with symmetrical shapes reflecting the ordered nature of the lattice. This is particularly the case with ionic materials.

See also BAND THEORY, COVALENT CRYSTAL, IONIC SOLID, POLYCRYSTALLINE.

solid angle An angle in three dimensions, as at the point of a cone. Solid angle is defined as the ratio of an area of a section of a sphere to the square of its distance from the point where the angle is measured. Solid angles are measured in STERADIAN.

solid solution A solid MIXTURE in which the atoms, ions or molecules of the constituents are entirely intermixed, rather than appearing as small crystals of each type of material. Certain alloys, such as those formed between gold and silver, are solid solutions. When such a solution is heated, it does not have a single melting point, but melts over a range of temperatures.

solid-state (*adj.*) Describing an electronic system that does not use any THERMIONIC VALVES.

solid-state detector A PN JUNCTION DIODE used to detect IONIZING RADIATION from the ELECTRON-HOLE PAIRS produced in the DEPLETION LAYER by the passage of the ionizing radiation. See RADIATION DETECTORS.

solid-state physics The physics of solid materials, in particular their thermal, electrical and magnetic properties. The creation and investigation of new semiconducting materials, and their use for electronic devices, such as INTEGRATED CIRCUITS, is an important area of solid-state physics.

soliton A pulse-like disturbance that is able to move through a DISPERSIVE material without itself spreading. A soliton was first observed as a water wave on a canal, but subsequent attempts to exploit this property of dispersive materials, for example as a way of sending signals down optical fibres without dispersion,

have met with only limited success.

sonar A system similar in principle to RADAR, but which locates objects using the reflection of SOUND. Sonar is used in ships to measure the depth of the water and to detect fish, submarines, etc.

sonic boom The sudden loud sound heard when a SHOCK WAVE from a SUPERSONIC aircraft reaches the ear.

s-orbital The lowest energy ORBITAL for a given PRINCIPAL QUANTUM NUMBER. There is only a single s-orbital for each principal quantum number and the electron WAVEFUNCTION for an s-orbital is spherical in shape.

sound A LONGITUDINAL WAVE motion that can be heard by the human ear. Sound waves need a supporting medium: they can travel through solids, liquids and gases, but not through a vacuum.

The human ear uses a thin layer of skin called the eardrum to convert the pressure changes in the air into the movement of small bones which then stimulate nerves producing the sensation of sound. The human ear will respond to sound waves in a frequency range of roughly 16 Hz to 15 kHz; the sensitivity is less towards the end of this range and the upper limit decreases with age. Other animals can hear higher frequency sounds and bats use sound waves to locate their prey, sending out pulses of high frequency sound and listening for the reflections. The reflection of a sound wave is called an echo.

See also INFRASOUND, SONAR, SPEED OF SOUND, ULTRASOUND.

source In electronics, the ELECTRODE from which CHARGE CARRIERS enter the CHANNEL in a FIELD EFFECT TRANSISTOR.

source slit In a SPECTROMETER or SPECTROSCOPE, a narrow slit from which the light to be examined is diffracted. The SPECTRUM is then a series of images of this slit in different wavelengths. See also DIFFRACTION, SPECTROSCOPY.

space-time Space and time considered as a single entity, mathematically described by a VECTOR in four dimensions. By considering space and time together, it is possible to simplify the equations of the SPECIAL THEORY OF RELATIVITY, which treat both distances and times as dependent on the motion of the observer. The GENERAL THEORY OF RELATIVITY describes space-time as being distorted by the effects of GRAVITY.

spark The unstable flow of electric currents through gases at high pressures (greater than about 0.1 ATMOSPHERES), with the emission of light and sound. Sparks occur as a result of AVALANCHE BREAKDOWN.

spark chamber A PARTICLE DETECTOR which comprises a volume of gas through which parallel wires are run in a criss-cross arrangement, with alternate sets of wires at right angles to one another and connected to opposite sides of a high-voltage supply. Sparks between the wires form first along any IONIZATION trails left by fast-moving charged particles passing through the chamber. The spark trails can be photographed or the information fed to a computer by detecting the pulses of current in those wires that experienced a spark. *See also* DRIFT CHAMBER.

spark counter A RADIATION DETECTOR that detects IONIZING RADIATION by producing a spark from a high voltage supply in the air between two ELECTRODES through which the ionizing radiation passes.

spark plug A pair of ELECTRODES used to produce a SPARK that ignites the fuel/air mixture in a PETROL ENGINE. They are built to withstand high pressures and temperatures.

special theory of relativity Proposed by Albert Einstein (1879–1955) in 1905, this theory takes as its starting point the idea that there is no one viewpoint for any physical system that is more valid than any other. Thus any observer who is in an non-accelerating FRAME OF REFERENCE (*see* INERTIAL REFERENCE FRAME), will agree with a second such observer about the laws of physics. Thus physics experiments performed in a train that is moving (but not accelerating) are just as valid, and should lead to the same conclusions as those performed by an observer at rest on the surface of the Earth, which is in any case moving through space.

When James Maxwell (1831–1879) put forward the electromagnetic theory of light in 1864, it became clear that light waves travelled at a constant speed irrespective of any relative motion between the source of the light and the observer measuring its speed. This was confirmed experimentally in the MICHELSON–MORLEY EXPERIMENT. Einstein took the incompatibility between the constant SPEED OF LIGHT and Newton's laws and made the bold step of devising a new system of mechanics in which

NEWTON'S LAWS OF MOTION were only approximately valid, with the approximation becoming increasingly good at speeds much less than the speed of light ($3 \times 10^8\,\mathrm{ms^{-1}}$). Objects travelling at speeds sufficiently close to the speed of light for the differences between relativity and NEWTONIAN MECHANICS to be important are said to be RELATIVISTIC.

Consequences of the special theory of relativity include the idea that distances and times between events are different for different observers; the result that nothing can travel faster than the speed of light; and that mass appears to increase for a moving particle. The increase in mass with KINETIC ENERGY turns out to be just one example of another idea that comes from the special theory of relativity, the EQUIVALENCE OF MASS AND ENERGY.

The predictions of the special theory of relativity have subsequently been verified experimentally with fast-moving ELEMENTARY PARTICLES; for example, PI-MESONS moving at speeds very close to the speed of light are able to reach the end of a tube that, according to Newtonian mechanics, is far too long for them to pass along without undergoing radioactive decay. As seen by an observer in the laboratory, this is explained by the fast-moving particles decaying more slowly (TIME DILATION). An observer moving along with the particles would see them decaying at the normal rate, but would see the tube as being shorter (LORENTZ CONTRACTION).

If an observer at rest with respect to an object observes a mass m_0, then an observer moving at a speed v relative to the first will observe mass m, length l and time t with

$$m = m_0/\sqrt{(1 - v^2/c^2)}$$

where c is the speed of light.

See also GALILEAN RELATIVITY, GENERAL THEORY OF RELATIVITY, QUANTUM ELECTRODYNAMICS, POLE AND BARN PARADOX, TWIN PARADOX.

specific charge The CHARGE carried by an object, especially a proton, electron or ion, divided by its mass.

specific gravity *See* RELATIVE DENSITY.

specific heat capacity The HEAT CAPACITY per unit mass of a substance. Specific heat capacity is measured in $\mathrm{Jkg^{-1}K^{-1}}$.

Energy flow = mass × specific heat capacity
× temperature change

The specific heat capacity of water is unusually large at 4200 $J kg^{-1} K^{-1}$. This is due to the HYDROGEN BONDS in water, which absorb energy as the water is heated. The consequence of this is that water heats up and cools down more slowly than most other substances.

For solids and liquids it makes little difference whether the heat capacity is measured under conditions of constant pressure or constant volume, but gases expand substantially when they are heated under a constant pressure. The work then done in pushing back the atmosphere makes the specific heat capacity at constant pressure greater.

Specific heat capacities can be measured by finding the amount of heat needed to change the temperature of a certain quantity of the material by a measured amount. If the temperature of the container in which the specific heat capacity is being measured also changes during the experiment, it is important to take account of the energy involved in heating the container rather than the material under test. This is done by performing the measurement in a CALORIMETER.

See also CONSTANT FLOW METHOD, COOLING CORRECTION, EQUIPARTITION OF ENERGY, METHOD OF MIXTURES, MOLAR HEAT CAPACITY, RATIO OF SPECIFIC HEATS.

specific latent heat of fusion The energy needed to turn one kilogram of solid material at its melting point into liquid at the same temperature.

specific latent heat of vaporization The energy needed to convert one kilogram of liquid at its boiling point to gas at the same temperature.

specific volume The reciprocal of DENSITY, the volume occupied by unit mass of a material, usually at a specified temperature and pressure.

spectral class A classification of STARS according to their surface temperature. The temperature of a star determines the features of its BLACK BODY spectrum and of the LINE SPECTRUM which is seen superimposed on this. This enables stars to be classified in a sequence O, B, A, F, G, K, M, in order of decreasing temperature.

spectral line A narrow range of wavelengths present in an EMISSION SPECTRUM or absent from an ABSORPTION SPECTRUM. The wavelength of a spectral line corresponds to the energy of a transition between two ENERGY LEVELS in the

atom or ion which produced the line. *See also* SPECTROSCOPY, SPECTRUM.

spectrometer An instrument for forming and recording a SPECTRUM. *See* SPECTROSCOPY.

spectroscope An instrument for forming a SPECTRUM and viewing it directly. *See* SPECTROSCOPY.

spectroscopic binary *See* BINARY STAR.

spectroscopy The study of the ELECTROMAGNETIC RADIATION produced by a sample, usually in the INFRARED, visible and ULTRAVIOLET regions of the ELECTROMAGNETIC SPECTRUM. Spectroscopy is a powerful tool for chemical analysis. INFRARED SPECTROSCOPY gives information about the chemical bonds in organic molecules, whilst visible and ultraviolet spectroscopy provides information about which elements are present and their IONIZATION states. Visible spectroscopy is also used in astronomy to provide information about the surface temperatures of stars and, via the DOPPLER EFFECT, about their motion.

A SPECTROMETER is a device used to produce a record of a SPECTRUM, whilst a SPECTROSCOPE enables a visible spectrum to be viewed directly. In either instrument, light from the sample is used to illuminate a slit. Light diffracting (*see* DIFFRACTION) from this slit (the SOURCE SLIT) is focused into a parallel beam by a CONVERGING LENS. The lens and slit assembly is called a collimator. The light is separated into its different WAVELENGTHS by a DIFFRACTION GRATING.

See also ATOMIC ABSORPTION SPECTROSCOPY, ATOMIC EMISSION SPECTROSCOPY, MICROWAVE SPECTROSCOPY.

spectrum (*pl.* *spectra*) The arrangement of ELECTROMAGNETIC RADIATION in order of wavelength or frequency. White light separated into its component wavelengths by a PRISM or DIFFRACTION GRATING gives a characteristic spectrum of colours.

When a sample is heated, or bombarded with ions or electrons, or absorbs photons of electromagnetic radiation, it emits radiation of wavelengths characteristic to the sample. This type of spectrum is called an emission spectrum. If radiation of a continuous range of wavelengths is passed through a sample, the sample absorbs certain characteristic wavelengths. When the transmitted radiation is viewed by a SPECTROSCOPE, the absorbed wavelengths show up as dark bands or lines. This is called an ABSORPTION SPECTRUM. A line spectrum is one in which only certain wavelengths appear, while a continuous spectrum is one in which all the wavelengths in a certain range appear. The absorption or emission spectrum of an element is unique to that element, and details can also be obtained about the temperature and speed of motion of a light source by accurate analysis of the wavelengths it produces. This information is particularly useful in cases where it is not possible to obtain a sample for chemical analysis, in astrophysics for example, or where only very small amounts of material are available.

The term 'spectrum' can also apply to any distribution of entities or properties arranged in order of increasing (or decreasing value). For example, a mass spectrum is an arrangement of molecules, ions or isotopes by mass (*see* MASS SPECTROSCOPY).

See also ELECTROMAGNETIC SPECTRUM, HYDROGEN SPECTRUM, RYDBERG EQUATION, SPECTRAL LINE, WAVE NATURE OF PARTICLES.

specular, *regular* (*adj.*) In physics, describing a REFLECTION in which light is all reflected in one direction rather than being scattered in all directions.

speed The distance an object travels divided by the time it takes to travel through that distance. If the direction of travel is specified, the quantity is a VELOCITY, which is a vector (a quantity with both size and direction), as opposed to a scalar (a quantity with no associated direction). *See also* SPEEDOMETER.

speed of light (*c*) All ELECTROMAGNETIC WAVES, including light, travel through empty space at the same speed, 3.00×10^8 ms^{-1}. In the SI system, the speed of light is now fixed by the definition of the metre as being 299,792,458 ms^{-1}.

The first measurement of the speed of light was made by the Danish astronomer Olaus Römer in 1674 and involved studying the orbits of the satellites of the planet Jupiter. These appear to be completed more frequently as the Earth moves nearer to Jupiter, and less frequently as it moves away, due to the differing times taken for the light to reach the Earth. Later methods involved chopping a light beam into a series of pulses, using a toothed wheel (as did the French scientist H.L. Fizeau in 1849), or a rotating octagonal mirror (as did the American physicist Albert Michelson in 1878). In each of these two cases the pulsed light beam is reflected from a distant point and

returns to the apparatus. It can only re-enter the optical system if the wheel or mirror has now moved far enough for the light pulse to pass through the next gap between the teeth, or to be reflected off the next face of the mirror. In this way, the speed of the light is related to the rotation of the wheel or mirror.

In modern methods of measuring the speed of electromagnetic waves, radio waves form a STANDING WAVE pattern and their frequency is measured by counting the waves over a known time interval and the wavelength from the separation of nodes in the standing wave pattern.

speed of sound The speed at which sound, and other LONGITUDINAL WAVES, pass through a specified medium, usually air. The speed of sound in air is 331 ms^{-1} at 0°C, but increases proportionally to the square root of the ABSOLUTE TEMPERATURE. The speed of sound in a gas is proportional to the ROOT MEAN SQUARE speed of the molecules, so sound travels faster in hydrogen and helium than in air, and more slowly in carbon dioxide. The speed of sound in liquids and solids is generally faster than in air. The far greater rigidity of these materials compared to gases greatly outweighs the increased INERTIA resulting from their greater density, thus any disturbance in one part of the material rapidly spreads to other regions.

In a gas, the speed of sound is $(\gamma p/\rho)^{1/2}$ where p is the pressure, γ the RATIO OF SPECIFIC HEATS and ρ the DENSITY. In a solid the speed is $(E/\rho)^{1/2}$, where E is YOUNG'S MODULUS and ρ the density.

See also MACH NUMBER.

speedometer A speed-measuring instrument. The speedometer commonly used in cars exploits the effect of EDDY CURRENTS set up in a metal disc by a magnet rotating at the same rate as the wheels of the car. The DOPPLER EFFECT can also be used to measure the speed of objects ranging from cars in police speed-traps to stars (*see* HUBBLE'S LAW).

spherical aberration *See* ABERRATION.

spherical co-ordinates *See* POLAR CO-ORDINATES.

spherical mirror A MIRROR with a reflecting surface that is part of a sphere. *See* CURVED MIRROR.

spin Attempts to produce a theory of QUANTUM MECHANICS that also incorporated the ideas of the SPECIAL THEORY OF RELATIVITY led to the discovery that some types of particle possess an inherent ANGULAR MOMENTUM rather as if they were spinning. The spin of a particle is always a whole number times $h/4\pi$, where h is PLANCK'S CONSTANT. Particles with half-integral spin (odd number times this basic unit) are called FERMIONS, of which the electron is an example. Particles with no spin or integral spin (an even number times $h/4\pi$), such as photons, are called BOSONS. *See also* PAULI EXCLUSION PRINCIPLE, STERN– GERLACH EXPERIMENT.

split-ring commutator A metal ring, split into segments that are connected to the coils in a D.C. MOTOR or DYNAMO. As the commutator rotates with the coils, brushes carry current in and out through those segments of the commutator that are best placed to produce a turning effect in the desired direction (in a motor) or are generating the largest VOLTAGE (in a dynamo).

spontaneous emission The change of an atom from one ENERGY LEVEL to another with the emission of a PHOTON, taking place without the influence of any external electric field. *Compare* STIMULATED EMISSION.

spontaneous fission NUCLEAR FISSION by radioactive decay, without having been struck by a neutron. *See also* INDUCED FISSION, RADIOACTIVITY.

spring constant *See* HOOKE'S LAW.

spring pendulum A mass oscillating on the end of a spring. A spring pendulum is isochronous (period of oscillation is independent of amplitude) provided the spring obeys HOOKE'S LAW. For a spring pendulum with spring constant k, and a mass m, the period is

$$T = 2\pi(k/m)^{1/2}$$

spring tide The TIDE produced when the gravitational effects of the Moon and Sun combine, producing the greatest rise and fall in tide level.

SPS (Super Proton Synchrotron) Currently the largest proton SYNCHROTRON, located at CERN (European Centre for Nuclear Research) in Geneva. The SPS is 7 km in circumference and can accelerate electrons to 270 GeV. It can accelerate antiprotons to the same energy in the other direction, and was used in the COLLIDING BEAM EXPERIMENTS that led to the discovery of the W BOSON and Z BOSON in 1984.

square 1. A geometrical figure having four straight sides of equal length and at right angles to one another. If the length of the sides is a, the area of the square is a^2.

2. A number multiplied by itself, for example, the square of 3 is 9.

square wave An electrical signal that switches suddenly and regularly between two values, generally remaining at each of the two values for equal periods of time.

ssb *See* SINGLE SIDEBAND.

stable equilibrium An EQUILIBRIUM state where a small displacement from the equilibrium position will cause the system to create forces tending to return it to its equilibrium position. A simple example of this is a ball resting in the bottom of a bowl.

stall (*vb.*) In engineering, to cease to function, particularly of an AEROFOIL, ELECTRIC MOTOR or INTERNAL COMBUSTION ENGINE. An aerofoil will stall when the ANGLE OF ATTACK exceeds a certain value (typically 14°) and the airflow towards the rear of the wing becomes TURBULENT. The result is a loss of LIFT and an increase in DRAG (*see* AERODYNAMICS). An electric motor stalls when the load exceeds the force produced by the motor. In some motors this can result in overheating and damage, since the wire in the motor is only designed to cope with a current which is limited by the BACK E.M.F. created when the motor is turning. A stall in an internal combustion engine occurs when the load causes the motor to slow down too much in the interval between one POWER STROKE and the next.

standard atmosphere *See* ATMOSPHERE.

standard deviation In statistics, a measure of how much the values recorded depart from the MEAN. The standard deviation is the square root of the mean value of the square of the difference of each value from the mean. Thus if $<x>$ denotes the mean of x, the standard deviation, σ, is

$$\sigma = \sqrt{<(x - <x>)^2>}$$

See also NORMAL DISTRIBUTION.

standard electrode potential The POTENTIAL DIFFERENCE in an electrochemical CELL with a concentration of one MOLE of metal ions per decimetre cubed, and a temperature of 25°C, with an ANODE made of the same metal, measured relative to a HYDROGEN ELECTRODE.

standard form A way of expressing a number as a number with one digit before the decimal point multiplied by a power of 10. For example, 15,479 is written in standard form as 1.5479×10^4, and 0.0001471 as 1.471×10^{-4}.

Standard form is particularly useful in expressing numbers that are so large or so small that they would otherwise contain a large number of zeros.

standard model In elementary particle physics, the picture of the FOUR FORCES OF NATURE unified to produce the ELECTROWEAK FORCE, the STRONG NUCLEAR FORCE and GRAVITY, with these forces acting on three pairs of LEPTONS (ELECTRON, MUON and TAU LEPTON and their NEUTRINOS) and three pairs of QUARKS (up, down, charm, strange, top and bottom). This model has firm experimental support, whilst anything beyond this is widely regarded as unsupported by experimental evidence. *See also* HIGGS BOSON, QUANTUM THEORY, UNIFIED FIELD THEORY.

standard temperature and pressure (s.t.p.) Conditions where the pressure is one ATMOSPHERE (1.01×10^5 Pa) and the temperature is 0°C.

standing wave An oscillating motion along the length of some object, such as a stretched string or air in a pipe. To support a standing wave, the object must be capable of supporting the motion of a PROGRESSIVE WAVE along its length and have ends at which this wave will be reflected. The standing wave is formed by the superposition of waves travelling in both directions along the length of the system, each being reflected when it reaches the end. A standing wave can exist only at those frequencies where the wave is IN PHASE with itself after it has travelled along the system, been reflected at one end, travelled back along the system and been reflected at the other end. On reflection, there may be a 180° phase change. If this occurs at both ends, or at neither end, then standing waves will be supported if they have a wavelength such that the length of the system is a whole number of half-wavelengths. If there is a phase reversal on reflection at one end only, the wavelength must be such that the length of the system is a whole number of half-wavelengths plus an odd quarter wavelength.

Standing waves are also important in QUANTUM THEORY, as the ENERGY LEVELS in an atom are determined by the patterns of standing electron waves that can be supported within the electric field around the nucleus (*see* WAVE-PARTICLE DUALITY). *See also* ANTINODE, FUNDAMENTAL, HARMONIC, OVERTONE, NODE.

Stanford Linear Accelerator Centre (SLAC) California, USA. The home of what is currently the largest LINEAR ACCELERATOR. This machine is 3 km long and can accelerate electrons to an energy of 30 GeV.

star A self-luminous celestial body. A star is a large, roughly spherical mass of gas, mostly hydrogen and helium held together by its own gravity and releasing energy by NUCLEAR FUSION. The Sun is a fairly typical star, though stars with masses between 0.1 and 50 times the mass of the sun are fairly common.

The formation of a star begins with the collapse of a cloud of hydrogen and helium to form a PROTOSTAR. Provided this cloud exceeds a minimum mass of about 0.08 solar masses, the core will reach a temperature high enough to initiate nuclear fusion. Hydrogen nuclei are combined to form helium nuclei in a process known as the PP CHAIN, with the release of energy. This stabilises the star against further collapse. The star is now on the MAIN SEQUENCE, and will produce energy steadily for about 90 per cent of its life.

The lifetime of a star depends on its mass. Heavier stars have more fuel, but are far brighter, so use up their energy more rapidly; the Sun is estimated to be roughly halfway through its 10 billion year life, but the most massive stars probably burn out in about 10 million years. The fact that we see such massive stars today suggests that star formation is an ongoing process.

When most of the hydrogen is consumed, a light star, less than about 0.5 solar masses, will shrink to form a WHITE DWARF. More massive stars expand and cool, forming RED GIANTS, where further nuclear fusion takes place. The medium-sized stars then loose their outer layers in the form of a PLANETARY NEBULA, revealing a white dwarf at the core. In the largest stars, the red giant stage becomes unstable as the nuclear processes become ENDOTHERMIC at higher MASS NUMBERS. This leads to a collapse of the core to form a NEUTRON STAR, whilst the outer layers are blown off in an explosion called a SUPERNOVA. In the case of the most massive stars, the core may collapse without limit, leading to a BLACK HOLE.

The material released in a supernova explosion may eventually recollapse to form a new star, containing mostly hydrogen and helium, but also a small proportion of heavier elements. The Sun and the Solar System are made of material that was processed inside a previous star before being ejected in a supernova explosion.

See also GALAXY, HERTZSPRUNG–RUSSELL DIAGRAM, PULSAR, SPECTRAL CLASS, STELLAR PARALLAX, WEINS' LAW.

state function In THERMODYNAMICS, a quantity, such as TEMPERATURE, PRESSURE or INTERNAL ENERGY, that depends only on the current state of the system and not on how it arrived at that state. Thus the WORK done on a system is not a state function as the system could reach a particular state from a given initial state by various paths involving different amounts of work. Since a simple system carries no memory of how it arrived in a particular state, the laws of thermodynamics are most powerful when formulated in terms of state functions that require knowledge only of the initial and final states of a system.

states of matter The three physical forms in which a substance can usually exist: SOLID, LIQUID and GAS. The different properties of these three states are explained by the KINETIC THEORY of matter. PLASMA is sometimes regarded as a fourth state of matter.

static electricity Effects produced by the separation of opposite electric CHARGES when these charges are at rest. Such effects are described as electrostatic. An object with no overall charge is said to be electrically neutral. When two objects are rubbed together some electrons may be transferred from one object to the other, leaving one with a negative charge and the other with an equal positive charge. This process is called CHARGING BY FRICTION.

static equilibrium See EQUILIBRIUM.

static friction The FRICTION between two objects that are not moving relative to one another. See also LIMITING FRICTION.

statics The branch of mechanics that deals with objects at rest. For an object to remain at rest the vector sum of all the FORCES acting on it must be zero, and the total MOMENT produced by all the forces, taken about any point, must also be zero. These conditions for EQUILIBRIUM are the starting point for all statics problems. See also DYNAMICS.

stationary point A point at which the GRADIENT of a function is zero, but where the sign of the gradient on either side of the function has the

same sign, so that this does not represent a local MAXIMUM or MINIMUM value for the function.

statistical mechanics A development of KINETIC THEORY that applies the laws of MECHANICS to systems containing large numbers of particles in order to calculate quantities such as TEMPERATURE and PRESSURE from the average values of the ENERGY and MOMENTUM of these particles. See also FERMI DISTRIBUTION, MAXWELL–BOLTZMANN DISTRIBUTION.

statistics The branch of mathematics that deals with the collection and interpretation of large numbers of results, and the prediction of the outcome of sampling processes, where a small number of measurements are taken on a much larger group. See also NORMAL DISTRIBUTION, STANDARD DEVIATION.

steady-state theory The cosmological theory that the universe has existed in a steady state throughout time and will continue to do so, with a constant average density. The theory does not explain certain observable events, and most cosmologists now favour the BIG BANG theory. See also COSMOLOGY.

steel An ALLOY in which iron is the predominant component.

Stefan–Boltzmann constant, *Stefan's constant* (σ) The constant σ in STEFAN'S LAW, equal to $5.67 \times 10^{-8}\,\mathrm{Wm^{-2}K^{-4}}$. Early work on the QUANTUM MECHANICS of BLACK-BODY RADIATION showed that the Stefan–Boltzmann constant could be expressed in terms of other fundamental constants by the relationship

$$\sigma = 2\pi^4 k^5 / 15h^3 c^2$$

where k is the BOLTZMANN CONSTANT, h is PLANCK'S CONSTANT and c is the SPEED OF LIGHT.

Stefan-Boltzmann law See STEFAN'S LAW.

Stefan's constant See STEFAN–BOLTZMANN CONSTANT.

Stefan's law, *Stefan-Boltzmann law* That the total amount of BLACK-BODY RADIATION emitted per unit area from a BLACK BODY at an ABSOLUTE TEMPERATURE T is σT^4, where σ is the STEFAN–BOLTZMANN CONSTANT.

stellar parallax A method of measuring the distances of nearby stars. A nearby star will appear to change its position slightly against the background of more distant stars as the Earth moves around its orbit. This leads to a unit of distance called the PARSEC.

step-down Describing a TRANSFORMER that has a TURNS RATIO less than 1, so the voltage across the SECONDARY COIL is less than that across the PRIMARY COIL.

step index Describing an optical fibre which is made from an outer layer surrounding a core made of material which has a constant REFRACTIVE INDEX across its cross-section. *Compare* GRADED INDEX.

step-up Describing a TRANSFORMER with a TURNS RATIO greater than 1, so the voltage across the SECONDARY COIL is greater than that across the PRIMARY COIL.

steradian The unit of SOLID ANGLE. The size of the solid angle at the point of a cone is equal to the area of the section of a sphere at the base of the cone divided by the square of the distance of this surface from the point of the cone. A complete sphere encloses a solid angle of 4π steradian at its centre.

Stern–Gerlach experiment The experiment in which SPIN was first observed experimentally. In this experiment a beam of neutral silver atoms was split into two when passed through an inhomogeneous (non-uniform) magnetic field.

stimulated emission The emission of a PHOTON by an atom in the presence of ELECTROMAGNETIC RADIATION. When a photon of a certain energy interacts with an excited atom, the atom drops from its excited ENERGY LEVEL to a lower one, with the emission of a photon of energy equal to the difference in energy between the two levels. For stimulated emission to take place, the interacting photon must have the same energy as the emitted photon. Stimulated emission in a RESONANT CAVITY is responsible for the operation of the LASER. *Compare* SPONTANEOUS EMISSION.

Stokes' law A law predicting the amount of DRAG produced by a spherical object of radius r, moving with LAMINAR FLOW through a liquid of VISCOSITY η at a speed v. Stokes' law states the viscous drag will be F, where

$$F = 6\pi\eta rv$$

stopping potential In the PHOTOELECTRIC EFFECT, the voltage needed to prevent all the electrons released from reaching a nearby negative ELECTRODE.

strain A measure of the extent to which an object has been deformed. The TENSILE strain in an object is the amount by which it has been extended divided by its original length.

strain energy *See* ELASTIC ENERGY.

strain gauge A device for measuring STRAIN in a machine or other structure. A strain gauge usually consists of metal or semiconductor filament that is attached to the object to be measured and which receives the same strain. The electrical resistance of filament increases as the filament is deformed. By making the filament a component in a WHEATSTONE BRIDGE circuit, the change in resistance can be measured, and thus the strain calculated.

strange A FLAVOUR of QUARK, with charge $-\frac{1}{3}$ in units of the electron charge.

strange attractor In CHAOS theory, a point in PHASE SPACE close to which the path of a chaotic system always passes. If a system is released from a phase space point close to the strange attractor, then small variations in the point of release will lead to large variations in the subsequent behaviour of the system.

strangeness A QUANTUM NUMBER carried by the strange QUARK. It is +1 for the strange quark and −1 for the strange antiquark, but zero for all other particles. Strangeness is conserved in all interactions except the WEAK NUCLEAR FORCE, thus strange particles decay into non-strange particles relatively slowly.

stratopause The boundary between the STRATOSPHERE and the MESOSPHERE. *See* ATMOSPHERE.

stratosphere The layer of the upper ATMOSPHERE in which temperature rises with increasing height due to the absorption of ultraviolet radiation.

stratus Cloud that takes the form of an even layer, usually covering most or all of the sky. *See also* CUMULUS, WEATHER SYSTEMS.

streamlined (*adj.*) Having a smooth shape designed to minimize DRAG forces and minimize the amount of TURBULENT flow.

stress The FORCE on a solid object divided by the cross-sectional area over which that force acts. Stress is a useful way of measuring the force on an object regardless of its size. The point at which a material will break, for example, can be expressed as a stress, regardless of the size of the sample being considered.

stroboscope A device that produces flashes of light at regular intervals, used in physics experiments and as a special effect in discotheques.

stroboscopic photograph A series of pictures recorded on a single piece of film by illuminating the moving object with light from a STROBOSCOPE.

strong nuclear force The force that holds QUARKS together inside PROTONS, NEUTRONS and other HADRONS. The strong nuclear force was first thought to be the force responsible for holding protons and neutrons together in the atomic nucleus. It has since been recognized that this effect is a left-over from the far stronger forces between quarks. The interactions between quarks are described by the exchange of particles called GLUONS, and a full QUANTUM MECHANICAL theory of the strong interaction has been produced. This theory is called QUANTUM CHROMODYNAMICS and the quality that quarks possess that has the same role as CHARGE in QUANTUM ELECTRODYNAMICS is called COLOUR, though it has nothing to do with colour in the visual sense.

subatomic particle Any LEPTON or HADRON, including in particular the PROTON, NEUTRON and ELECTRON, from which atoms are made.

subduction zone A place on the Earth's surface where one TECTONIC PLATE slides under another, with the lower plate re-melting whilst the upper plate may be forced up to form mountain ranges. Subduction zones are areas where EARTHQUAKES occur most frequently as the plates grate past one another.

sublime(*vb.*) To change directly from a solid to a gas without passing through a liquid phase.

sublimation The direct change from solid to gas without passing through a liquid phase when a material is heated. Solid carbon dioxide (dry ice) and iodine are two examples of materials that sublime rather than melt at ATMOSPHERIC PRESSURE.

submarine An underwater vessel that can float or sink by adjusting its average density by flooding tanks with water or using compressed air to force that water out of the tanks. Submarines are used by the military, as they can move quickly and are difficult to detect since sea water conducts electricity and so is virtually opaque to electromagnetic radiation. SONAR systems can be used to detect submarines, but must be immersed in water to operate, so cannot be carried on fast aircraft or satellites.

subsidiary maximum In a DIFFRACTION pattern, a maximum of INTENSITY other than the central maximum.

subsidiary quantum number The QUANTUM NUMBER used to label an ORBITAL giving a measure of the ANGULAR MOMENTUM of an electron in the orbital, and describing the shape of the orbital. The subsidiary quantum number is usually expressed as a letter in the sequence s, p, d, f, from an early mistaken categorization of lines in SPECTRA as sharp, principal, diffuse and fine.

summing amplifier An AMPLIFIER with two or more inputs and an output equal to the sum of these inputs, each multiplied by a constant.

Sun The STAR at the centre of our SOLAR SYSTEM. It has a mean diameter of 1,392,000 km, a mass of 2.0×10^{30} kg and lies at a distance of 149,500,000 km from the Earth. It is composed of about 75 per cent hydrogen, 24 per cent helium and about 1 per cent other elements. THERMONUCLEAR REACTIONS take place in the core of the Sun, converting hydrogen into helium and generating vast amounts of energy. Temperatures here are thought to reach 15,000,000°C, but are closer to 6,000°C at the surface. The Sun is classified as a MAIN SEQUENCE STAR.

superatom *See* BOSE–EINSTEIN CONDENSATION.

superconductivity The disappearance of electrical RESISTANCE, exhibited by some materials, called SUPERCONDUCTORS, at low temperatures. The theory of superconductivity is complex, but it involves electrons forming pairs, called Cooper pairs, with equal and opposite momenta (*see* MOMENTUM). Whenever one member of the pair is scattered, the other receives an equal and opposite momentum change, hence there is no overall change in the rate at which charge is carried through the material. At low temperatures, the Cooper pairs behave like BOSONS and experience a BOSE–EINSTEIN CONDENSATION, so they are almost all in the lowest ENERGY LEVEL. An applied electric field then causes all the pairs to accelerate together, without any scattering.

It has been suggested that the phenomenon of superconductivity could be used to build highly efficient electrical machines. However, so far applications have been limited, owing to the high costs involved in cooling the materials to the low temperatures required. Liquid helium (boiling point 4 K) is often used. Some HIGH-TEMPERATURE SUPERCONDUCTORS are known, which may be cooled with liquid nitrogen (boiling point 77 K), but these materials do not have good mechanical properties. Another problem is that many electrical machines rely on magnetic fields, and magnetic fields have a tendency to destroy superconductivity. Despite this, superconducting magnets have been used in medical MAGNETIC RESONANCE IMAGING machines and in some PARTICLE ACCELERATORS.

superconductor Any material that exhibits SUPERCONDUCTIVITY. Many metals become superconductors at low enough temperatures, but the temperatures below which superconductivity occurs are often very low, only a few KELVIN. One of the best metallic superconductors is Niobium-tin alloy, which retains its superconductivity up to 22 K in zero magnetic field. *See also* HIGH-TEMPERATURE SUPERCONDUCTOR.

supercooled (*adj.*) Describing a material that has cooled below its melting point, but remained liquid.

superfluidity The disappearance of all VISCOSITY in liquid helium–4 below 2.2 K. Superfluid helium–4 also displays other unexpected properties, such as the ability to escape from an open vessel by forming a thin film and climbing up the inside of the vessel and down the outside, to collect in droplets at the base of the vessel.

Superfluidity is a consequence of a BOSE–EINSTEIN CONDENSATION in the atoms of helium–4 at low temperatures. At low temperatures, almost all the atoms are in the lowest energy state, and under the influence of a force they gain or lose momentum together, behaving as a single entity without any viscous forces resulting from collisions.

supergravity A theory derived from the combination of the ideas of GAUGE THEORIES and SUPERSYMMETRY that reproduces the GENERAL THEORY OF RELATIVITY. Unfortunately the pattern of ELEMENTARY PARTICLES predicted cannot be made to fit the observed pattern of the STANDARD MODEL. Whilst it is possible that QUARKS and LEPTONS are composite objects made up from the particles predicted by supergravity, most scientists now accept supergravity as being of theoretical interest only.

superheated The state of a liquid heated above its boiling point without boiling. Such a state can be achieved provided the liquid is clean and the walls of its container smooth, since extra energy is needed to form small bubbles unless CONDENSATION NUCLEI are present. These are normally provided by dirt or imperfections in the walls of the container. When a superheated liquid does boil, it will do so rather violently – an effect called bumping.

supernova A sudden brightening of a star as the outer layers are blown off when the core collapses to form a NEUTRON STAR. The brightness of a star may increase by a factor of several million in a period of a few minutes, before decaying over a period of a few weeks.

superpose (*vb.*) To add together two or more WAVES. *See* INTERFERENCE. *See also* PRINCIPLE OF SUPERPOSITION.

superposition The adding together of two or more WAVES. *See* PRINCIPLE OF SUPERPOSITION.

Super Proton Synchrotron *See* SPS.

supersaturated vapour The state of a VAPOUR with a PARTIAL PRESSURE higher than its SATURATED VAPOUR PRESSURE. This is an unstable state, and droplets of liquid will form on any small particles or any irregularities in the surface of the container – these are called CONDENSATION NUCLEI.

supersonic (*adj.*) Faster than the SPEED OF SOUND.

superstring theory A theory of particle physics that postulates that ELEMENTARY PARTICLES are not point-like objects, but extremely small strings in a multidimensional space, where all but four of the dimensions (SPACE-TIME) have collapsed to very small distances. Despite the attractiveness of these theories, there is no experimental evidence to support them.

supersymmetry A SYMMETRY that links ordinary algebraic variables with anticommuting variables such that $ab = -ba$. The use of this symmetry in physics leads to a prediction that for every type of FERMION there is a corresponding BOSON and vice versa. Unfortunately, it is not possible to match the known fermions and bosons of the STANDARD MODEL in this way, so supersymmetric theories predict the existence of many new particles that have not yet been discovered. Whilst it has been suggested that such particles may account for the MISSING MASS of the Universe, the failure to detect such particles in other experiments has led to growing scepticism about the validity of simple supersymmetric theories, though the ideas have been extended to other theories such as SUPERGRAVITY and SUPERSTRING THEORY.

surface tension The force that appears at the surface of a liquid and tends to pull the liquid into spherical droplets. It results from the imbalance of INTERMOLECULAR FORCES acting on a molecule near the surface of a liquid. Such molecules are attracted by all the molecules below it in the liquid, whereas a molecule inside the liquid is attracted by molecules from all sides. The result is physically the same as if the surface were made of an elastic sheet with a constant force across any imaginary line on that surface, proportional to the length this line. The surface tension – which is temperature dependent, falling to zero at the CRITICAL TEMPERATURE of the liquid – is equal to the force per unit length along such a line.

The effect of surface tension allows small objects, such as some insects, to rest on a water surface. Another result is that the pressure inside a bubble is higher than that outside, by an extent which depends on the surface tension of the liquid and the radius of the bubble. It is highest when the bubble is smallest; for similar reasons, a rubber balloon requires the most effort to blow it up when it is small.

For a liquid with surface tension T, in a bubble of radius r, the pressure inside the bubble exceeds that outside by Δp, where

$$\Delta p = 2T/r$$

If the bubble has two surfaces, i.e. is made of a thin film of liquid with gas inside and outside, the difference is then

$$\Delta p = 4T/r$$

susceptance The reciprocal of REACTANCE. The SI UNIT of susceptance is the SIEMENS.

susceptibility The extent to which a material increases (if FERROMAGNETIC or paramagnetic; see PARAMAGNETISM) or decreases (if diamagnetic; see DIAMAGNETISM) the strength of a MAGNETIC FIELD in which it is placed.

Susceptibility = relative permeability – 1

See also RELATIVE PERMEABILITY.

s-wave In SEISMOLOGY, the transverse waves that arrive after the P-WAVES in an EARTHQUAKE. See also SEISMIC WAVE.

switch In electronics, any device for making or breaking a connection in an electric circuit, or for diverting a current from one part of a circuit to another.

symmetry The property of a system or object remaining unchanged when certain changes are made. In particular, it is the property of certain shapes that remain unchanged under specified transformations.

synchrocyclotron A development of the CYCLOTRON that used a varying frequency for the accelerating electric field to overcome the changing period of particle orbits as they reached RELATIVISTIC speeds. As particle accelerators became larger, the provision of a magnetic field over a large area became increasingly problematic and the synchrocyclotron was superseded by the SYNCHROTRON.

synchronization pulse A signal transmitted, in TELEVISION for example, to keep the transmitting and receiving apparatus in step. In television these are transmitted with the picture information to ensure that the scanning at the receiving end remains in step with the transmitted information.

synchrotron A PARTICLE ACCELERATOR that overcomes the limitations imposed on the CYCLOTRON by the RELATIVISTIC speeds of particles causing a change in orbital period. Unlike the SYNCHROCYCLOTRON, the synchrotron does not require a magnetic field to be maintained over a large area. The synchrotron uses a steadily increasing magnetic field and an increasing frequency of alternating current to accelerate charged particles around a circular evacuated tube, called the beam tube. ELECTRODES placed at various locations around the tube accelerate the particles, whilst ELECTROMAGNETS focus the beam of particles and bend it into a roughly circular path. High energy synchrotrons are generally very large machines as the amount of energy lost by SYNCHROTRON RADIATION decreases rapidly with increasing radius. See also COLLIDING BEAM EXPERIMENT, SPS.

synchrotron radiation A characteristic pattern of ELECTROMAGNETIC RADIATION given off by the charged particles in a SYNCHROTRON, which are accelerating by virtue of their CIRCULAR MOTION. This is an important source of X-rays in medicine and other applications, but it places a limit on the energy that can be achieved by a given size of synchrotron.

T

tachyon A hypothetical particle travelling faster than the speed of light. This is not forbidden by the SPECIAL THEORY OF RELATIVITY, which only forbids objects from accelerating past the speed of light. Tachyons would have some very distinctive properties if they did exist, but as yet there is no experimental evidence for them.

tangent A function of angle. In a right-angled triangle, the tangent of an angle is equal to the length of the side opposite the angle divided by the side adjacent to the angle. The tangent of any angle is equal to its SINE divided by its COSINE.

tau lepton, *tauon* (τ) The heaviest known LEPTON, with a mass more than 3,000 times that of the electron, but a HALF-LIFE of only 3×10^{-13} s.

tauon *See* TAU LEPTON.

tectonic plate A region of solid rock in the Earth's crust. The crust is composed of several such plates which drift around on the surface of the Earth, moving at speeds up to a few centimetres per year. This motion, called continental drift, is driven by CONVECTION CURRENTS in the mantle. The plates effectively float on the mantle, with removal of material from one part of the plate causing the plate to rise up, whilst other plates sink – a process called isostasy. In some places, such as the Mid-Atlantic ridge, plates are moving apart and hot mantle material solidifies. Volcanoes are found in such areas as magma forces its way to the surface in a semi-molten form called lava. As the magma cools and solidifies, it stores a record of the MAGNETIC FIELD OF THE EARTH, which has reversed many times since the Earth was

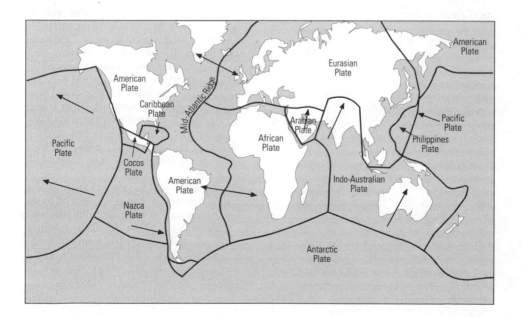

Tectonic plates.

formed. This enables the rate of spreading to be compared at different places on the Earth. It is believed that these plates originally supported a single large continent called Pangaea.

tectonics The branch of GEOLOGY concerned with the study of the structure and motion of TECTONIC PLATES.

telecommunications The conveying of information from one place to another, by wire or radio.

telephone A TELECOMMUNICATIONS system for the transmission of voice. The ANALOGUE speech signal is converted into an electrical signal by a microphone in the mouthpiece. This signal is transmitted via a series of switching stations, called exchanges, by wire, FIBRE OPTIC cables, microwave or satellite, to the receiving telephone, where it is converted back into speech by a loudspeaker in the earpiece. A dialling system and the exchange enables each telephone to be connected to any other telephone in the system. Telephone is an example of a DUPLEX system (two-way communication; in a SIMPLEX system communication can only be achieved in one direction).

Telephone signals are often converted to DIGITAL form if they are to be sent over long distances. In addition to the improved quality and freedom from INTERFERENCE (unwanted signals added to the signal being sent) that can be achieved with digital techniques, the individual digits take less time to send than the speech they represent. This means that several signals can be sent along the same wire, a process called MULTIPLEXING. One form of multiplexing involves sending digital signals representing several different telephone calls in turn down the same wire – this technique is called TIME-DIVISION MULTIPLEXING.

telescope A device that produces a magnified image of a distant object. Telescopes are divided into two classes, REFLECTING TELESCOPES, which use a CONCAVE mirror to gather the light from the distant object, and refracting telescopes, which use a CONVERGING LENS (the object lens). For visual use, the real image produced is then viewed through an eyepiece – a converging lens that acts as a magnifying glass. Most modern large astronomical telescopes are not used directly, however, and photographic film, a CCD or other detector is placed at the FOCUS of the mirror (modern astronomical telescopes are always reflecting telescopes).

To prevent disturbance from light pollution (man-made light sources scattering off particles in the atmosphere) and to minimize REFRACTION from layers of air at different temperatures, large astronomical telescopes are now located on mountain tops or launched into space, as was done with the HUBBLE SPACE TELESCOPE. For use in astronomy, telescopes are built with as large a mirror as possible, this is done both to collect more light, allowing fainter objects to be detected, and to minimize the effects of DIFFRACTION (the WAVEFRONTS of the light entering the telescope are limited by the size of the mirror; the larger the mirror, the less diffraction this will produce).

See also RADIO TELESCOPE.

television A system for the transmission and reception of pictures using RADIO WAVES. Television receivers rely on the CATHODE RAY TUBE. In a television set, voltages are applied to coils around the neck of the tube to produce a scanning pattern called a raster, in which the electron beam moves across the screen from left to right in a series of lines, each line just below the previous one, before returning to the top left-hand corner. As the beam moves, its brightness is controlled by the transmitted signal and a picture is built up on the screen. *See also* COLOUR TELEVISION.

temperature A measure of the hotness or coldness of an object. If two objects are placed in contact with one another, HEAT will flow from the one at the higher temperature to the one at the lower temperature. If they are at the same temperature, there will be no heat flow. The SI UNIT of temperature is the KELVIN. *See also* ABSOLUTE TEMPERATURE, TEMPERATURE SCALES, THERMOCOUPLE, THERMOMETER, PYROMETER.

temperature coefficient of resistivity The fractional change in RESISTIVITY for each degree CELSIUS of temperature change. The temperature coefficient of resistance is positive for metals, which have resistances that increase with increasing temperature, but negative for semiconductors, where the resistance decreases with increasing temperature. If there is a change in temperature $\delta\theta$, there will be a change in resistivity $\delta\rho$, given by

$$\delta\rho = \rho\,\alpha\,\delta\theta$$

where α is the temperature coefficient of resistivity.

Thermionic emission is an ACTIVATION PROCESS with an ACTIVATION ENERGY equal to the WORK FUNCTION of the metal. *See also* CATHODE RAY TUBE, ELECTRON GUN, THERMIONIC VALVE.

thermionic valve Any device in which electrons are emitted from a heated CATHODE by the process THERMIONIC EMISSION, and are attracted to an ANODE at which they are collected. A number of other ELECTRODES may also be present to modify the electron flow. Apart from the CATHODE RAY TUBE, X-RAY TUBE and MAGNETRON, thermionic valves are largely obsolete, having been replaced by solid-state semiconductor devices, which are smaller, more robust and consume less power.

thermistor An electronic device designed to exploit the decrease in resistance of a semiconductor, such as germanium, with increasing temperature. Thermistors are used in electronic circuits that are required to respond to changing temperatures.

thermocouple A pair of junctions between two different metals. If there is a temperature difference between these junctions, the different thermal motion of the electrons in the two metals will cause an ELECTROMOTIVE FORCE by the THERMOELECTRIC EFFECT. This can be used to measure temperature.

thermodynamic equilibrium The state in KINETIC THEORY where individual molecules are exchanging quantities such as energy and MOMENTUM, or reacting chemically, but the total amount of any chemical present, or the total energy, is unchanging. Thus the system can be meaningfully described by quantities such as temperature or the chemical concentration of its constituents.

thermodynamics The study of thermal ENERGY changes and ENTROPY. Whilst thermodynamics makes some complex predictions about the behaviour of systems, its attractiveness lies in the fact that it makes only very general assumptions, so these predictions are applicable to a wide range of systems regardless of the detailed way in which a given system operates. The principles of thermodynamics can be applied to objects as diverse as black holes and living organisms.

Thermodynamics is based on four laws. The ZEROTH LAW defines the concept of TEMPERATURE, by stating that two objects are at the same temperature if there is no net heat flow between them when they are in thermal contact. The FIRST LAW encapsulates the LAW OF CONSERVATION OF ENERGY including INTERNAL ENERGY and the recognition that heat is a form of energy. The SECOND LAW defines the concept of entropy, a measure of the degree of disorder in a system, and states that the entropy of a closed system can never decrease. The consequence of the third law, the NERNST HEAT THEOREM, is that it is impossssible to reach ABSOLUTE ZERO in a finite number of steps.

See also ARROW OF TIME, CARNOT ENGINE.

thermoelectric effect, Seebeck effect The production of a voltage across a pair of junctions between two different metals, when the junctions are kept at different temperatures. This is a way of converting heat energy directly to electricity, but the EFFICIENCY of energy conversion in the thermoelectric effect is low, and it is usually used as a way of measuring temperature differences rather than as an energy source. Some spacecraft, however, have used a THERMOPILE and heat from a radioactive source to provide electricity. *See also* THERMOCOUPLE.

thermoluminescence LUMINESCENCE of an object when it is heated. This luminescence, which is usually very faint, is caused by electrons becoming trapped within defects in a crystal lattice. When heated, lattice vibrations release the electrons. These EXCITED STATES arise as a result of exposure to IONIZING RADIATION, so increase over time. In this way thermoluminescence can be used for the dating of archaeological artefacts, especially pottery.

thermometer Any device for measuring TEMPERATURE. Thermometers are generally classified according to the THERMOMETRIC PROPERTY they exploit. Liquid in glass thermometers are based on the THERMAL EXPANSION of a liquid, usually alcohol or mercury. RESISTANCE THERMOMETERS use the change in resistance of an electric CONDUCTOR, often platinum. The CONSTANT VOLUME GAS THERMOMETER uses the variation in pressure of a fixed mass of gas held in a constant volume.

thermometric property Any property that varies with temperature, such as the volume of a liquid, the pressure of a fixed volume of gas, or the resistance of an electric conductor, and which can be used to measure temperature.

thermonuclear reaction A CHAIN REACTION in NUCLEAR FISSION maintained by slow moving NEUTRONS. The neutrons are described as thermal because collisions with the MODERATOR in a nuclear reactor give them a distribution of energies which can be characterized by a temperature in a similar way to distribution of energies of the molecules in a gas.

thermopile A BATTERY of THERMOCOUPLES.

thermosphere *See* ATMOSPHERE.

thermostat A temperature-controlled switch, usually connected to a heater or refrigerator for the purpose of maintaining a constant temperature. Most thermostats are based on a pair of contacts, one of which is attached to a BIMETALLIC STRIP, whilst the other can be moved on a screw thread to select the temperature at which the thermostat operates. The contacts are often magnetized so that they attract one another in order to produce a switch that opens and closes sharply to reduce damage and electrical interference caused by sparks.

thin-film interference The INTERFERENCE effect in which light is reflected off the front and back surfaces of a thin layer and interferes when brought to a focus in the eye. The two reflected beams have travelled different distances before reaching the eye and interfere constructively or destructively (*see* INTERFERENCE) depending on their wavelength and the thickness of the film. The fact that different colours have different wavelengths means that some colours in the white light interfere constructively, others destructively, producing the coloured bands that are seen in soap bubbles and thin layers of oil.

threshold The smallest value of energy, or an energy-related quantity such as frequency, that is required to produce a given effect.

threshold of hearing The quietest sound that can be heard. This varies with FREQUENCY and from person to person, but for a person with normal hearing, this is assigned the level 0 dBA. *See also* DECIBEL.

thrust A pushing or propelling force, as produced by a JET ENGINE, for example.

thunder The sound produced by the sudden heating and consequent rapid expansion of the air when LIGHTNING is formed.

thyristor, *silicon controlled rectifier* A SEMICONDUCTOR device that allows current to flow in one direction (from the ANODE to the CATHODE) on the application of a voltage to a third ELECTRODE, called the gate. Once a current is flowing, it will continue to flow regardless of the gate voltage provided the current remains above a threshold value.

tidal forces Forces producing a STRESS within an object in a GRAVITATIONAL FIELD resulting from the difference in the strength of gravity between one side of an object and the opposite side. This arises due to the differing distances of different parts of the object from the source of the gravitational field. Such forces are responsible for TIDES on the Earth, as a result of the differing effect on the Moon's gravity on different parts of the Earth. Tidal forces are also believed to be responsible for the formation of RING SYSTEMS around the GAS GIANTS, and would cause the destruction of any solid object as it approached an extremely strong source of gravity, such as a BLACK HOLE.

tidal power The generation of electricity on a commercial scale from the tidal rise and fall of water levels. A tidal power station is operating at La Rance in France and one is planned for the Severn Estuary in the UK. Problems arise from the fact that the times of peak generation do not coincide with peaks in demand, the large building costs and the environmental impact on the area flooded. There are relatively few areas that have sufficiently strong tides or the natural basin needed to build such a scheme, but where they can be constructed they seem economically worthwhile.

tide A change in sea level caused by the different gravitational attraction of the Moon on that part of the Earth which is closest to the Moon at any time compared to that part which is furthest away. In effect, one part of the Earth is pulled towards the Moon more strongly than the average whilst the opposite part is left behind. As the Earth rotates, two high and two low tides are seen each day.

Tides are also affected to a lesser extent by the gravitational pull of the Sun. When the effects of the Sun and Moon combine, large tides, called spring tides, are seen. When the Sun and the Moon are at right angles as seen from the Earth, their effects tend to cancel, producing smaller tides, called neap tides.

Tidal effects are observed elsewhere in the Solar System and are believed to be responsible for the formation of RING SYSTEMS around the GAS GIANTS.

timbre The property of a sound that the ear distinguishes in two sounds having the same amplitude and frequency, but produced by two different instruments. This is due to the differing shape of the sound wave, caused by the different HARMONIC content in the sounds. A SINUSOIDAL sound wave will produce a pure sound.

time The continuous passage of existence, originally marked by the rising and setting of the Sun, but now recorded by a wide range of oscillating devices. The SI UNIT of time is the second. The SPECIAL THEORY OF RELATIVITY recognizes that the finite speed of light means that the passage of time is not the same for all observers. *See also* SPACE-TIME, TIME DILATION.

timebase A voltage applied to the X-PLATES of the CATHODE RAY TUBE in a CATHODE RAY OSCIL-LOSCOPE to move the electron beam across the screen at a steady rate, to enable a graph of voltage against time to be produced.

time constant For any process where a quantity changes exponentially with time, the time constant is the time taken for that quantity to fall to 1/e times its initial value if decreasing or to rise to e times its initial value if increasing, where e is the EXPONENTIAL constant. Used particularly of circuits containing a RESISTANCE R and a CAPACITANCE C, when the time constant is RC.

time dilation The effect that results in any moving clock running more slowly than when at rest according to the SPECIAL THEORY OF RELA-TIVITY. If an observer at rest with respect to an object observes a time interval t_0 then an observer moving at a speed v relative to the first will observe time interval t where

$$t \doteq t_0/\sqrt{(1 - v^2/c^2)}$$

where c is the SPEED OF LIGHT. *See also* TWIN PARADOX.

time-division multiplexing A system for sending several pieces of information (such as telephone conversations) down a single link (such as a FIBRE OPTIC). This is achieved by encoding the information digitally and then sending each set of information very rapidly, one after another.

Titan The largest satellite of SATURN, notable as one of the few satellites large enough to retain its own atmosphere.

tokamak A device for producing energy from NUCLEAR FUSION. In a tokamak, a PLASMA is contained in a torus or ring shape, with the plasma effectively forming the SECONDARY COIL of a TRANSFORMER. Magnetic fields are used to

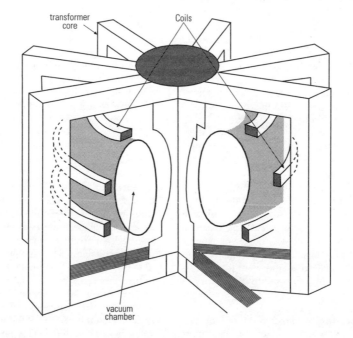

Tokamak.

confine the plasma and to induce currents in it, which heat it to the temperatures required for fusion. No tokamak has yet been built that produces a greater energy output than input, but larger machines are proposed, which should be more efficient than the small experimental machines currently operating.

tomography A MEDICAL IMAGING technique that overcomes the limitations of RADIOGRAPHS due to problems in their interpretation and the reduction in sharpness caused by the scattering of X-rays within the patient. In tomography, the amount of X-rays able to penetrate the patient is measured over a wide range of directions. A computer then reconstructs this information into a measurement of the density of the patients body tissues at each point within the patient. The information is then displayed on a screen as a series of images of slices through the body.

toner Finely powdered ink used in electrostatic printing systems such as the PHOTOCOPIER. The toner is transferred to a piece of paper by a roller and then attached by passing the paper over a heater or heated roller, which melts the toner powder.

tonne A unit of MASS. One tonne is equivalent to 1,000 kg.

top In particle physics, a heavy QUARK with charge $+\frac{2}{3}$ of the electron charge. The top quark was long predicted by the STANDARD MODEL of particle physics, and discovered in 1993.

topology The study of shapes and the properties that they retain even when they are distorted.

top-pan balance See BALANCE.

Toricellian vacuum The space above the mercury in a MERCURY BAROMETER. It contains mercury vapour at its SATURATED VAPOUR PRESSURE, which is very low.

toroidal Having the shape of a TORUS.

torque The turning MOMENT produced by a FORCE. Torque is a measure of the ability to produce acceleration in a rotating machine such as an INTERNAL COMBUSTION ENGINE.

torr A unit of PRESSURE used in high-vacuum technology. One torr is equivalent to the pressure produced by 1 mm depth of mercury in the Earth's GRAVITATIONAL FIELD (approximately 133 Pa).

torsion The twisting effect produced by two opposing FORCES acting on an object at different points and having different LINES OF

ACTION. The size of this effect is measured by the COUPLE produced by the forces.

torus A doughnut shape, formed by bending the ends of a cylinder round until they meet.

total internal reflection The situation where a wave travelling in a dense material meets a boundary with a less dense material at such an ANGLE OF INCIDENCE that the ANGLE OF REFRACTION predicted by SNELL'S LAW would be greater than 90°. In this case there will be no REFRACTION, but only REFLECTION. The smallest angle of incidence at which total internal reflection takes place is called the critical angle. The critical angle for glass is about 42°, which means that totally internally reflecting PRISMS can be made which will change the direction of light by 90° or 180° without any need for a reflective coating.

tracer technique The use of RADIOACTIVITY to determine the path of a particular element in some system. Since radioactive ISOTOPES are chemically no different from stable isotopes of the same element, they can be used to see how a system handles a particular element by introducing a radioactive isotope and monitoring the flow of radioactivity through the system. Examples of the use of tracer techniques include medical diagnosis and the study of the rate of wear in engines.

trajectory The path of an object, especially a PROJECTILE or other object, moving solely under the influence of gravity.

transducer Any device for converting changes in a non-electrical signal (such as pressure, temperature, sound, etc.) to electrical signals, and vice versa.

transformation A mapping that shifts all points in a space, or all values of a variable, to new points or values according to a specified procedure.

transformer A device for changing the voltage of ALTERNATING CURRENT signals and power supplies. A transformer consists of two coils wound on a LAMINATED SOFT iron core. An alternating current in one of these coils (the primary coil) causes a changing magnetic field in the core that induces an alternating current in the other coil (the secondary coil). By altering the ratio of the number of turns in the secondary to the primary (called the TURNS RATIO), the ratio of the secondary to primary voltages can be altered. This ratio is equal to the turns ratio. A transformer with a turns ratio greater than one is called a step-up transformer and will produce an output voltage higher than the input, though at a lower current. Step-down transformers are those with a turns ratio less than one – they produce low voltage outputs but are capable of delivering higher currents. Step-down transformers are needed to generate the high currents required in arc welding, for example.

Transformers are never 100 per cent efficient. Energy losses are divided into two classes – COPPER LOSSES, caused by the electrical RESISTANCE of the transformer coils, and IRON LOSSES, resulting from hysteresis and EDDY CURRENTS in the core.

Transformers are used in electrical power transmission systems. Lower currents are needed at high voltages to transmit a given amount of power, which makes for cheaper cables. Transformers are used to step-up the voltage where the electricity is generated and then to step it down when it reaches the towns where it is used.

If the number of turns in the primary coil of a transformer is n_p and in the secondary n_s, the primary and secondary voltages and currents V_p, V_s and I_p, I_s, then

$$V_s/V_p = n_s/n_p$$

and if the transformer is 100 per cent efficient,

$$V_p I_p = V_s I_s$$

transistor A three-terminal SEMICONDUCTOR device exhibiting the property of GAIN, so that a small current or voltage can be used to control a larger current, enabling transistors to be used for amplification and switching, for example in the manufacture of LOGIC GATES. There are two common types, the JUNCTION TRANSISTOR and the FIELD EFFECT TRANSISTOR. Each type operates on a different set of principles. First manufactured in 1947, transistors rapidly became common as an alternative to THERMIONIC VALVES, since they required less power and operated at lower voltages. In the 1960s, the development of integrated circuits containing many transistors led to the manufacture of many electronic devices which would have been impossibly large, expensive and unreliable had they been based on valves.

translucent (*adj.*) Describing a material through which light can pass, but with the light being scattered, so that an object cannot be seen clearly through the material. *Compare* OPAQUE, TRANSPARENT.

transmittance The proportion of incident ELECTROMAGNETIC RADIATION transmitted through a partially TRANSPARENT material, rather than being absorbed or reflected. For a perfectly transparent material the transmittance is 1, for a perfectly OPAQUE material, it is 0.

transparent (*adj.*) Describing a material through which light, or some other specified form of electromagnetic radiation, can pass without scattering, so that an object can be seen through the material.

transport coefficient Any quantity that describes some property of a gas that depends on intermolecular collisions to transfer some quantity from one part of the fluid to another. Examples include THERMAL CONDUCTIVITY, in which energy is transferred from one molecule to another by collisions; VISCOSITY, which relies on molecular collisions to transfer MOMENTUM; and DIFFUSION, where collisions limit the spread of molecules from one place to another. *See also* KNUDSEN REGIME, MEAN FREE PATH.

transverse wave A WAVE in which the motion of the particles in the wave is at right angles to the direction of propagation. The wave can be thought of as a series of crests and troughs. Examples are waves on a water surface and ELECTROMAGNETIC WAVES. Transverse waves can be further described by their POLARIZATION. *Compare* LONGITUDINAL WAVE.

travelling wave *See* PROGRESSIVE WAVE.

triac A semiconductor device widely used in the control of the power of ALTERNATING CURRENTS. A triac does not conduct until a voltage is applied to an ELECTRODE called the GATE. It will then conduct in either direction, regardless of the gate voltage, until the current through it falls below a THRESHOLD value.

tribology The study of FRICTION and its effects, especially the reduction of friction with lubricating oils.

triclinic (*adj.*) Describing a CRYSTAL structure where none of the faces are at right angles to one another. The size and shape of the UNIT CELL is characterized by three lengths and three angles.

triple point The one combination of temperature and pressure at which the solid, liquid and gas PHASES of a substance can exist together in equilibrium. The triple point of water is at a temperature of 0.01°C and a pressure of 600 Pa.

tritium The ISOTOPE hydrogen–3. It is widely used in NUCLEAR FUSION processes and is unstable, with a HALF-LIFE of 12 years.

tropical (*adj.*) Relating to the tropics. A tropical AIR MASS is one that has come from equatorial regions.

tropopause The boundary between the TROPOSPHERE and the STRATOSPHERE. *See* ATMOSPHERE.

troposphere The lowest level of the Earth's ATMOSPHERE, extending up to about 17 km above the Earth's surface. Most of the WEATHER SYSTEMS occur in this layer.

tuned circuit, *resonant circuit* A circuit containing an INDUCTANCE and a CAPACITANCE. These act in a way analogous to a mass on a spring and have a RESONANT FREQUENCY f given by

$$f = \frac{1}{2}\pi(LC)^{1/2}$$

where L is the inductance and C the capacitance.

turbine A machine for converting the KINETIC ENERGY of a fluid (usually water or air heated by burning fuel). A turbine consists of a rotating set of blades surrounded by fixed blades to channel the fluid flow into the path that produces the largest force on the rotating blades. *See also* GAS TURBINE.

turbocharging A system used in some INTERNAL COMBUSTION ENGINES in which the energy of the expanding exhaust gases is used to drive a small TURBINE which compresses the fuel/air mixture entering the engine, enabling more fuel to be burnt by an engine of a given size, increasing its power output.

turbofan *See* HIGH-BYPASS ENGINE.

turboprop A GAS TURBINE engine used to drive an aircraft propeller.

turbulent (*adj.*) Describing irregular motion in a fluid with sudden unpredictable changes in flow direction over time or from one part of the fluid to the next. *Compare* STREAMLINED. *See also* BOUNDARY LAYER, DRAG, VORTEX.

turning point A point at which a mathematical FUNCTION reaches a local MAXIMUM or MINIMUM; in other words, a point where the gradient of the function is zero, with the sign of the gradient being different on either side of the point.

turns ratio In a TRANSFORMER, the number of turns in the SECONDARY COIL divided by the number of turns in the PRIMARY COIL.

twin paradox A RELATIVISTIC PARADOX in which one of a pair of identical twins sets off on a long space journey. When he returns he will have aged less than his twin (TIME DILATION). However, if only relative motion is important, why is it not possible to think of the other twin as being the one who made the journey? The resolution of this paradox lies in careful consideration of what happens when the twin turns round to begin his return journey; up to that point the position of the two twins is symmetrical, but when the travelling twin starts his return journey, he must either jump to a different NON-INERTIAL FRAME OF REFERENCE, or decelerate, making his frame non-inertial. Only when the twins are brought back together is it possible to compare their ages unambiguously.

two-stroke cycle A system of operation used by some smaller PETROL ENGINES, such as those used on small motorcycles. In the two-stroke engine, the space below the cylinder is used to draw in a fresh load of fuel/air mixture as the previous load is leaving the top of the engine. This provides more even power but has disadvantages in that oil must be dissolved in the petrol to provide adequate lubrication of the piston. This oil is partially burnt and expelled in the exhaust.

typhoon *See* HURRICANE.

U

UHF (ultra high frequency) RADIO WAVES with wavelength from about 10 cm to 1 m. They are used for many forms of RADIO communication and also for television, where the high rate at which picture information has to be transmitted requires each signal to have a large BANDWIDTH (spread of frequencies). This bandwidth requirement is less serious at high frequencies. As with VHF, only line-of-sight communication is possible.

ultracentrifuge A CENTRIFUGE that operates at very high speeds.

ultra high frequency *See* UHF.

ultrasound SOUND waves of too high a frequency for the human ear to hear. The PIEZO-ELECTRIC effect can be used to produce sound waves with frequencies of several megahertz (MHz). Such high frequencies mean that the wavelengths are very short, so narrow beams of sound can be produced without problems from DIFFRACTION. Such ultrasonic beams can then be fired into objects and the reflections studied to give a picture of the structure of the object being studied. *See also* ULTRASOUND IMAGING.

ultrasound imaging A technique for studying the interior of opaque structures, widely used in medicine. ULTRASOUND is especially used in the routine examination of human foetuses. The technique relies on the reflection of ultrasound from boundaries between materials of differing densities. The DOPPLER EFFECT also enables motion, particularly blood flow rates, to be measured by ultrasonic techniques – the shift in frequency of the reflected wave gives a measure of the motion of the reflecting particles.

Whilst ultrasound is non-ionizing, so cannot produce mutations, high levels can produce tissue changes, and more intense ultrasound is used in physiotherapy and to break up kidney stones.

ultraviolet ELECTROMAGNETIC WAVES with wavelengths in the range 4×10^{-7} m to about 10^{-9} m. They are produced by the more energetic changes in energy in atomic electrons. Ultraviolet radiation from the Sun is mostly absorbed in the upper layers of the atmosphere (the OZONE LAYER), so relatively little reaches the Earth. That which does reach ground level is responsible for the tanning and burning effect of exposure to sunlight and, with prolonged exposure, is believed to be responsible for skin cancer. Ultraviolet radiation can be detected by PHOTOGRAPHIC FILM and can be made visible by FLUORESCENCE.

ultraviolet catastrophe A failure in the attempt to calculate the SPECTRUM of BLACK-BODY RADIATION based on the physics of ELECTROMAGNETIC WAVES. These calculations suggest that the amount of radiation given off would increase without limit as the wavelength became shorter, leading to high levels of ULTRAVIOLET radiation which is not observed in practice. The solution to this problem lies in treating electromagnetic radiation as PHOTONS, with the shorter wavelengths requiring higher energies for their production. This solution, together with the PHOTOELECTRIC EFFECT, was the first step in the development of quantum physics.

uncertainty principle *See* HEISENBERG'S UNCERTAINTY PRINCIPLE.

underdamped (*adj.*) Describing an oscillating system where the DAMPING is less than that required to prevent OSCILLATION. *See also* OVERDAMPED.

unified field theory A theory in particle physics that describes at least two of the FOUR FORCES OF NATURE as being aspects of a single force. The STANDARD MODEL, which brings together the ELECTROMAGNETIC FORCE and WEAK NUCLEAR FORCE is the only truly successful example of such a theory, but physicists have been inspired by this success to apply the idea to the STRONG NUCLEAR FORCE and the gravitational force as well, to create a GRAND UNIFIED THEORY or a THEORY OF EVERYTHING.

unit 1. A fixed size of some quantity, such as length, in terms of which all other examples of that quantity can be expressed. Thus, once the length of 1 METRE is fixed, all other distances can be measured in terms of the metre. It is possible to fix units for all quantities independently, but in practice once a few units, called base units, have been fixed, other quantities can be defined in terms of these. Speed for example can be measured in terms of metres and seconds, units of time and distance. *See also* C.G.S. UNITS, METRIC SYSTEM, SI UNIT.

2. Relating to the number 1 – a digit immediately to the left of the decimal point, or at the right-hand end of a number with no decimal point, representing the number of 1's in the number, as opposed to the number of 10's for example.

unit cell, *primitive cell,* The smallest part of a CRYSTAL lattice that is needed to describe the structure of a crystal. A crystal can be thought of as being made up of repeated unit cells. For example, the unit cell for sodium chloride can be regarded as a cube containing a central sodium ion with half a chlorine ion in the middle of each face and one eighth of a sodium ion at each corner.

universal gas constant *See* MOLAR GAS CONSTANT.

universal gas equation *See* IDEAL GAS EQUATION.

unstable equilibrium The situation in which a small displacement from an EQUILIBRIUM state produces forces tending to move the system further from its equilibrium position. A simple example of this is a pencil balanced on its point.

up A FLAVOUR of QUARK, with charge $+\frac{2}{3}$ in units of the electron charge.

upthrust The upward force on an object immersed (totally or partially) in a fluid. *See* FLOTATION.

Uranus The seventh planet in order from the Sun, with an orbital radius of 19.2 AU (2.9 billion km). Uranus is a GAS GIANT, with a diameter of 50,000 km (3.9 times that of the Earth) and a mass of 8.7×10^{25} kg (14 times that of the Earth). Like the other gas giants it has an atmosphere composed mostly of hydrogen and shows evidence of active weather systems. It has a RING SYSTEM similar to that of SATURN, but with much finer rings, made of a much darker material. Uranus is unique in having its axis tilted to lie almost in the plane of its orbit. Fifteen satellites are known, nine of which were discovered following the visit of the VOYAGER spacecraft in 1986. Uranus orbits the Sun every 84 years, and rotates on its axis every 17 hours.

U-value A quantity often used in the study of heat flow, particularly in the design of buildings. The U-value relates to a sheet of material of a specified thickness. It is the heat flow in each square metre of the sheet produced by a temperature difference of 1 KELVIN, expressed in $WK^{-1}m^{-2}$.

V

vacuum A region containing no matter of any kind. In particle physics, the true nature of the vacuum state is a matter of some interest. In this context, the vacuum is defined as being the lowest energy state, and will contain VIRTUAL PARTICLE/ANTIPARTICLE pairs.

valence band In the BAND THEORY of solids, the energy band occupied by the VALENCE ELECTRONS. In metals, this band is either only partially full or overlaps with an empty band. In nonmetals it is completely full, so electrons in the valence band cannot gain energy to conduct heat or electricity. *See also* CONDUCTION BAND.

valence electron An electron in the outer SHELL of electrons of an atom, which may be involved in a bonding process. *See also* BAND THEORY.

valence shell The outer electron SHELL of an atom, containing the VALENCE ELECTRONS.

value In mathematics, the particular number that an algebraic quantity represents at a particular moment. For example the value of x is 4.

valve 1. In general, any device that can be closed or opened to allow a fluid to pass through it.

2. *non-return valve* A device through which fluid can flow in only one direction. Such a valve generally comprises an opening closed by a flap or ball held in place by a light spring. Fluid pressure on one side of the opening overcomes the force from the spring and opens the valve. Fluid pressure from the other side holds the valve tightly closed.

3. *safety valve, pressure relief valve* Similar in structure to a non-return valve, but held firmly closed by a weight or stiff spring. The closing force is only overcome when the fluid pressure exceeds a certain value.

4. An electronic device based on THERMIONIC EMISSION. *See* THERMIONIC VALVE.

Van Allen belt One of two doughnut-shaped regions around the Earth in which radiation consisting of high-energy charged particles is trapped by the MAGNETIC FIELD OF THE EARTH.

The lower belt, comprising electrons and protons extends from 1,000 to 5,000 km from the Earth's surface, while the upper belt, which contains mostly electrons, is at a distance 15,000 to 25,000 km from the surface.

van de Graaf generator A machine for creating electrostatic charges. The charges are generated by friction in smaller machines and by a beam of charged particles in larger machines. The charges are transferred to an insulating belt, which carries them to the top of the machine, where they are transferred to a large metal dome-shaped ELECTRODE, insulated from its surroundings. *See also* CHARGING BY FRICTION.

van der Waals' bond A very weak bond that holds separate molecules together in molecular solids such as solid carbon dioxide. The bond originates from the VAN DER WAALS' FORCE between neutral molecules. The very weak nature of this bond is reflected in the low melting points of most molecular materials.

van der Waals' equation An EQUATION OF STATE for a REAL GAS. For N MOLES of gas at an ABSOLUTE TEMPERATURE T and pressure p, in a volume V, van der Waals' equation is

$$(p + a/V^2)(V - b) = NRT$$

where R is the MOLAR GAS CONSTANT and a and b are constants that represent the attractive INTERMOLECULAR FORCES and the volume of the gas molecules respectively.

van der Waals' force The attractive force between neutral molecules or atoms. The force is strongest if both molecules are POLAR, but a POLAR MOLECULE will create an induced DIPOLE in a non-polar molecule. There is also a weak induced-dipole to induced-dipole force between non-polar molecules. This arises because at any instant the centres of positive and negative charge may not coincide, even though the molecule may be non-polar on average. These forces are relatively weak and give rise to substances that have melting and boiling points below room temperature.

As with all INTERMOLECULAR FORCES, there is a repulsion at shorter distances, due to the effect of the PAULI EXCLUSION PRINCIPLE on the overlapping electron clouds of the molecules. Van der Waals' forces between non-polar molecules can be described by a POTENTIAL ENERGY with a repulsive core that varies like the inverse twelfth power of the separation and an attractive part, dominant at larger separations, but also falling off with distance, which varies like the inverse sixth power of separation. This is called the LENNARD–JONES 6-12 POTENTIAL.

See also REAL GAS.

vapour A term used to describe the gaseous state of a substance below its BOILING POINT.

vapour density A measure of the density of a gas, found by dividing the mass of a sample of the gas by the mass that the same volume of some reference gas (generally hydrogen) would have at the same temperature and pressure.

vapour pressure The PARTIAL PRESSURE exerted by a VAPOUR, especially the vapour found above the surface of a liquid. If the liquid and its vapour are held in a closed container, the vapour pressure will rise until it reaches the SATURATED VAPOUR PRESSURE.

variable An algebraic quantity that may take on a different value at different times. Variables are usually regarded either as independent variables, which can take on any value freely, or dependent variables, whose values are a FUNCTION of an independent variable.

variable capacitor A CAPACITOR whose CAPACITANCE can be changed. This can be constructed by arranging two sets of plates, one fixed and the other moveable, so that the degree of overlap can be altered. *See also* VARICAP DIODE.

variable resistor A RESISTOR whose RESISTANCE can be changed, made from a layer of resistance material with a sliding contact. When just the sliding contact and one end of the resistance are used in a circuit, to control the brightness of a lamp for example, variable resistors are sometimes called rheostats. When used in a POTENTIAL DIVIDER circuit, with a voltage applied between the two fixed ends, a fraction of which is taken off between the sliding contact and one of the fixed contacts, they are called potentiometers.

varicap diode A PN JUNCTION DIODE made with a particularly wide junction to enhance its CAPACITANCE when REVERSE BIASED. The capacitance decreases as the reverse bias voltage is increased. Varicap diodes are often used for the electronic control of tuning circuits in radios, etc.

vector A quantity that is described by both its size, or magnitude, and its direction in space. Examples of vector quantities include velocity and force. Vectors can be added by drawing them nose to tail and forming a single vector that joins the ends of the composite vector thus formed. *See also* COMPONENT, RESOLVE.

velocity The rate of change of displacement of a body. Velocity is the VECTOR quantity related to SPEED; that is, the speed plus the direction of motion.

velocity ratio The factor by which the speed at which the effort force applied to a machine must move is greater than the speed of movement of the LOAD.

$$\text{Velocity ratio} = \frac{\text{distance moved by effort}}{\text{distance moved by load}}$$

velocity selector A mechanism, usually based on electric and magnetic fields, for selecting charged particles moving at a particular speed. *See also* MASS SPECTROMETER.

Venus The second planet in order from the Sun, with an orbital radius of 0.72 AU (108 million km). Venus is similar in size to the Earth, with a diameter of 12,100 km (0.94 times that of the Earth) and a mass of 4.9 x 10^{24} kg (0.81 times that of the Earth).

Venus has a dense atmosphere with high levels of carbon dioxide, which have caused an extreme form of the GREENHOUSE EFFECT, leading to high temperatures on the surface. Nothing can be seen of the surface from Earth, due to a constant cover of dense white clouds that are now known to be made of sulphuric acid. The surface of Venus has been mapped using radar and its geology is believed to be similar to that of the other TERRESTRIAL PLANETS. Venus orbits the sun every 225 days and rotates on its axis, from west to east, every 243 days.

vernier A device for fine adjustment or precise measurement. A vernier scale typically consists of a small scale sliding along a main scale. The small scale contains 10 divisions in the space occupied by nine divisions on the main scale. By seeing which pair of scale marks align most accurately, it is possible to read the main scale to one tenth of a division.

very high frequency *See* VHF.

VHF (very high frequency) RADIO WAVES with wavelengths from about 1 m to 10 cm. They are used for many forms of RADIO communication, but because the waves are not reflected by the IONOSPHERE, they can only be used where there is an uninterrupted path from the transmitter to the receiver (this is called line-of-sight communication). This problem can be overcome by transmitting signals from communications satellites or by using RELAY STATIONS.

virtual earth A point in a circuit that, though not connected to EARTH, is maintained at a ELECTRIC POTENTIAL close to earth potential. *See also* OPERATIONAL AMPLIFIER.

virtual focus An imaginary point from which light rays appear to have come. *See* FOCUS.

virtual image An optical image where rays of light from a given point on the object only appear to be coming from a corresponding point on the image (as when an object is viewed in a PLANE MIRROR for example). A virtual image cannot be formed on a screen as the light rays do not actually arrive at the point where the image appears to be located. *See also* REAL IMAGE.

virtual particle In QUANTUM THEORY, a particle whose permanent existence is forbidden by the LAW OF CONSERVATION OF ENERGY, but which can exist temporarily according to HEISENBERG'S UNCERTAINTY PRINCIPLE. In particle physics, the exchanged GAUGE BOSONS responsible for the forces between particles of matter are virtual particles; they can also exist as real particles, though in the case of the W BOSONS and Z BOSONS, large amounts of energy are required.

viscosity A measure of the resistance to flow in a fluid. Forces between molecules in a fluid mean that MOMENTUM given to one part of the fluid tends to be transferred to nearby regions. This creates a force opposing any tendency for different parts of the fluid to move at different speeds, as happens at a boundary with a solid surface for example.

The viscosity of a fluid is measured as the force F per unit area A between two surfaces in the fluid that are moving at unit velocity v relative to one another and are separated by unit distance x.

$$F = \eta A (\mathrm{d}v/\mathrm{d}x)$$

where η is a constant of proportionality, called the coefficient of viscosity. *See also* POISEULLE'S EQUATION, SUPERFLUIDITY.

volatile (*adj.*) Easily evaporated.

volt (V) The SI UNIT of ELECTRIC POTENTIAL, POTENTIAL DIFFERENCE or ELECTROMOTIVE FORCE. One volt is an energy of one JOULE per COULOMB of CHARGE.

voltage The measure of the amount of ELECTRICAL POTENTIAL ENERGY carried by each unit of charge. The SI UNIT of voltage is the VOLT. *See also* ELECTRIC POTENTIAL, ELECTROMOTIVE FORCE, POTENTIAL DIFFERENCE.

voltaic cell *See* CELL.

voltammeter A container for ELECTROLYSIS experiments with ELECTRODES that can be removed to find their mass. By measuring the change of the mass of the electrodes in an electrolysis experiment, the ELECTROCHEMICAL EQUIVALENT of the element concerned can be found. *See also* HOFMANN VOLTAMMETER.

voltmeter An instrument for measuring VOLTAGE. A voltmeter is connected between the two points between which the POTENTIAL DIFFERENCE is to be measured. Voltmeters should have a very high resistance so that little current flows through them, so as not to affect the circuit to which they are connected. Modern voltmeters are usually based on OPERATIONAL AMPLIFIERS, which have a very high input resistance. The output of the operational amplifier drives a DIGITAL display. ·

volume 1. The amount of space occupied by an object. For a cube, its volume is equal to the third power of the length of the sides of the cube. For other shapes, the volume can be calculated by imagining them to be built up from a large number of small cubes. Volume is expressed in SI UNITS in metres cubed.

2. The loudness of a sound, related to the amplitude of the sound waves. Sound levels are often measured in DECIBELS.

vortex (*pl. vortices*) A TURBULENT flow characterized by a rotational movement of the fluid. Vortices are produced in aircraft wings, for example, as high pressure air from beneath the wing spills over at the wing tip into the low pressure area above the wing. *See also* DRAG.

Voyager The name given to each of the two unmanned spacecraft that were launched in 1976 to visit the GAS GIANTS. They sent back photographs and other data that greatly increased our knowledge of these planets. The craft took several years to travel from one planet to another, and used the GRAVITY of each planet to accelerate it to the next.

W

warm front A WEATHER SYSTEM in which a body of warm air meets and rises above a layer of cooler air. The arrival of a warm front produces thickening cloud and rain. *See also* COLD FRONT.

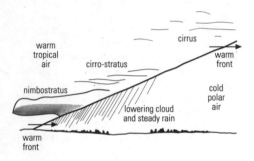

Formation of a warm front.

watt The SI UNIT of POWER. One watt represents WORK done, or ENERGY converted from one form to another at a rate of one JOULE per second.

wattmeter A device for measuring electrical POWER. One form of wattmeter operates on a principle similar to the MOVING-COIL GALVANOMETER, but whilst the voltage of the supply is fed to a moving coil, the current fed to the load flows through an ELECTROMAGNET, which provides the magnetic field in which the coil moves. In this way both current and voltage control the movement of the pointer.

wave A periodic variation in some quantity across a region of space; in particular, a progressive or travelling wave – a motion whereby ENERGY is transmitted through some medium (or empty space in the case of an ELECTROMAGNETIC WAVE) without the particles of the medium moving far from their equilibrium positions.

To support a wave motion, the material through which the wave travels must have INERTIA (so that any given part of the material tends to continue to move after it has been displaced) and a stiffness, tending to restore the material to its original shape. Another consequence of the stiffness of the material is that the displacement of one part of the material tends to cause displacement of nearby regions and in this way a disturbance in one part of the material can spread out, or propagate, through the material.

Waves are divided into two classes: transverse, where the motion is at right angles to the direction in which the wave is travelling; and longitudinal, where the wave motion is to-and-fro along the direction of travel.

Whilst waves can have any shape, SINUSOIDAL waves are the easiest to study, and any more complicated shape can be constructed by adding together sinusoidal waves of various frequencies. In a sinusoidal wave, the wave motion at any fixed point will be SIMPLE HARMONIC MOTION. The AMPLITUDE, PERIOD and FREQUENCY of the wave as a whole are defined as being the same as they are for this simple harmonic motion. Each point along the wave motion performs simple harmonic motion with a phase slightly behind the phase of the previous point. The frequency f of a wave is related to its speed v and WAVELENGTH λ by the equation

$$v = f\lambda$$

See also DIFFRACTION, INTERFERENCE, LONGITUDINAL WAVE, PHASE VELOCITY, REFLECTION, REFRACTION, STANDING WAVE, TRANSVERSE WAVE, WAVEFRONT.

wavefront A line drawn on a diagram that connects all points that are at the same PHASE, such as lines showing the positions of all the crests at a given instant. *See also* HUYGENS' CONSTRUCTION.

wavefunction In QUANTUM MECHANICS, a PROBABILITY AMPLITUDE, the square of which measures the probability of finding the particle at the point specified or the system in the state described by the particular wavefunction. *See also* SCHRÖDINGER'S EQUATION.

waveguide A metal tube, usually rectangular in cross-section, along which MICROWAVES can travel, being reflected from the inside walls of the tube. Waveguides are used to convey microwaves over relatively short distances, from a transmitter to an AERIAL for example.

wavelength (λ) The closest distance between two points in a WAVE that are moving IN PHASE, for example the distance between two adjacent peaks or two adjacent troughs.

wave mechanics The branch of QUANTUM THEORY that deals with the WAVE NATURE OF PARTICLES such as electrons, and the WAVEFUNCTIONS that describe them.

wavemeter A device for measuring the WAVELENGTH of a RADIO WAVE or MICROWAVE, often based on measuring the distance between NODES in a STANDING WAVE.

wave nature of light One aspect of light in the framework of WAVE-PARTICLE DUALITY. The wave nature of light was first demonstrated in 1801 by Thomas Young (1773–1829), in what is now known as YOUNG'S DOUBLE SLIT EXPERIMENT.

wave nature of particles In 1924, Prince Louis de Broglie (1892–1987) suggested that particles would behave like waves with a wavelength now called the DE BROGLIE WAVELENGTH. Three years later this property was confirmed experimentally as ELECTRON DIFFRACTION by Clinton Davisson (1881–1958) and Lester Germer (1896–1971). In electron diffraction experiments, an ELECTRON GUN fires electrons through a thin film of metal or graphite. The regular lattice of atoms acts like a DIFFRACTION GRATING for the electron waves and a diffraction pattern is observed on a PHOSPHOR screen.

A consequence of the wave nature of electrons was the explanation of the experimental fact that atoms absorb and emit light of certain wavelengths only (*see* SPECTRUM). The explanation for this is that atoms can exist only in certain energy states in much the same way that STANDING WAVES can be produced with only certain frequencies – the permitted states correspond to standing wave patterns of the electrons in the atom. The lowest energy state of any atom is called its ground state. Other, higher energy states are called excited states. In a transition from the ground state to an excited state, an atom will absorb a PHOTON of energy equal to the energy difference between the two states. An atom in an excited state will lose energy by giving off a photon and moving to a lower energy excited state, or to the ground state. In each case, the energy of the photon will be equal to the difference in energy of the two states. The different energy states of different atoms account for the characteristic spectra produced by different elements, which is a useful tool for the analysis of chemical composition, in stars for example.

See also ENERGY LEVEL, HEISENBERG'S UNCERTAINTY PRINCIPLE, NEUTRON DIFFRACTION, SCHRÖDINGER'S EQUATION.

wavenumber The reciprocal of WAVELENGTH, the number of WAVES in one unit of length.

wave-particle duality The name given to the dual behaviour of objects originally thought of as either waves or particles. Thus light, traditionally a wave, can behave as a particle (*see* PHOTON, PHOTOELECTRIC EFFECT), whilst electrons, traditionally thought of as particles, have wave-like properties (*see* ELECTRON DIFFRACTION). *See also* WAVE NATURE OF PARTICLES.

wave power The commercial extraction of energy from waves to generate electricity. Whilst a number of experimental schemes have been tried, there are problems with the high cost of installation and the very large size of any commercial wave power station. The strongest waves are often found far from the places where the most energy is consumed and are also in sites where the installation is most vulnerable to storm damage.

W boson, *W particle* The charged particle that, together with the neutral Z BOSON are the GAUGE BOSONS responsible for the WEAK NUCLEAR FORCE. There are two oppositely charged W particles, W^+ and W^-. They have a mass of about 80 times that of the proton.

weak intermediate vector boson The W BOSON or Z BOSON.

weak nuclear force One of the FOUR FORCES OF NATURE, the INTERACTION responsible for BETA DECAY. It is mediated by two GAUGE BOSONS called the W BOSON and Z BOSON. The weak force is the only way in which NEUTRINOS can interact with matter. It is also vital in the conversion of hydrogen to helium releasing energy in STARS. *See also* ELECTROWEAK FORCE.

weather systems Regions where AIR MASSES lose water by condensation to form clouds. Water may then fall from the clouds as rain or snow.

Weather systems occur when different air masses meet. When TROPICAL MARITIME air meets cooler air, a WARM FRONT is formed. The warm air is less dense, so rises over the cooler air. The result is the formation of cloud as the air cools and becomes SATURATED. This happens first at high altitudes producing wispy clouds composed of ice crystals. These are called cirrus clouds. As the height of the saturated layer moves lower, stratus (layer) clouds are formed and rain falls, or snow if the temperature is below freezing. Sleet is snow that melts as it falls into warmer air.

If cold air runs into warmer air, it tries to force its way underneath, forming a COLD FRONT. The result is a disturbed area in which large heaped (cumulus) clouds form, producing showers. Cumulus clouds can also be formed by uneven solar heating of the ground. Dark areas warm up faster than paler areas and rising air cools and becomes saturated at a certain level, producing flat-bottomed cumulus clouds. As the water vapour condenses in the cloud, the LATENT HEAT released slows the rate of cooling, so the air in the cloud can remain warmer than its surroundings, causing rapidly rising currents that reach the STRATOSPHERE. Within such a cloud (called CUMULONIMBUS) water droplets are lifted to great heights and freeze. As they fall, further water freezes on the surface, forming a layered ball of ice, which falls as a hailstone. Collisions between hailstones and water droplets within the cloud produce electrostatic charges, with the top of the cloud positive and the base negative. This charge can become large enough to produce large sparks. The flash of light from the spark is seen as LIGHTNING, whilst the sound is heard as thunder.

The greater HEAT CAPACITY of the sea than the land and the mixing with cooler layers below the surface, which can take place in water but not on dry land, mean that the oceans tend to remain at fairly constant temperatures from one day to the next. The land, on the other hand, heats up during the day and radiates that heat away at night, particularly if the sky is clear. This may set up CONVECTION CURRENTS which cause winds blowing from sea to land by day (sea breezes) and in the opposite direction at night. The cooling of the land on a clear night also leads to condensation which may appear as dew or frost if there is no wind, or as mist or fog if the wind is sufficient to form a layer of cold air just above the cold ground.

See also ANTICYCLONE, ATMOSPHERE, CLIMATE, DEPRESSION.

weber (Wb) The SI UNIT of MAGNETIC FLUX. One weber is the flux produced by a MAGNETIC FIELD of one TESLA over an area of one square metre.

weight The force of GRAVITY on an object.

weightlessness The apparent absence of GRAVITY experienced in an orbiting spacecraft, or any other object falling freely in a GRAVITATIONAL FIELD. *See* FREE FALL.

wet and dry bulb hygrometer A HYGROMETER in which the temperature of the air is measured by one thermometer whilst a second thermometer surrounded by a piece of damp cloth measures a temperature that is lower by an amount called the WET BULB DEPRESSION. The wet bulb depression will be greatest when the air is dry, falling to zero when the air is SATURATED.

wet bulb depression The amount by which the wet bulb temperature is below the dry bulb temperature in a WET AND DRY BULB HYGROMETER. When the wet bulb depression falls to zero, the air is completely saturated with water vapour; that is, the RELATIVE HUMIDITY is 100 per cent.

wet cell An obsolete term for an electrolytic CELL in which the ELECTROLYTE is a liquid rather than a paste, so must be kept upright to avoid spillage. An example is the LEAD-ACID CELL. Unlike a DRY CELL, a wet cell is RECHARGEABLE.

Wheatstone bridge A circuit containing two pairs of RESISTORS, each pair forming one arm of the bridge, and acting as a POTENTIAL DIVIDER, connected across a single power supply. A GALVANOMETER is connected between the junction of the resistors in each pair, and the bridge is said to be balanced when no current flows through this galvanometer. This will be the case when the POTENTIAL DIFFERENCE across the galvanometer is zero; that is, when the ratios of the two resistance in each potential divider are equal. This balance condition does not depend on the voltage of the supply.

The circuit can be used to compare one resistance with another. The two resistors to be compared form one of the potential dividers, whilst the other takes the form of a POTENTIOMETER, traditionally in the form of a metre length of RESISTANCE WIRE with a sliding contact. The Wheatstone bridge is very sensitive to small changes in resistance, a fact that is exploited in some of its applications (*see* RESISTANCE THERMOMETER).

white dwarf A very small hot STAR. As the surface area is small the LUMINOSITY is low. Such stars lie at the bottom left of the HERTZSPRUNG–RUSSELL DIAGRAM. White dwarfs are supported by ELECTRON DEGENERACY PRESSURE – the PAULI EXCLUSION PRINCIPLE places limits on the extent to which the electrons can overlap and the resulting material, similar in many ways to a metal, is able to support itself against further gravitational collapse. Stars of about 0.5 SOLAR MASSES or less simply continue to collapse at the end of their main sequence lifetime. Heavier stars pass through a RED GIANT stage first. *See also* PLANETARY NEBULA.

white noise NOISE that has a relatively wide range of frequencies.

Wien's law A BLACK BODY will emit the greatest amount of radiation at a wavelength inversely proportional to the ABSOLUTE TEMPERATURE. The colour of the light from stars can be used as a measure of their surface temperature, with hotter stars appearing more blue whilst cooler stars are orange or red.

WIMP (weakly interacting massive particle) In COSMOLOGY, one hypothetical form of MISSING MASS in the Universe.

wind power The installation of wind-driven electrical generators; one of the more econom-ically effective sources of renewable energy, though only some areas have sufficient wind and enough land available. In Europe, there has been increasing opposition to the visual impact of large 'wind farms'. The costs of installation are low compared to some forms of renewable energy, but no power is produced on still days.

wire wound resistor A RESISTOR made from a coil of RESISTANCE WIRE.

work The effect of a FORCE moving through a distance and converting one form of ENERGY to another.

$$\text{Work done} = \text{force} \times \text{distance moved in the direction of the force}$$

So if a force F moves through a distance d, with the angle between the direction of the force and the direction of motion being θ, the work done is

$$Fd\cos\theta$$

The SI UNIT of work is the JOULE.

work function The minimum amount of energy needed to remove an electron from the surface of a metal in the PHOTOELECTRIC EFFECT.

work hardening The increase in hardness and brittleness of a material (usually a metal) when it undergoes PLASTIC deformation. *See also* ANNEALING.

W particle *See* W BOSON.

X

x-plates One of the two pairs of plates fitted to a CATHODE RAY TUBE to control the direction of the electron beam. The other pair are the y-plates.

X-ray An ELECTROMAGNETIC WAVE with a wavelength shorter than about 10^{-9} m, produced by the most energetic energy changes of atomic electrons. Such waves are called gamma rays (*see* GAMMA RADIATION) if they are produced by changes within a NUCLEUS. X-rays are a form of IONIZING RADIATION and the shorter wavelengths (hard X-rays) are highly penetrating. They can be detected by photographic film or with a fluorescent screen or by the IONIZATION they produce in a GEIGER COUNTER or other detector of ionizing radiation. The penetrating quality of X-rays has lead to their use in examining the internal structure of various objects, including the human body (*see* MEDICAL IMAGING).

X-ray crystallography The study of the structure of crystals using X-RAY DIFFRACTION. In a simple experiment, the crystal is placed in the path of an X-RAY beam and surrounded by PHOTOGRAPHIC FILM. When the film is developed, spots appear in the directions of CONSTRUCTIVE INTERFERENCE. Since a crystal is a three-dimensional structure, it is possible to identify many crystal planes, or layers of atoms. The result is complex, but with modern computational techniques it is possible to work back to the structure that produced the pattern.

Some samples cannot be obtained as single crystals, but only as a powder or POLYCRYSTALLINE sample, which effectively contains crystals aligned in all possible directions. The result is that the pattern of spots on the X-ray film becomes a pattern of rings. In the same way that the width of the individual slits in a DIFFRACTION GRATING affects the overall shape of the pattern obtained from the grating, the distribution of the electrons in the atoms affects the overall intensity distribution of the X-rays. It is the electrons that diffract the X-ray beam and thus information can be obtained about the distribution of the electrons in the atoms or molecules from which the crystal is made. It was in this way that the structure of DNA was first uncovered by British biologists Francis Crick (b. 1916) and James Watson (b. 1928) in 1953. Similar techniques have more recently been used to unravel the structure of HIGH-TEMPERATURE SUPERCONDUCTORS.

X-ray diffraction The DIFFRACTION of an X-RAY beam off the atoms in a crystal. Because the atoms in a crystalline lattice are arranged regularly, they act rather like a DIFFRACTION GRATING for the X-rays, and strong CONSTRUCTIVE INTERFERENCE is seen in certain directions.

For a beam of X-rays of wavelength λ, striking a crystal at an angle θ to the planes of the crystal, which have a separation d between one plane and the next, constructive interference will occur if

$$2d\sin\theta = n\lambda$$

where n is a whole number. *See also* BRAGG'S LAW.

X-ray tube A device used to produce X-RAYS, similar in structure to the THERMIONIC DIODE but operating at very high voltages (typically 100 kV). At these voltages, electrons gain a great deal of energy from the electric field. This energy is suddenly lost when they hit the metal anode, and some of it is given out as X-rays. A good deal of heat is also produced, so the anode is made of a good thermal conductor, such as copper, and may be cooled by passing water through it.

Y

yagi A type of radio AERIAL. In a yagi, several elements re-radiate received signals in such a way that they constructively interfere (*see* CONSTRUCTIVE INTERFERENCE) at an element connected to a radio receiver or TV set. Thus a stronger signal is produced than would be achieved by a simple aerial.

yield point For a DUCTILE material, the point beyond which a further small increase in TENSILE STRESS produces a large increase in STRAIN. Beyond the yield point, the material has effectively lost all its strength; it may extend a great deal more before it breaks, but this extra extension will require little more force.

Young's double slit experiment The experiment that first showed the WAVE NATURE OF LIGHT. In this experiment a COHERENT light source (light from a laser, or light diffracted from a single slit) is shone onto a pair of parallel slits. The light from these slits spreads out by DIFFRACTION and as the beams overlap they interfere (*see* INTERFERENCE), producing areas of constructive and destructive interference, which are seen as bright and dark bands called fringes. *See also* FRESNEL'S BIPRISM.

Young's fringes The bright and dark bands seen as a result of constructive and destructive interference of light in YOUNG'S DOUBLE SLIT EXPERIMENT.

Young modulus A measure of the stiffness of a stretched rod or wire under tension or compression. It is equal to the STRESS of the wire or rod divided by the longitudinal STRAIN.

y-plates One of the two pairs of plates fitted to a CATHODE RAY TUBE to control the direction of the electron beam. The other pair are the x-plates.

Z boson, *Z particle* The neutral particle that together with the charged W BOSON are the GAUGE BOSONS responsible for the WEAK NUCLEAR FORCE. The Z boson has a mass about 90 times that of the proton.

zeroth law of thermodynamics A definition of the concept of TEMPERATURE, which states that two objects are at the same temperature if there is no net flow of ENERGY between them when they are placed in contact. *See also* THERMODYNAMICS.

zinc-carbon cell, *dry cell* A common type of electrolytic CELL. It is not RECHARGEABLE, but is cheap to manufacture and gives an ELECTROMOTIVE FORCE of about 1.5 V. The ANODE takes the form of a carbon rod surrounded by a powdered mixture of carbon and manganese dioxide which acts to remove the waste products of the reaction, which would otherwise build up around the anode and increase the INTERNAL RESISTANCE of the cell. The ELECTROLYTE is ammonium chloride solution made into a paste with flour so it will not easily leak from the cell. The CATHODE is zinc made in the form of a cylinder, open at one end, and surrounded in more expensive versions by a steel outer shell since the zinc reacts to form zinc chloride as the cell is discharged, and an old cell may leak corrosive electrolyte. The gap between the anode and the open end of the cathode is sealed with an insulating material, usually plastic.

zodiac The region of the sky extending 8° each side of the ECLIPTIC within which the Sun and planets appear to move.

zone plate A transparent piece of material carrying a series of concentric opaque circles arranged such that the FRESNEL DIFFRACTION resulting from this pattern will have the effect of bringing light to a focus at a single point in much the same way as a lens. The FOCAL LENGTH of such a plate depends on the relationship between the wavelength of the light and the scale of the pattern. Zone plates are used as an alternative to lenses in some optical systems. They have the advantage that they are flat, but because the effective focal length varies with wavelength they can only be used with MONOCHROMATIC light.

Z particle *See* Z BOSON.

APPENDICES

Appendix I: SI units

Base units

Physical quantity	Name	Symbol	Definition
length	metre	m	the length equal to the length of path travelled by light in a vacuum in 1/(299,792,458) seconds
mass	kilogram	kg	the mass equal to that of the international prototype kilogram kept at Sèvres, France
time	second	s	the duration of 9,192,631,770 oscillations of the electromagnetic radiation corresponding to the electron transition between two hyperfine levels of the ground state of the caesium–133 atom
electric current	ampere	A	the constant electric current which, if maintained in two straight parallel conductors of infinite length and negligible cross-section, placed 1 metre apart in a vacuum, would produce a force between these conductors equal to 2×10^{-7} metres
thermodynamic temperature	kelvin	K	the fraction 1/273.16 of the thermodynamic temperature of the triple point of water
luminous intensity	candela	cd	the luminous intensity, in a given direction, of a source of monochromatic radiation of frequency 5.4×10^{14} Hz and has a radiant intensity in that direction of 1/683 watt per steradian
amount of substance	mole	mol	the amount of substance containing as many atoms (or molecules or ions or electrons) as there are carbon atoms in 12 g of carbon–12.

Supplementary units

plane angle	radian	rad	the plane angle subtended at the centre of a circle by an arc of equal length to the circle radius
solid angle	steradian	sr	the solid angle that encloses a surface on a sphere equal in area to the square of the radius of the sphere

Derived SI units

Physical quantity	Name	Symbol	SI equivalent
activity	becquerel	Bq	
dose	gray	Gy	
dose equivalent	sievert	Sv	
electric capacitance	farad	F	A sV^{-1}
electric charge	coulomb	C	As
electric conductance	siemens	S	
electric potential difference	volt	V	WA^{-1}
electric resistance	ohm	Ω	VA^{-1}
energy	joule	J	Nm
force	newton	N	kg ms^{-2}
frequency	hertz	Hz	s^{-1}
illuminance	lux	lx	
inductance	henry	H	V sA^{-1}
luminous flux	lumen	lm	
magnetic flux	weber	Wb	
magnetic flux density	tesla	T	
power	watt	W	J s^{-1}
pressure	pascal	Pa	Nm^{-2}

Appendix II: SI prefixes

Submultiple	Prefix	Symbol	Multiple	Prefix	Symbol
10^{-1}	deci	d	10^{1}	deca	da
10^{-2}	centi	c	10^{2}	hecto	h
10^{-3}	milli	m	10^{3}	kilo	k
10^{-6}	micro	μ	10^{6}	mega	M
10^{-9}	nano	n	10^{9}	giga	G
10^{-12}	pico	p	10^{12}	tera	T
10^{-15}	femto	f	10^{15}	peta	P
10^{-18}	atto	a	10^{18}	exa	E

Appendix III: SI conversion factors

Unit name	Symbol	Quantity	SI equivalent	Unit
acre		area	0.405	hm²
ångstrom	Å	length	0.1	nm
astronomical unit	AU	length	0.150	Tm
atomic mass unit	amu	mass	1.661×10^{-27}	kg
bar		pressure	0.1	MPa
barn	b	area	100	fm²
British thermal unit	btu	energy	1.055	kJ
calorie	cal	energy	4.187	J
cubic foot	cu ft	volume	0.028	m³
cubic inch	cu in	volume	16.387	cm³
cubic yard	cu yd	volume	0.765	m³
curie	Ci	activity	37	GBq
degree (angle)	°	plane angle	π/180	rad
degree Celsius	°C	temperature	1	K
degree Fahrenheit	°F	temperature	5/9	K
degree Rankine	°R	temperature	5/9	K
dyne	dyn	force	10	μN
electronvolt	eV	energy	0.160	aJ
erg		energy	0.1	μJ
fathom (6 ft)		length	1.829	m
foot	ft	length	30.48	cm
foot per second	ft s⁻¹	velocity	0.305	ft s⁻¹
gallon (UK)	gal	volume	4.546	dm³
gallon (US)	gal	volume	3.785	dm³
gallon (UK) per mile		consumption	2.825	dm³km⁻¹
gauss	Gs, G	magnetic flux density	100	μT
hectare	ha	area	1	hm²
horsepower	hp	power	0.746	kW
inch	in	length	2.54	cm
knot		velocity	1.852	km h⁻¹
light year	ly	length	9.461×10^{15}	m
litre	l	volume	1	dm³
Mach number	Ma	velocity	1193.3	km h⁻¹

SI conversion factors (continued)

Unit name	Symbol	Quantity	SI equivalent	Unit
maxwell	Mx	magnetic flux	10	nWb
micron	μ	length	1	μm
mile (nautical)		length	1.852	km
mile (statute)		length	1.609	km
miles per hour	mph	velocity	1,609	km h^{-1}
minute = (1/60)°	′	plane angle	π/10800	rad
oersted	Oe	magnetic field strength	1/(4π)	kA m^{-1}
ounce (avoirdupois)	oz	mass	0.2	g
ounce (troy)		mass	31.103	g
parsec	pc	length	30857	Tm
pint (UK)	pt	volume	0.568	dm^3
pint (US)		volume	0.473	dm^3
pound	lb	mass	0.454	kg
poundal	pdl	force	0.138	N
pound-force	lbf	force	4.448	N
pound-force per inch		pressure	6.895	kPa
pounds per square inch	psi	pressure	6.895×10^3	kPa
rad		absorbed dose	0.01	Gy
rem		dose equivalent	0.01	Sv
röntgen	R	exposure	0.258	mC kg^{-1}
second = (1/60)′	″	plane angle	π/648	mrad
solar mass	M	mass	1.989×10^{30}	kg
square foot	sq ft	area	9.290	dm^3
square inch	sq in	area	6.452	cm^2
square mile (statute)	sq mi	area	2.590	km^2
square yard	sq yd	area	0.836	m^2
standard atmosphere	atm	pressure	1.101	MPa
stokes	St	viscosity	1	cm^{2}s^{-1}
therm		energy	0.105	GJ
ton		mass	1.016	Mg
tonne	t	mass	1	Mg
torr, or mm Hg		pressure	0.133	kPa
yard	yd	length	0.914	m

Appendix IV: Common measures

Length

Metric units	Imperial equivalent	Imperial units		Metric equivalent
1 millimetre (mm)	0.03937 in			
10 mm = 1 centimetre (cm)	0.39 in	1 inch		2.54 cm
10 cm = 1 decimetre (dm)	3.94 in	1 foot	12 in	30.48 cm
100 cm = 1 metre (m)	39.37 in	1 yard	3 ft	0.9144 m
1000 m = 1 kilometre (km)	0.62 mi	1 mile	1760 yd	1.6093 km

Area

Metric units	Imperial equivalent	Imperial units		Metric equivalent
1 square millimetre	0.0016 sq in			
100 cm² = 1 square centimetre	0.155 sq in	1 square inch		6.45 cm²
10,000 cm² = 1 square decimetre	15.5 sq in	1 square foot	144 sq in	0.0929 m²
10,000 m² = 1 square metre	10.76 sq ft	1 square yard	9 sq ft	0.836 m²
1 hectare	2.47 acres	1 acre	4,840 sq yd	0.405 ha
		1 square mile	640 acres	259 ha

Volume

Metric units	Imperial equivalent	Imperial units		Metric equivalent
1 cubic centimetre	0.016 cu in	1 cubic inch		16.3871 cm³
1,000 cm³ = 1 cubic decimetre	61.024 cu in	1 cubic foot	1,728 cu in	0.028 m³
1,000 dm³ = 1 cubic metre	35.31 cu ft	1 cubic yard	27 cu ft	0.765 m³
	1.308 cu yd			

Common measures (continued)

Liquid volume

Metric units	Imperial equivalent	Imperial units	Metric equivalent
1 litre	1.76 pt	1 pint	0.57 l
1 hectolitre	22 gal	2 pt	1 quart	1.14 l
100 l		4 qt	1 gallon	4.55 l

Weight

Metric units	Imperial equivalent	Imperial units	Metric equivalent
1 gram	0.035 oz		1 ounce	28.3495 g
1 kilogram	2.2046 lb	16 oz	1 pound	0.4536 kg
1 tonne	0.0842 ton cm^2	14 lb	1 stone	6.35 kg
1,000 g		8 st	1 hundredweight	50.8 kg
1,000 kg		20 cwt	1 ton	1.016 t

Appendix V: Physical concepts in SI units

Concept	Symbol	SI unit	SI symbol	Defining equation
Area	A, a	square metre	m^2	$a = l^2$
Volume	V, v	cubic metre	m^3	$V = l^3$
Velocity	v, u	metre/second	ms^{-1}	$v = dl/dt$
Acceleration	a	metre/second^2	ms^{-2}	$a = d^2l/dt^2$
Density	ρ	kilogram/metre^2	$kg\,m^{-2}$	$\rho = m/v$
Moment of inertia	I	kilogram metre^2	$kg\,m^2$	$I = Mk^2$
Momentum	p	kilogram metre/second	$kgms^{-1}$	$p = mv$
Angular momentum	$I\omega$	kilogram metre^2/second	kgm^2s^{-1}	$I\omega$
Force	F	newton	N	$F = ma$
Torque (moment of force)	$T, (M)$	newton metre	Nm	
Work (energy, heat)	$W, (E)$	joule	J	$W = \int Fdl$
Kinetic energy	$T, (W)$	joule	J	$T = \tfrac{1}{2}mv^2$
Potential energy	V	joule	J	$V = \int Fdl$
Heat (enthalpy)	$Q, (H)$	joule	J	$H = U + pV$
Power	P	watt	W	$P = dW/dt$
Pressure (stress)	$p, (\sigma, \phi)$	newton/metre^2	Nm^{-2}	$p = F/A$
Surface tension	$\gamma, (\sigma)$	newton/metre	Nm^{-1}	$\gamma = F/l$
Viscosity, dynamic	η, μ	newton second/metre^2	Nsm^{-2}	$F/A = \eta dv/dl$
Viscosity, kinematic	ν	metre^2/second	m^2s^{-1}	$v = \eta/\rho$

Physical concepts in SI units (continued)

Concept	Symbol	SI unit	SI symbol	Defining equation
Temperature	θ, T	degree Celsius, kelvin	°C, K	$TK = (\theta \pm 273.15)°C$
Electric charge	Q	coulomb	C	$F = (Q_1Q_2)/(4\pi\varepsilon_0r^2)$
Electric potential (potential difference)	V	volt	V	$V = \int_\infty Edl$ $(V_{ab} = -\int_\beta Edl)$
Electric field strength	E	volt/metre	Vm^{-1}	$E = -dV/dl$
Electric resistance	R	ohm	Ω	$R = V/I$
Conductance	G	siemens	S	$G = 1/R$
Electric flux	ψ	coulomb	C	$\psi = Q$
Electric flux density	D	coulomb/metre²	Cm^{-2}	$D = d\psi/dA$
Frequency	f	hertz	Hz	cycles per second
Permittivity	ε	farad/metre	Fm^{-1}	$\varepsilon = D/E$
Relative permittivity	ε_r	farad/metre	Fm^{-1}	$\varepsilon_r = \varepsilon\varepsilon_0$
Magnetic field strength	H	ampere turn/metre	Atm^{-1}	$dH = idl\sin\theta/4\pi r^2$
Magnetic flux	Φ	weber	Wb	$\Phi = -\int edt$
Magnetic flux density	B	tesla	T	$B = d\Phi/dA$
Permeability	μ	henry/metre	Hm^{-1}	$\mu = B/H$
Relative permeability	μ_r	henry/metre	Hm^{-1}	$\mu_r = \mu/\mu_0$
Mutual inductance	M	henry	H	$e_2 = Mdi_1/dt$
Self inductance	L	henry	H	$e = Ldi/dt$
Capacitance	C	farad	F	$C = Q/V$

Physical concepts in SI units (continued)

Concept	Symbol	SI unit	SI symbol	Defining equation
Reactance	X	ohm	Ω	$X = \omega L$ or $1/\omega C$
Impedance	Z	ohm	Ω	$X = \sqrt{R^2 = X^2}$
Susceptance	B	siemens	S	$B = 1/X$
Admittance	Y	siemens	S	$Y = 1/Z$
Luminous flux	Φ	lumen	lm	lm = cd sr
Illuminance	E	lux	lx	lx = lm m^{-2}

Appendix VI: Fundamental constants

Constant	Symbol	Value	Units
acceleration of free fall	g	9.80665	ms^{-2}
speed of light in vacuum	c	299,792,458	ms^{-1}
elementary charge	e	1.60217733	10^{-19} C
electron rest mass	m_e	9.1093897	10^{-31} kg
permeability of free space	μ_0	$4\pi \times 10^{-7}$ $= 12.566370614....$	$N\,A^{-2}$ $10^{-7}\,N\,A^{-2}$
permittivity of free space	ε_0	8.854187187....	$10^{-12}\,F\,m^{-1}$
electronic radius	r_e	2.18794092	10^{-15} m
neutron rest mass	m_n	1.6749286	10^{-27} kg
proton rest mass	m_p	1.6726231	10^{-27} kg
Avogadro constant	L, N_A	6.0221367	$10^{23}\,mol^{-1}$
Boltzmann constant	$k = R/N_A$	1.380658	$10^{-23}\,JK^{-1}$
Molar volume of ideal gas at STP	V_m	2.241409	$10^{-2}\,m^3mol^{-1}$
Faraday constant	F	96,485.309	$C\,mol^{-1}$
Stefan–Boltzmann constant	s	5.67051	$10^{-8}\,Wm^{-2}K^{-4}$
gas constant	R	8.314510	$J\,mol^{-1}K^{-1}$
gravitational constant	G	6.67259	$10^{-11}\,m^3kg^{-1}s^{-2}$
Planck constant	h	6.6260755	$10^{-34}Js$
Bohr radius	r_B	5.29177249	10^{-11} m
Bohr magneton	μ_B	9.2740154	$10^{-24}\,JT^{-1}$
Nuclear magneton	μ_N	5.0507866	$10^{-27}JT^{-1}$

Appendix VII: Electromagnetic spectrum

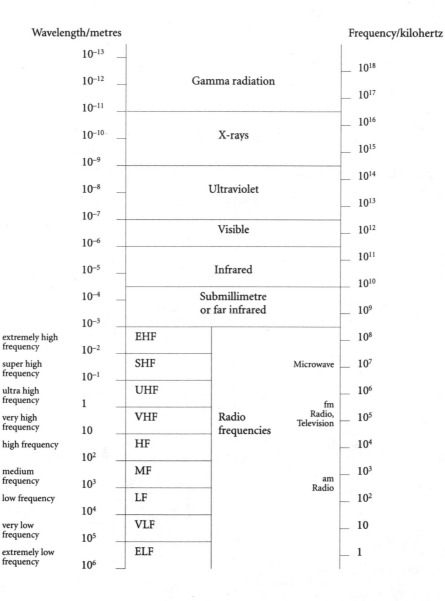

Appendix VIII: Periodic table of the elements

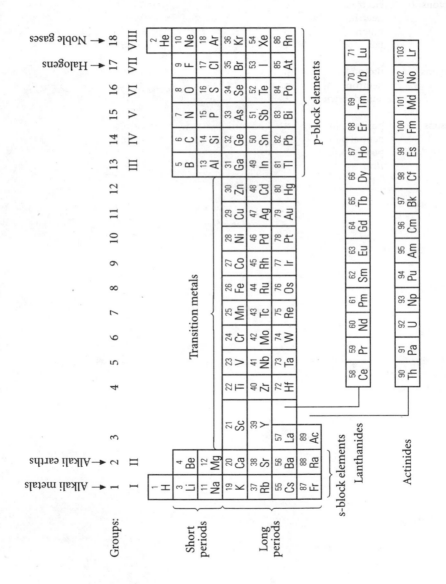

Appendix IX: Elementary particles

Family	Particle	Symbol	Mass (MeV)	Mean life
Bosons	photon	$^{\star}\gamma$	0	stable
Leptons	electron	e	0.5110	stable
	neutrino	ν	0	stable
	muon	μ	106	2.20×10^{-6}
	tau	τ	1784	$<2 \times 10^{-12}$
Baryons	proton	p	938	stable ($>10^{33}$ years)
	neutron	n	939.6	917
	lambda particle	λ	1116	2.6×10^{-10}
Mesons	positive pion	π^+	140	2.6×10^{-8}
	neutral pion	π°	135	8.4×10^{-17}
	positive kaon	K^+	494	1.24×10^{-8}
	J/psi	J/ψ	3097	10^{-20}

Appendix X: The Solar System

Planet	Mean distance from Sun (million km)	Mass (Earth masses)	Sidereal period	Axial rotation (equatorial)	Diameter (equatorial) (km)	Satellites	Atmosphere
Mercury	57.91	0.054	88 d	58 d 16 h	4,878	0	hydrogen, helium, neon
Venus	108.21	0.815	224.7 d	243 d	12,104	0	carbon dioxide
Earth	149.60	1.000*	365.26 d	23 h 56 m 4 s	12,756	1	nitrogen, oxygen
Mars	227.94	0.107	687 d	24 h 37 m 23 s	6,794	2	carbon, dioxide
Jupiter	778.34	317.89	11.86 y	9 h 50 m 30 s**	142,800	16	hydrogen, methane
Saturn	1430	95.14	29.46 y	10 h 14 m**	120,000	18	hydrogen, helium
Uranus	2869.6	14.52	164.79 y	16–28 h**	51,000	15	methane, helium, hydrogen
Neptune	4496.7	17.46	164.79 y	18–20 h**	49,500	2	methane, hydrogen
Pluto	5900	0.1 (approx)	247.7 y	6 d 9 h	2,300	1	methane

d = days
y = years
h = hours
km = kilometres
*The mass of the Earth is 5.976×10^{24} kilogram
**Different latitudes rotate at different speeds